WHAT EVERY ENGINEER SHOULD KNOW ABOUT
MATLAB®
and Simulink®

WHAT EVERY ENGINEER SHOULD KNOW
A Series

Series Editor*

Phillip A. Laplante
Pennsylvania State University

*Founding Series Editor: **William H. Middendorf**

WHAT EVERY ENGINEER SHOULD KNOW ABOUT
MATLAB®
and Simulink®

Adrian B. Biran

With contributions by Moshe Breiner

CRC Press
Taylor & Francis Group
Boca Raton London New York

CRC Press is an imprint of the
Taylor & Francis Group, an **informa** business

MATLAB® and Simulink® are trademarks of The MathWorks, Inc. and are used with permission. The Math-Works does not warrant the accuracy of the text of exercises in this book. This book's use or discussion of MATLAB® and Simulink® software or related products does not constitute endorsement or sponsorship by The MathWorks of a particular pedagogical approach or particular use of the MATLAB® and Simulink® software.

CRC Press
Taylor & Francis Group
6000 Broken Sound Parkway NW, Suite 300
Boca Raton, FL 33487-2742

© 2011 by Taylor and Francis Group, LLC
CRC Press is an imprint of Taylor & Francis Group, an Informa business

No claim to original U.S. Government works

Printed in the United States of America on acid-free paper
10 9 8 7 6 5 4 3 2 1

International Standard Book Number: 978-1-4398-1020-0 (Paperback)

Library of Congress Cataloging-in-Publication Data

Biran, Adrian.
    What every engineer should know about MATLAB and Simulink / Adrian B. Biran.
        p. cm. -- (What every engineer should know ; 45)
    Includes bibliographical references and index.
    ISBN 978-1-4398-1020-0 (pbk. : alk. paper)
    1. Engineering mathematics--Data processing. 2. MATLAB. 3. SIMULINK. I. Title.
II. Series.

TA345.B4855 2011
620.001'51--dc22
                                                        2010019524

Visit the Taylor & Francis Web site at
http://www.taylorandfrancis.com

and the CRC Press Web site at
http://www.crcpress.com

Dedicated to Suzi, Mihal, Paul and Zohar

Contents

Preface

Target audience

This book is primarily intended for students of mechanical and electrical engineering; it can also be useful to students of civil engineering, to students of computer sciences, and to practicing engineers who want to learn MATLAB® and Simulink®. Moreover, the book can be used by students of physics, and in some European countries by high-school students in their last year before the maturity examinations, and by technicians.

Objectives and general approach

The general objective is to teach MATLAB by showing that it is good for engineering students and for practicing engineers. We do this by examples in which MATLAB yields easy and elegant solutions of problems in mathematics and mechanical and electrical engineering. In each group of applications we start with geometrical examples because we consider geometry as essential in mechanical and civil engineering. It is easier to understand many mathematical concepts by showing their geometrical interpretation. Moreover, to understand and experimentally solve a problem of geometry one may need as few tools as a ruler and compass, while experiments in mechanics or electricity require a laboratory. Geometry is the basis of computer graphics that are extensively used in the design of machine parts, of car, airplane and ship surfaces, and of building and architectural structures. Last, but not least, we often appeal to geometry to visualize phenomena that belong to various fields of engineering or science. MATLAB was initially developed as software for linear algebra. There is a one-to-one correspondence between objects of linear algebra and those of geometry. Showing this can give an important insight into the nature of problems and their solutions.

The geometrical applications are followed by examples in those basic fields of mechanical engineering that are taught in the first university courses. The examples belong to mechanics, that is, statics, kinematics, dynamics and mechanisms, where MATLAB data types, operations, and functions fit in a most natural manner.

We end each group of applications with examples in electrical engineering. Again, the examples come from fields that are taught in the first basic courses.

Often we deliberately introduce the same MATLAB concept more than once, for instance, the first time by a geometrical example, the second time time, by a mechanical example, and the third time, by an electrical example. In this way we allow students to concentrate on those examples that belong to their own sphere of interest. We also show in this way that the same MATLAB facilities can be used in many branches of engineering or science.

We use examples to introduce MATLAB data types, operations, and functions, but we do not limit ourselves to isolated applications. We also try to teach the reader general methods and algorithms for solving problems, how to solve them in an efficient way, and how to avoid or minimize computer errors.

In the third part of the book we introduce the reader to Simulink®. Throughout the book we prefer applications in which MATLAB and Simulink simplify the formulation and make computing more elegant, and applications for which MATLAB provides spectacular visualization. These applications belong to disciplines taught in basic courses.

The contents of the book

The book is divided into three parts. In Part I, *Introducing MATLAB*, we teach the reader how to start using MATLAB for solving problems and how to process and present calculation and experimental results. As this part contains the basics of working in MATLAB, its chapters are more detailed and extensive than those of other parts. Part II, *Programming in MATLAB*, shows that the computer has its limitations that one must take into account when programming and when evaluating results. Part II also contains an introduction to object-oriented programming (OOP). In Part III, *Progressing in MATLAB*, we introduce the reader to more advanced features of the software. By the end of this part the reader should be able to solve most problems encountered in basic engineering courses. The knowledge acquired in this stage will also allow the reader to go further to specialized MATLAB toolboxes. In this part we introduce Simulink and a few very elementary applications of MATLAB in control and signal processing.

Chapter 1, *Introduction*, shows that anything a calculator can do, MATLAB can do better.

Chapter 2, *Vectors and matrices*, shows how MATLAB arrays can be used to represent in a most natural and elegant way geometrical, mechanical and electrical objects, and to perform meaningful operations on them. In this chapter the reader also discovers that programming can help in solving tasks that otherwise would be so tedious and time consuming that many users would even choose not to deal with them. Beginning with this chapter we introduce the readers to the subjects of program structuring and checking.

Chapter 3, *Equations*, shows how to treat this ubiquitous subject in MAT-LAB and what the limitations of this treatment are. To better understand the subject we use concrete examples of intersecting lines and planes. Doing so we can also give geometric interpretations of the notions of ill-conditioned systems and of least-squares solutions. Given a system of linear equations, MATLAB always yields an answer. This answer, however, can be a unique solution, a particular solution, a solution in the least-squares sense, or no solution at all. We show how to distinguish among these cases. This chapter continues with polynomial equations. The reader is warned that the function `root` can yield erroneous results and we give some hints on how to detect and correct errors. The chapter ends by describing an iterative method for solving transcendental equations.

Students and engineers must also deliver reports of calculations. The results of laboratory experiments must be processed, visualized in graphs, interpreted, and presented in reports. Scientists publish papers describing such results. In Chapter 4, *Processing and publishing the results*, we describe the `diary` option that allows the user to record full working sessions. As these records are in plain ASCII code, they can be further modified in a word processor or a typesetter. The data can be processed by interpolation and curve fitting and presented in 3-D plots, histograms or other plots. MATLAB provides functions for connection with Excel® and for producing figures that can be readily inserted into Word® or Tex files.

Part II starts with Chapter 5, *Some facts about numerical computing*. Not a few students and engineers believe that once they use a computer they need not care much about how to supply the input data, how to write a program or how reliable are the results. The computer is supposed to be infallible and capable of solving any problem. To prove that such beliefs are misconceptions we describe the sources of numerical errors, we define computing complexity and show how to improve it. Part of this chapter is based on Chapter 2 of Biran and Breiner (2002); the rest is new.

Chapter 6, *Data types and object-oriented programming*, defines structures, cell arrays and classes as MATLAB building blocks. This chapter includes an elementary introduction to object-oriented programming (OOP) with an application to calculating with units. We reuse in this chapter material from Chapter 18 of Biran and Breiner (2002).

Part III starts with Chapter 7, *Complex numbers*. Once considered impossible, later a mathematical curiosity, complex numbers are now an indispensable tool in the study of oscillations, electrical circuits and control engineering. We show that in MATLAB complex numbers can be dealt with as easily as real numbers can. The chapter ends with applications to a simple mechanism.

Chapter 8 is about *Numerical integration*. We reuse part of Chapter 11 in Biran and Breiner (2002) and add some newer features introduced with MATLAB 7. In this book we continue our policy of referring only to the basic MATLAB package and not to toolboxes. However, in this chapter we make an exception and give a few simple examples of symbolic integration with the help of the *Symbolic Math Toolbox*^{TM}. We do this to show the reader one possible direction of going beyond the treatment exposed in this book.

Chapter 9, *Ordinary differential equations*, discusses the numerical integration of ordinary differential equations (ODEs). Several methods of integration are described as well as the corresponding MATLAB functions. The material is based on Chapter 14 of Biran and Breiner (2002).

Chapter 10, *More graphics*, continues what we have begun in the previous chapters. The last part of the chapter is an introduction to graphical user interfaces (GUIs).

In Chapter 11, *An introduction to Simulink*, we begin the description of this important toolbox that is also part of the *MATLAB for Students* package. We start with the simple example of the addition of two waves and continue with examples requiring the integration of differential equations. Progressively we go from first- to second-order linear equations and end with nonlinear equations. The chapter contains examples in both mechanical and electrical engineering.

Chapter 12, *Applications in the frequency domain*, is dedicated to some basic methods of signal processing and contains also an introduction to the Bode diagram. The chapter contains slightly updated versions of several sections of Biran and Breiner (2002).

The philosophy of the book

Let us begin by quoting from the Preface to Biran and Breiner (2002).

"The third edition of our book is meant to include some of the powerful improvements introduced in MATLAB 6. Additionally, we are aware that with this release the software grew to such an extent that the danger appears of not being able to see the forest because of the trees. Often, MATLAB 6 provides several possibilities of performing the same task and the beginner may get lost when faced with such a wide choice. Therefore, we think that an important task of the book is to guide the reader through the MATLAB 6 forest and choose a sufficient set of commands and functions that enable the completion of most engineering tasks.

The help facilities of MATLAB 6 contain excellent reference material. As the reader can easily access that help, we feel no necessity to compete in that direction, but leave our book as a tutorial, as it was conceived from the beginning. We also continue our policy of introducing notions in small portions, dispersed throughout the book."

If the above paragraphs were true for MATLAB 6, *a fortiori* they are valid for the MATLAB versions that followed. There is no need to copy the manuals and the help provided by the MathWorks^{TM}. Further, we do not want to write a reference book, because the reference manuals provided by The MathWorks are excellent. In reference books each chapter is dedicated to one subject and exhausts it. Then, for example, one has to learn everything on matrices before learning how to do a simple plot. Our philosophy is to teach the reader a bit of A, then a bit of B, and so on, and return to more of A, more of B and so on. This is the difference between a book on the grammar of a foreign language, and a book for learning the same language. In the latter, the reader learns a bit about nouns, then a bit about verbs, enough to build a very simple sentence. Returning to learn more about nouns, more about verbs, and adding some knowledge about adjectives and adverbs, the reader can now build a more complex sentence. This is the philosophy that led us when we wrote our previous books, and this is the philosophy that guides us in this book.

It is common knowledge that students learn easier when they are motivated. In our book we explain the engineering importance, significance and implications of the calculations performed. Seeing that MATLAB can help them in their current, immediate tasks can enhance the students' wish to learn and use the software.

Notations

To distinguish between explanations and what should appear on the computer screen, we present the latter in gray frames, for example

```
≫ 4 - 2
ans =
   2
```

Emboldened words indicate a key term being defined for the first time in the text, for example **array of numbers**. Italics are used to emphasize, for instance *command history*. Boldface letters are also used to name vectors or matrices, where it is usual to write so:

$$\mathbf{P} = \begin{bmatrix} 5 \\ 7 \end{bmatrix}$$

Typewriter characters are used for the names of MATLAB and Simulink functions and commands (e.g., `Delete`), program and function listings and names of files (e.g., F2C). Sometimes we also use boldface letters for the same purpose.

MATLAB®and Simulink®are registered trademarks of The MathWorks, Inc. For product information please contact:

The MathWorks, Inc.
3 Apple Hill Drive
Natick, MA, 01760-2098 USA
Tel: 508-647-7000
Fax: 508-647-7001
E-mail: info@mathworks.com
Web: www.mathworks.com

Companion software

Companion software can found at a site provided by The MathWorks
   http://www.mathworks.com/matlabcentral/fileexchange/
and one provided by the publisher
   http://crcpress.com/product/isbn/

Acknowledgments

We thank Simon Lake, Tracey Cummins, Nikki Bamister, Owen King, and Laurence Gaillard of Pearson Education, UK, for permission to reuse extensive parts of Biran and Breiner (2002).

For our first books, Courtney Esposito, of *The MathWorks*, and Baruch and Abigail Pekelman of *Omikron Delta* provided us with the newest versions of MATLAB and advice. Their work was continued by Naomi Fernandez and the MathWorks Book Program.

We acknowledge the professional help of Tim Pitts and Karen Mosman who were the editors of the first edition, *MATLAB for Engineers*, in 1995.

Irina Abramovici, of Technion's Taub Computer Center, provided help in LATEX for our previous books and continued to do so for this book.

We thank Allison Shatkin, Kari Budyk, Karen Simon, Shashi Kumar, James Miller, and Joel Schwarz of Taylor & Francis for their contributions in editing and producing the book.

Finally, many thanks to my wife, Suzi, for her patience, encouragement and continuous support.

Part I

Introducing MATLAB®

1

Introduction to MATLAB®

We begin this chapter by showing how to start MATLAB and use it as a calculator. Next, we show how to use MATLAB as a scientific calculator. We begin to discover the superiority of MATLAB by storing numerical values under the names of constants, by storing several numerical values in arrays and by calling these values by name, for use in various operations. We further learn how to plot simple functions, built in to MATLAB, and how to program simple functions.

By the end of this chapter the reader will be able to perform in MATLAB all the calculations previously carried on with the hand-held computer, and will know how to plot the results and how to program simple functions that may speed up repetitive calculations.

1.1 Starting MATLAB

We assume that you installed MATLAB according to the instructions you received and that you are familiar with your computer and operating system. To start, double-click on the MATLAB icon, in Windows, or type the instruction matlab, in Unix. One or more windows appear on your screen. The configuration depends on your MATLAB version and on the configuration you used when closing the last working session. The default desktop corresponding to version 7.7 (R2008b), is shown in Figure 1.1.

The various windows are identified by the names appearing in their title bars: *Command Window*, marked 1 in Figure 1.1, *Workspace*, marked 2, *Command History*, marked 3, and *Current Directory*, marked 4. For the moment we are interested only in the *Command Window*. You may close the other windows by clicking on the *Close* icon, that is on the 'X' located in the upper, right-hand corner of each window. You can change the size of any window by dragging the bars that separate it from other windows. Also try the icons situated in the toolbar of each window, at the left of the *Close* icon. Finally, you may click on *Desktop* in the main toolbar and choose the option you are interested in. To restore the screen to the configuration shown in Figure 1.1 click *Desktop* → *Desktop Layout* → *Default*. If necessary, it is possible to enlarge the command window by clicking on the square situated

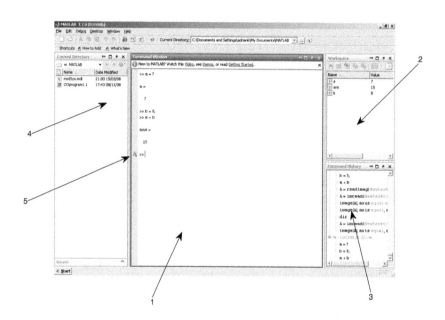

FIGURE 1.1: The MATLAB desktop

two places at the left of the *Close* icon.

For multimedia introductions to MATLAB click on one of the blue items listed in the information line situated at the top of the *Command Window*. Another help, introduced in 2008, is the *fx* icon marked 5 in Figure 1.1; clicking on it opens a box in which you can browse lists of the many built-in functions provided by MATLAB.

The symbol ≫ appears in the *Command Window*; it *prompts* you to enter a command. In the next subsection you'll learn the simplest ones. For the moment try

```
≫ ver
```

to get details about the MATLAB version, the computer and the operating system you are using. The display will also include a list of the MATLAB-related toolboxes installed on the computer.

1.2 Using MATLAB as a simple calculator

We can use MATLAB to carry on the same arithmetical operations that we perform on a simple hand-held calculator. Thus, for addition, we use the '+' key

```
≫ 2 + 4
```

and pressing the **Enter** key we obtain

```
ans =
  6
```

For subtraction we use the minus key, '-', and press the **Enter** key

```
≫ 4 - 2
ans =
  2
```

We perform multiplication with the symbol '*' (press simultaneously the **8** and **Shift** keys)

```
≫ 2*3
ans =
  6
```

and division with the aid of the **slash** key, '/',

```
≫ 4/2
ans =
  2
```

We can 'chain' operations. Try for yourself

```
≫ 4*5 + 2*8
```

Usually, hand-held calculators have a key for canceling an erroneous input; it is marked 'CE'. In MATLAB we use the **Backspace** arrow, ←, to return the cursor over the number or operation symbol we want to correct. Then, we can delete that number or symbol and type the correct input. As an example try the sequence

```
≫ 4 + 2
≫ 4
≫ 4 + 5
```

As expected, MATLAB allows for more corrections than a hand-held calculator. Suppose, for example, that we write

```
≫ 4*5 + 2*8
```

while we intend

```
≫ 4*4 + 2*8
```

We can use the left arrow key, '←', to return over 5, next the **Delete** key
to erase 5, and finally write 4 and press **Enter**:

```
≫ 4*5 + 2*8
≫ 4*4 + 2*8
```

We can do even more. Suppose that we carried on the wrong operation
and we realized this immediately afterwards. We do not have to type again
the whole input, but we can use the **Up arrow**, ↑, to retrieve our command,
correct it and execute again the operation. This way of performing corrections
is called *last-line editing*. In the newest versions of the software and with
present-day computers it is possible to return more than one line back.

To raise a number to an integer power, for example 2^4, on some hand-held
computers we press in sequence 2, ×, and three times '='. In MATLAB we
use the '^' symbol (press simultaneously the '**6**' and **Shift** keys)

```
≫ 2^4
ans =
   16
```

Some simple calculators have also a key for extracting square roots; it is
marked $\sqrt{\ }$. In MATLAB we use for this

```
≫ sqrt(4)
ans =
   2
```

Hand-held calculators also have a memory key, or textbfM+ and **M-** keys
for adding to or subtracting from a stored number. In general, on those
calculators we are limited to one number. In MATLAB we can store as many
numbers as we need and distinguish between them by allocating different

names. Try, for example,

```
≫ A = 5
A =
    5
≫ B = 2
B =
    2
≫ C = 3
C =
    3
```

While on hand-held calculators we have a special key, usually marked **MRC** (memory recall) to retrieve one stored number, in MATLAB you retrieve the 'memorized' values by *calling them by name*. If you entered the three numbers indicated above, try for yourself

```
≫ A, B, C
```

and see the results.

MATLAB is *case sensitive*, which means that $A \neq a$. Enter, for example

```
≫ A = 3
A =
    3
≫ a = 2
a =
    2
```

and type

```
≫ A, a
```

Mind the difference.

Above, when we stored a named number, or wrote an expression to be calculated, the computer returned our input. This takes some time that can be negligible for short operations, but can become annoying for many, longer or repetitive operations. In most cases we do not need this 'confirmation' and can spare time and keep a cleaner screen. To do so we end the command by a semicolon, ';',

```
≫ A = 5; B = 2; C = 3;
```

Try this for yourself and compare the difference.

What about the **M+** and **M-** keys? Let us suppose that we stored the value 5 under the name *A*, as we did above, and we want to add to it the number 2. We do this by the simple command

```
≫ A = A + 2
A =
   7
```

The computer stores the updated value. If we want to add more numbers we do not need to use several times the **M+** and **M-** keys. We can proceed in one command, using even combined operations, as below

```
≫ A = 5;
A = A + 2*3 + 5*4
A =
   31
```

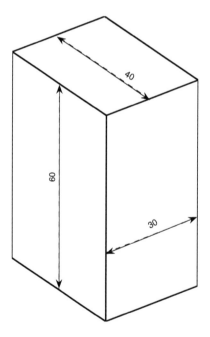

FIGURE 1.2: A parallelepiped

It is here that the superiority of MATLAB, and of computer languages in general, begins to show. If we have to use several times a certain value, we do not have to enter it again and again. We just *allocate* that value to an appropriately named constant and then *call it by name*. As an example, consider the parallelepiped shown in Figure 1.2. Its breadth is 40 mm, depth, 30 mm, and height, 60 mm. We use the initials of these dimensions as names of constants and allocate them the above-mentioned values. Next, we calculate the area of the base, the volume of the parallelepiped and the area of the developed surface as follows:

```
≫ B = 40;
≫ D = 30;
≫ H = 60;
≫ BaseArea = B*D
BaseArea =
  1200
≫ Volume = BaseArea*H
Volume =
  72000
≫ Developed = 2*BaseArea + 2*B*H + 2*D*H
Developed =
  10800
```

The developed surface is the plane figure that, when conveniently folded, yields the surface of the parallelepiped.

In the calculations shown above we used the value of B twice by calling it directly and twice by calling the intermediate result **BaseArea**. We defined, however, the value of B only once, in the first line. Thus, not only are we spared work, but we also avoided *input errors*. More specifically, we avoided the possibility of entering different values by defining B several times.

To see what variables are stored now in your computer type **who**. To obtain more details use the 'long form of **who**', that is the command **whos**. The resulting display show the names, the size, the number of bytes and the class of the stored variables. The same information is contained in the *Workspace* window. In later chapters we shall explain the notions of *bytes* and *class*.

1.3 How to quit MATLAB

If you want to close the work session and quit the MATLAB environment, you may use one of the following options;

- type **quit** or **exit** on the command line;

- click on the **File** pull-down menu, then on **Exit MATLAB**;

- click on the MATLAB icon, in the upper, left-hand corner of the window, then on `Close`.

1.4   Using MATLAB as a scientific calculator

1.4.1   Trigonometric functions

Scientific calculators have keys for many functions, for example for trigonometric functions. Thus, on some calculators, to obtain the sine of 30°, we enter 30 and press the key marked `sin`. On other calculators we have to press first the `sin` key, enter afterwards 30, and finally press the = key. There is also one key that defines the units of angles as degrees, radians, or even grades. In MATLAB we have to write explicitly the name 'sin' followed by its **argument**, in radians, between parentheses:

```
≫ sin(pi/3)
ans =
   0.8660
sin(pi/2)
ans =
   1
```

In the first and fourth lines `pi` is the name under which MATLAB stores the value of the number π. Try for yourself

```
≫ pi
ans =
   3.1416
```

In engineering angles are usually measured in degrees. As there are 180 degrees in an angle of π radians, we can convert 30 degrees to $30\pi/180$ radians and calculate the sine of 30° by

```
≫ sin(30*pi/180)
ans =
   0.5000
```

In the newest MATLAB versions there is no need for such conversions; we can use trigonometric functions with the argument given in degrees. The names of the new functions end with the letter 'd'. For example, we can calculate sine and cosine values as follows

```
>> sind(0)
ans =
   0
sind(30)
ans =
   0.5000
cosd(30)
ans =
   0.8660
```

As an exercise, let us check a known trigonometric relationship

```
>> sind(30)^2 + cosd(30)^2
ans =
   1
```

To calculate tangents of angles measured in degrees we use the `tand` function

```
>> tand(0)
ans =
   0
tand(45)
ans =
   1.0000
tand(90)
ans =
   Inf
```

'Inf' is the MATLAB symbol for *infinity*. The tangent of an angle can be defined as the ratio of the sine to the cosine of that angle. Let us check this relationship

```
>> sind(30)/cosd(30) - tand(30)
ans =
   -1.1102e-016
```

The answer appears in *scientific notation* and should be read as -1.1102×10^{-16}. The correct result should have been 0. The very small number yielded by MATLAB is due to errors of *numerical calculations*. As we shall learn more in Chapter 5, numerical calculations are bound to produce errors. In fortunate cases such errors can be minimized by careful programming, in many cases to reduce errors we must use considerable computer resources. In engineering, however, in most cases we must and can be satisfied by a precision compatible with our technological needs.

Years ago, when hand-held computers that could be programmed appeared on the market, we could hear engineers say, 'Why should I use your computer in which I have to write so many letters, while I can use my calculator by pressing only one, two or at most three keys?' The complete answer appears obvious after discovering how powerful programmed computing can be. For the moment, let us be satisfied with a simpler answer. On many calculators we find 15 keys. With the aid of the '2ndF' (second function) key we may double the number of available functions, and on some calculators it is even possible to add a few more functions. We approach thus 40 functions. This is a very small number compared to the combinations that can be obtained even with a few letters. As we shall learn in this book, MATLAB provides us with many more functions whose utility in engineering will become progressively evident. Let us see, for example, a few more trigonometric functions available in MATLAB, but usually not on hand-held calculators. Thus, the *cotangent* is defined as the inverse of the *tangent*

```
≫ cotd(30)
ans =
  1.7321
1/tand(30)
ans =
  1.7321
```

The *secant* is the inverse of the *cosine*

```
≫ secd(30)
ans =
  1.1547
1/cosd(30)
ans =
  1.1547
```

and the *cosecant* is the inverse of the *sine*

```
≫ cscd(30)
ans =
  2.0000
1/sind(30)
ans =
  2.0000
```

Think also that for all the above functions we have in parallel the variant with the argument in radians and the variant with the argument in degrees. Table 1.2 summarizes the trigonometric functions provided by MATLAB.

Exercise 1.1 Elementary trigonometry
For $x = 15°$ verify that

$$\sin(-x) = -sinx$$
$$\cos(-x) = \cos x$$
$$\tan(-x) = -\tan x$$

Use in this exercise and the following the MATLAB functions that accept arguments in degrees.

Exercise 1.2 Trigonometric identities
For $\alpha = 30°, ; \beta = 15°$ verify that

$$\sin(\alpha + \beta) = \sin\alpha \cos\beta + \cos\alpha \sin\beta$$
$$\cos(\alpha + \beta) = \cos\alpha \cos\beta - \sin\alpha \sin\beta$$
$$\tan(\alpha + \beta) = \frac{\tan\alpha + \tan\beta}{1 - \tan\alpha \tan\beta}$$

Hints. First define the constants `alpha` and `beta`. Next, to calculate the right-hand side of the last equation, write the numerator and the denominator between parentheses:

```
(tan(alpha) + tan(beta))/(1 - tan(alpha)*tan(beta))
```

To enhance readability, we use spaces between different terms. Thus, there is a blank space before the '+' sign and one after it. Of course, this detail of style does not influence the execution of the statement, but it is highly recommended when we want to print a publishable report of our calculations.

1.4.2 Inverse trigonometric functions

We calculate inverse trigonometric functions on a hand-held calculator by pressing first the 2ndF key, and next the key with the name of the respective direct trigonometric function. Usually, above the latter key appears the name of the direct function with the exponent '-1'. In MATLAB we write the MATLAB name of the inverse function, immediately followed by the argument between parentheses. Thus, to obtain the *arc* whose sine equals 1, we use the function `asin`

```
≫ asin(1)
ans =
    1.5708
```

This arc equals, as expected, $\pi/2$ radians. To check this we note that the result 1.5708 appeared as a constant named **ans** (abbreviation of 'answer'). We use this fact to write

```
≫ ans - pi/2
ans =
    0
```

We see that the constant **ans** stores only the value yielded by the last executed operation.

To obtain the angle, in degrees, whose sine equals 1 we call the function **asind**

```
≫ asind(1)
ans =
   90
```

To obtain the arc whose cosine equals 0, we call the function **acos**

```
≫ acos(0)
ans =
    1.5708
```

The result is in radians. To calculate the angle, in degrees, we call the function **acosd**

```
≫ acosd(0)
ans =
   90
```

More inverse trigonometric functions are listed in Table 1.2. To learn more about a certain function enter **help** followed by the name of the function, for example

```
≫help atan
ATAN Inverse tangent, result in radians.
   ATAN(X) is the arctangent of the elements of X.
   See also ...
   ...
```

Another possibility is to click on the **Help** icon on the menu bar (the rightmost icon), then on **Full Product Family Help**, on **Index**, and enter in the **Search** frame the name of the function we are interested in. The explanation will be displayed in the right-hand window. We can simplify this approach. For example, to learn more about the inverse cotangent function, $cotd^{-1}$, type on the command line `doc acotd`.

The function `atan2` deserves some explanations and an illustration. We shall do this in Chapter 2. For the moment let us note that, while a direct trigonometric function has only one value, inverse trigonometric functions are multi-valued. Thus, while $\sin \pi/2$ has only one value, 1, $\sin^{-1} 1$ has the values $\pi/2$, $\pi/2 + 2\pi$, $\pi/2 + 4\pi$, and so on. MATLAB yields only the *principal value*, that is the first value listed here.

Exercise 1.3 Pythagoras' theorem

Let the sides of the triangle ABC be

$$\overline{AB} = 30, \ \overline{BC} = 50, \ \overline{CA} = 40$$

The dimensions are in mm, as usual in mechanical drawings. Applying the cosine law, calculate in MATLAB the angle $\widehat{BAC}$, first in radians, then in degrees. As you will discover that the angle is right, show in MATLAB that the three sides are related by Pythagoras' theorem.

1.4.3 Other elementary functions

Scientific calculators have a key marked 'y^x' that allows us to raise a number to any power. To do the same thing in MATLAB, we use the symbol '^' (press simultaneously the 'Shift' and '6' keys). For instance, we carry on the operation $3^{2/3}$ as follows

```
≫ 3^(2/3)
ans =
   2.0801
≫ ans^(3/2)
ans =
   3
```

Scientific calculators have keys for calculating logarithms and exponentials. To calculate natural logarithms in MATLAB we use the function `log`

```
≫ log(1)
ans =
   0
≫ log(10)
ans =
   2.3026
```

The exponential function is calculated with the MATLAB function `exp`. If we continue the calculation shown above

```
≫ exp(ans)
ans =
   10.0000
```

as expected.

To calculate the common logarithm, that is the logarithm in base 10, we call the MATLAB function `log10`

```
≫ log10(1)
ans =
   0
≫ log10(10)
ans =
   1
```

Let us check that the logarithm of a product equals the sum of the logarithms of the multiplicands

```
≫ log(5) + log(7) - log(5*7)
ans =
   0
```

The corresponding rule for division is

```
≫ log(5) - log(7) - log(5/7)
ans =
   -5.5511e-017
```

Again we find a very small number where we should have obtained a zero. In Chapter 5 we are going to explain the source of this very small error.

1.5   Arrays of numbers

A feature specific to MATLAB is that one single command is sufficient to obtain the values of a trigonometric function for several arguments. This propriety is also valid for exponential and logarithmic functions. Thus, returning to the **tand** example, we could have obtained the three values by typing

```
≫ tand([ 0 45 90 ])
ans =
  0 1.0000 Inf
```

We grouped the three angle values between square brackets creating thus a data type called **array of numbers**. In this case the result is also an array. We can give an array a name and call it by name in various operations. Try for yourself

```
≫ A = [ 0 45 90 ];
≫ tand(A)
ans =
  0  1.0000  Inf
≫ B = [ 1 10 100 ];
≫ log10(B)
ans =
  0  1  2
```

The elements of an array can be separated by blank spaces, as above, or by commas, as below

```
≫ A = [ 0, 45, 90 ]
A =
  0 45 90
```

In our example the elements of the array are displayed as a single row. In many applications it is usual to call such an array **row vector**.

Exercise 1.4 Array of angles

Define an array whose elements are the angles

$$0°, 10°, 20°, 30°, 45°, 60°, 75°, 90°$$

and calculate the cosines of these angles. Let the name of the array of angle values be x, and that of the array of cosine values, y.

Warning. Do not try at this stage to use arrays of numbers for arithmetic operations. There is a catch to this application and we must wait until we learn more about arrays.

1.6 Using MATLAB for plotting

We know now that MATLAB can do what a hand-held calculator can do, and much more, and usually more easily. We are approaching the stage in which we can compare MATLAB with more advanced instruments, namely programmable hand-held computers. Today many of these have also *plotting* facilities. In this respect MATLAB is by far superior and we know already enough to give a simple example.

Let us **plot** the function $y = \sin x$ in the interval 0 to 2π radians. The sequence of commands is

```
≫ x = 0:  pi/60:  2*pi;
≫ y = sin(x);
≫ plot(x, y), grid
≫ title('Sine function')
≫ xlabel('x, radians')
≫ ylabel('sin(x)')
```

The command in the first line generates the sequence

$$0, \ \pi/60, \ 2\pi/60, \ \ldots, \ 2\pi$$

This sequence is stored under one name, x. We created an **array** of numbers, but, this time, instead of specifying each element in part, we used the fact that the elements of the array are equally spaced. The scheme used in the first line is

  array name = first element: increment: last element;

Mind the separation by colons, ':'.

The command in the second line, y = sin(x), produces for each number in x the corresponding sine value and stores it in the array y. The commands in the third line plot y against x, and draw a grid that allows us to read easier the graph. There are here two commands separated by a comma. The first command, **plot**, connects with straight-line segments the points with coordinates (x, y). The second command, **grid**, produces two sets of equally-spaced straight lines, one horizontal, the other vertical. MATLAB allows also for more sophisticated grids, some of which we are going to encounter later in this book.

The command in the fourth lines generates the title **Sine function**. The text of the title must be written between right quotes, ' ' (third row from bottom, at right, in the QWERTY keyboard), to tell MATLAB that it has to deal with a *string of characters*. A character string is an ordered set of symbols. In MATLAB these can be Latin or Greek letters and some symbols belonging to LaTeX, a high-quality document production system. While we can apply mathematical operations on numerical data, we cannot do the same with character strings. For the moment we shall use character strings only to display information, later we shall gradually learn that we can do with them more complex and intelligent things.

The last two commands label the x and the y axes. Again, the texts of the labels are written between quotes because they are character strings. The resulting graph is shown in Figure 1.3.

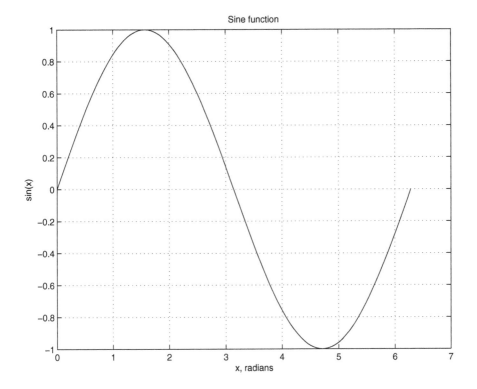

FIGURE 1.3: A plot of the sine function

Exercise 1.5 Plotting a cosine function

Instead of plotting the sine function as above, change the command to plot the cosine function.

Exercise 1.6 Plotting two functions

MATLAB allows us to plot two or more curves by one command. To exercise this facility, repeat the commands in Section 1.6, but instead of the third line write

```
≫ y = sin(x);
≫ z = cos(x);
≫ plot(x, y, x, z), grid
```

The two lines appear in the plot in two colours. We shall show immediately how to add annotations on the graph to make clear the meaning of each line.

Exercise 1.7 Plotting against angles measured in degrees

Repeat the plot described in Section 1.6 and completed in the previous exercise, but use now functions that receive the arguments in degrees. Obviously, the plot will cover the range $0°$ to $360°$.

1.6.1   Annotating a graph

If you executed the two plots proposed above, you obtained two curves; they have different colors, but you can know which one represents the sine and which one the cosine functions only because you know, for example, that $\sin 0 = 0$ and $\cos 0 = 1$. Obviously, this is inconvenient. To distinguish between different lines in more complex plots we may not wish to work like detectives. We do not need to because MATLAB provides for several possibilities to annotate graphs. First, let us tell MATLAB to use line colors and types of our choice. For example, let us plot the sine function as a black, solid line, and the cosine function as a red, dashed line. Next, let us add a *legend*. The necessary commands are

```
≫ x = 0:  pi/60:  2*pi;
≫ y = sin(x);
≫ z = cos(x);
≫ plot(x, y, 'k-', x, z, 'r--')
≫ xlabel('x, radian')
≫ legend('sinx', 'cosx')
```

In the fourth command, the string 'k-' means 'black, solid line'. Mind the use of the letter 'k', and not 'b', which is reserved for *blue*. The string 'r--' means 'red, dashed line'. In the command **legend** we supplied the two strings we want to display in the legend, in the order in which we plotted the curves. The resulting graph is shown, in black and white, in Figure 1.4. To see it in colors repeat the commands for yourself. To learn more on this subject type **doc print**.

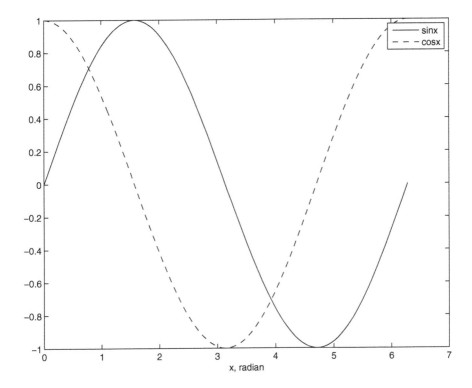

FIGURE 1.4: Annotated plots of two functions

1.7 Format

In the previous subsection we met the constant name `pi` and could read its value with four decimal digits. The *transcendental* number π has an infinite number of decimal digits. As we shall learn later, the computer cannot store an infinite number of digits. Moreover, we can neither write nor read an infinite number of digits. We can, however, ask for more *precision*. To do so in MATLAB we change the display format

```
≫ format long
pi
ans =
   3.14159265358979
```

We can return to the initial format as follows

```
≫ format short
pi
 ans =
 3.1416
```

Other MATLAB formats can be found by using the help facilities. Type, for example, **help format** or **doc format**. We shall discuss some of these possibilities when need will arise.

1.8   Arrays of numbers

1.8.1   Array elements

Let us repeat the first two commands in Section 1.6.

```
≫ x = 0:  pi/60:  2*pi;
≫ y = sin(x);
```

As already explained, we generated two *arrays* of numbers. Any element of an array is identified by its **index**. Indices are positive integers (*natural numbers*) and the first index in an array is always 1. MATLAB does not allow for other numbers as indices. To find the first elements of the arrays x and y we type

```
≫ x(1), y(1)
ans =
 0
ans =
 0
```

To find the fifteenth elements we enter

```
≫ x(15), y(15)
ans =
 0.7330
 ans =
 0.6691
```

and to find the last elements we type

```
≫ x(end), y(end)
ans =
  6.2832
ans =
 -2.4493e-016
```

The last answer should have been 0. Instead of this, due to numerical errors, we obtain a very small number.

The command **length** yields the number of elements in an array like those introduced in this chapter.

```
≫ length(x), length(y)
ans =
  121
ans =
  121
```

We can extract several elements of an array by specifying a range of indices. For example, below we retrieve the first five elements of the arrays x, y and store them in the arrays xi, yi

```
≫ xi = x(1:5), yi = y(1:5)
xi =
     0 0.0524 0.1047 0.1571 0.2094
yi =
     0 0.0523 0.1045 0.1564 0.2079
```

Exercise 1.8 Zooming a plot

To see only a part of the plot in Figure 1.3, for instance the first quarter of it, you may extract the first thirty-one elements of the arrays x and y,

```
≫ xi = x(1:31); yi = y(1:31)
```

and plot yi against xi.

1.8.2 Plotting resolution

The command **plot** does not produce a smooth curve; it draws only straight-line segments between the points indicated in the commands. In the examples shown above we noted the coordinates of the points by x and y. The distance between two points on the x-axis was $\pi/60$. The points were so close that we had the impression that the sine curve was smooth. If we increase the distance between points, we can see that the points of the resulting array y

are, in fact, connected by straight-line segments. Retry the plot for yourself, playing with intervals like $\pi/16$ and $\pi/8$.

1.8.3   Array operations

In the first part of this chapter we learned how to apply arithmetic operations over the set of numbers we work with on a computer. Later on, in Chapter 5, we are going to learn more about the nature of this set. Now we want to show how to extend to arrays the operations we applied to numbers. In this subsection we shall talk about operations applied to each element of an array in part; let us say, *element by element.*

The addition of two arrays produces a third array whose elements are the sums of the elements of the first two arrays. Formally, for two arrays

$$\mathbf{A} = [a_1, \ a_2, \ \ldots \ a_n]$$
$$\mathbf{B} = [b_1, \ b_2, \ \ldots \ b_n]$$

their sum is defined as

$$\mathbf{A} + \mathbf{B} = [a_1 + b_1, \ a_2 + b_2, \ \ldots \ a_n + b_n]$$

As an example, let us generate the array of the first nine natural numbers, in natural order, and the array of the same numbers in inverse order, and add the two arrays

```
≫ A = 1:9
A =
   1  2  3  4  5  6  7  8  9
≫ B = 9:  -1:  1
   B =  9  8  7  6  5  4  3  2  1
≫ A + B                                            ans =
   10  10  10  10  10  10  10  10  10
```

In the first command, `A = 1:9`, we specified no increment. Then, by default, MATLAB uses the increment 1. In the second command, `B = 9: -1:  1`, we specified a negative increment, that is a *decrement.*

The difference of the two vectors, A and B shown above, is by definition

$$\mathbf{A} - \mathbf{B} = [a_1 - b_1, \ a_2 - b_2, \ \ldots, \ a_n - b_n]$$

```
≫ B - A
ans =
   8   6   4   2   0   -2   -4   -6   -8
≫ A - A
ans =
   0   0   0   0   0   0   0   0   0
```

The *array multiplication*, or *multiplication element by element*, of the vectors A and B is defined as

$$\mathbf{A}.*\mathbf{B} = [a_1 b_1,\ a_2 b_2,\ \ldots,\ a_n b_n]$$

Note the *array operator* .*

```
≫ A.*B
ans =
    9   16   21   24   25   24   21   16   9
```

The *array division*, or *division element by element* of the vectors A and B is defined as

$$\mathbf{A}./\mathbf{B} = [a_1/b_1,\ a_2/b_2,\ \ldots,\ a_n/b_n]$$

```
≫ A./B
ans =
   0.1111   0.2500   0.4286   ...      9.0000
≫ B./A
ans =
   9.0000   4.0000   2.3333   1.5000 ...     0.1111
```

Note here the *array operator* '. /'.

To square the elements of an array we can use array multiplication. Thus, to obtain the table of squares of the first ten natural numbers we can proceed as follows

```
≫ C = 1:10
C =
   1   2   3   4   5   6   7   8   9   10
≫ C.*C
ans =
   1   4   9   16   25   36   49   64   81   100
```

MATLAB, however, provides the possibility of squaring element by element using an operator consisting of the point, '.', followed by the power symbol, '^',

```
≫ C.^2
ans =
    1   4   9   16   25   36   49   64   81   100
```

This option can be used with any power. As an illustration of squaring element by element we check below the relationship between squares of sines and cosines for five angle values

```
≫ alpha = 0:   45:   180
alpha =
    0   45   90   135   180
≫ sind(alpha).^2 + cosd(alpha).^2
ans =
    1   1   1   1   1
```

As an application of array division we check the relationship between tangents, sines and cosines for the same angle values

```
≫ D = tand(alpha)
D =
    0   1.0000   Inf   -1.0000   0
≫ E = sind(alpha)./cosd(alpha)
 Warning:  Divide by zero.
E =
    0   1.0000   -Inf   -1.0000   0
```

As the display above uses only four decimal digits, we cannot be sure that there are no surprises in the following digits. Of course, we can improve our check by changing to **format long**. Then, we still do not know what happens beyond the sixteenth decimal digit. Therefore, it is better to check the difference $D - E$

```
≫ D - E
ans =
    0   0   Inf   0   0
```

MATLAB provides many other functions that work on arrays. In Examples 1.1 and 1.2 we introduce the function sum.

1.9 Writing simple functions in MATLAB

By this time we know that MATLAB enables us to work with many built-in functions. One of the great features of MATLAB is that it also enables us to write our own functions and to use them whenever we need. In this section we give a very simple example.

Some time before 1720, the German physicist Daniel Gabriel Fahrenheit (1686-1736) chose as the zero of the temperature scale the temperature of an ice-salt mixture that corresponded to the lowest temperature that occurred in Danzig (today Gdansk) in 1709. In this scale the temperature of the melting ice corresponds to 32 degrees and is denoted by 32°. In the same scale the temperature of boiling water, at sea level, is 212°F.

In 1742 the Swedish astronomer Anders Celsius (1701-1744) proposed another temperature scale in which the melting point of ice corresponds to zero degrees, and the boiling point of water to 100. These temperatures are denoted by 0°C and 100°C.

In the USA the meteorological bulletin still refers to temperatures measured in degrees Fahrenheit, while over all Europe the temperatures are given in degrees Celsius. As temperature scales are linear, it is easy to check that the following formula corresponds to the data mentioned above

$$C = \frac{5}{9}(F - 32) \tag{1.1}$$

Here C is the temperature in degrees Celsius, and F, the temperature in degrees Fahrenheit.

We can use Formula 1.1 each time we have to convert from Fahrenheit to Celsius degrees. If we have to perform frequently the conversion, writing each time the commands defined by Equation 1.1 would be annoying and a waste time. MATLAB allows us to write a corresponding **MATLAB function** and call it each time we need. A possible programme code is

```
function    C = F2C(F)

%F2C    Converts degrees Fahrenheit to degrees Celsius
%       input:  temperature in degrees Fahrenheit
%       output: same temperature in degrees Celsius

C = 5*(F - 32)/9;
```

MATLAB functions are written on files with the extension 'M'. Therefore, in MATLAB jargon such functions are called **M-files**. MATLAB functions are, in fact, programmes that are called with **input arguments** and yield **output arguments**.

Let us analyze the syntax of the function F2C. First, the function has a name, in this case F2C, an abbreviation for *Fahrenheit to Celsius*. As we shall see throughout this book, many built-in MATLAB functions use the character '2' as a mnemonic way of saying 'to'. Names of MATLAB functions must begin with a letter. The line

```
function    C = F2C(F)
```

is the *function definition line*; it informs MATLAB that the file shall be interpreted as a function, that it should be called with one input argument, F, and that it yields one output argument, C.

The following line is

```
%F2C    Converts degrees Fahrenheit to degrees Celsius
```

This line is called *H1 line* because it is the first **help** line. The H1 line is not an *executable statement* as it has no effect on the execution of the function. To make the line not executable, we begin it with the percent sign, '%'. So the following lines are called *help lines*. They simply explain how to use the function. As we shall learn more in this book, such lines, known as **comments**, are helpful to the programmer. We shall see in this section how they are also helpful to the user.

The last line

```
C = 5*(F - 32)/9;
```

is the **body** of the function. In this simple case the body consists of one line only. In most cases the body has more lines.

To write the above listing on file, open the MATLAB **editor** by clicking on the leftmost symbol on the toolbar of the window, as shown by the arrow labeled '1' in Figure 1.5. Figure 1.6 shows the editor window with the complete listing. Each line receives a number that is useful when correcting the file. Later on we shall talk more about this. To save the file click on the **File** pull-down menu, then on **Save as**. A dialogue box opens; it allows you to define the file name and type. By default, the box shows the name of the function and the type m. This is correct, so click on **Save**. To close the file click on **File**, at the left of the menu bar, then on **Close Editor**. Alternatively, you may click on the **Editor** icon, in the left, upper corner of the window. In the pull-down menu that opens click on **Close**. In both cases you will be prompted to confirm that you want to save the last changes. If you want to open again the file, and you did not edit many other files in the meantime, use the pull-down menu **File** to find the name F2C. In general, you can open any file in the editor by clicking in the toolbar, on the second icon from the left. In Figure 1.6 it is marked 2.

Examples of calling the function F2C are

```
≫F2C(32)
ans =
 0
≫ F2C(212)
ans =
 100
≫ F2C(80)
ans =
 26.6667
```

Try now

```
≫ help F2C
 F2C Converts degrees Fahrenheit to
  degrees Celsius
   input:  temperature in degrees Fahrenheit
   output:  same temperature in degrees Celsius
```

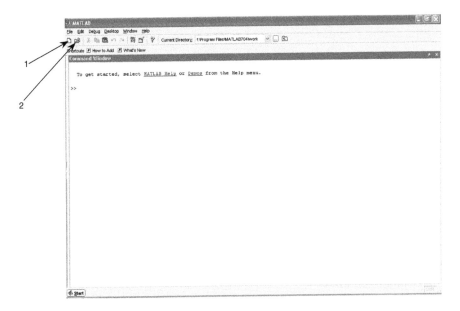

FIGURE 1.5: Opening a new file

We see that the help lines explain how to use the function. As to the H1 line, it has an additional function. Let us suppose, for example, that we do not know the name of the function F2C, but know that it refers to 'Fahrenheit'. We can use then the lookfor facility

```
≫lookfor Fahrenheit
F2C  Converts degrees Fahrenheit to degrees Celsius
```

After finding this result, MATLAB continues to search for additional occurrences of the word 'Fahrenheit'. To stop the search press Ctrl+C.

FIGURE 1.6: Writing the function F2C on file

We have given here a very simple example that can, however, show the advantage of building our own functions for jobs that we must repeat. The advantage in this case is limited. Gradually we are going to build more complex functions, for more complex problems, and then the advantage will be greater and more obvious.

Exercise 1.9 Celsius to Fahrenheit conversion

Write a function that converts temperatures measured in Celsius degrees to temperatures measured in Fahrenheit degrees.

1.10   Summary

Table 1.1: Arithmetic operations

Operation	Operator	Example
Addition	+	2 + 2
Subtraction	-	4 -2
Multiplication	*	2*2
Division	/	4/2
Square root	sqrt	sqrt(4)
Power	^	2^3

To perform arithmetical operations in MATLAB we use the symbols listed in Table 1.1. When the same number must be used several times, it is recommended to store it as a constant. Thus, we spare work and avoid input errors. We can retrieve the names of the stored constants by typing who, and obtain more details with whos. All this information can be also found in the *Workspace* window (1 in Figure 1.1).

Table 1.2: Trigonometric and inverse trigonometric functions

Function	Trigonometric		Inverse	
Argument	Degrees	Radians	Degrees	Radians
sine (sin)	sind	sin	asind	asin
cosine (cos)	cosd	cos	acosd	acos
tangent (tan)	tand	tan	atand	atan
				atan2
cotangent (cot)	cotd	cot	acotd	acot
secant (sec)	secd	sec	asecd	asec
cosecant (csc)	cscd	csc	acscd	acsc

The set of trigonometric and inverse trigonometric functions provided by MATLAB is listed in table 1.2. For each function there is a variant that accepts arguments in radians, and one that accepts arguments in degrees.

Exponential and logarithmic functions can be calculated with the functions listed in Table 1.3.

We use *arrays* to store several numerical values under one name. Thus, to

Table 1.3: Exponentials and logarithms

Operation	Command	Example	Meaning
Exponential function	exp	exp(7)	e^7
Natural logarithm	log	log(2)	$\log_e 7$
Common logarithm	log10	log10(10)	$\log_{10} 10$
Base 2 logarithm	log2	log2(8)	$\log2(8)$

Table 1.4: Array operations

Operation	Operator	Example
Addition	+	A + B
Subtraction	-	A - B
Multiplication	.*	A.*B
Division	./	A./B
Power	.^	A.^2

create in MATLAB the array of numbers

$$\mathbf{A} = [10,\ 20,\ 30,\ 45,\ 60,\ 75,\ 90]$$

we write

```
>> A = [ 10  20  30  45  60  75  90 ]
```

The array is displayed in one line. It is usual to call such an array *row vector*. When the array elements are equally spaced, we can generate the array with the command

    B = a: b: c;

where a is the first element, b, the spacing, and c, the last element. The spacing, b, can be any number, positive or negative. When $b = 1$, we can shorten the command to B = a: c;.

Any element of an array is identified by its *index* , which is the place of the element in the array. Thus, for the example given above

$$\mathbf{A}(1) = 10,\ \mathbf{A}(4) = 45,\ \mathbf{A}(end) = 90$$

The command length(A) yields the number of elements in the array $\mathbf{A}$. Given two arrays

$$\mathbf{A} = [a_1,\ a_2,\ \dots\ a_n]$$
$$\mathbf{B} = [b_1,\ b_2,\ \dots\ b_n]$$

their sum is defined as

$$\mathbf{A} + \mathbf{B} = [a_1 + b_1,\ a_2 + b_2,\ \ldots\ a_n + b_n]$$

Similarly, the difference of the two arrays is defined as

$$\mathbf{A} - \mathbf{B} = [a_1 - b_1,\ a_2 - b_2,\ \ldots,\ a_n - b_n]$$

their product element-by-element (*array multiplication*) as

$$\mathbf{A}.\ast\mathbf{B} = [a_1 b_1,\ a_2 b_2,\ \ldots,\ a_n b_n]$$

and their *array division*, or *division element by element*, as

$$\mathbf{A}./\mathbf{B} = [a_1/b_1,\ a_2/b_2,\ \ldots,\ a_n/b_n]$$

The symbols for array operations are listed in Table 1.4. The function **sum** introduced in examples 1.1 and 1.2 yields the sum of the elements of an array

$$sum(A) = a_1 + a_2\ \ldots,\ a_n$$

The commands introduced in this chapter are

exit terminates the MATLAB session.

format long after entering this command, the screen displays all results in scaled fix point format with 14 to 15 decimal digits.

format short after entering this command, the screen displays all results in scaled fix point format with four decimal digits.

help the command **help filename** causes the display of the comments written in the beginning of the file called **filename**.

legend the command **legend('string1', 'string2', ...)** displays a box showing the line types used in a given plot, followed each by the explanation contained in **'string1'**, **'string2'**, etc.

length given an array A, the command **length(A)** displays the number of its elements.

lookfor for example, **lookfor vector** looks for the word 'vector' in all the help lines of the existing M-files and displays the names of the files that contain the given keyword. Try this command for yourself; you will get a lot of answers.

plot **plot(x1, y1, 'LineSpec1', x2, y2, 'LineSpec2', ...** plots $x1$ against $y1$ with the line type and color specified in the string 'LineType1', $x2$ against $y2$ with the line type and color specified in the string 'Line-Type2', a.s.o.

quit terminates the MATLAB session.

sqrt `sqrt(A)` returns the square roots of the elements of the array A.

title `title('string')` displays 'string' at the plot top.

ver displays the version numbers of all MathWorks products installed on the computer.

xlabel `xlabel('xstring')` labels the $x-$axis with `xstring`.

ylabel `ylabel('ystring')` labels the $y-$axis with `ystring`.

1.11   Examples

EXAMPLE 1.1 Spring in series

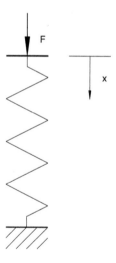

FIGURE 1.7: The sketch of a helicoidal spring

The helicoidal spring is a frequently used machine element. Figure 1.7 shows a schematic representation of a helicoidal spring. When the spring is subjected to a force F that acts as shown, the spring length is shortened by a distance x. If the force F acts in the opposite sense, the spring extends a length x. Within a certain range, the relationship between the force, F, and the compression or extension, x, is linear and can be represented by the equation

$$F = kx \qquad (1.2)$$

where k is called **spring constant**. In most cases helicoidal springs are designed to work in the linear range. If the force, F, is measured in N, and the displacement, x, in mm, the units of the spring constant are $rmN \ mm^{-1}$. Later in this book we are going to meet mechanical systems comprising a spring (see Section 11.5). For the moment we consider in Figure 1.8 three springs connected in series. We ask the question: is it possible to replace the three springs by one spring that behaves equivalently? The answer is 'Yes'. Let us remark that:

1. the force acting on each spring is F;

2. the total compression, x, is the sum of the individual compressions of the three springs:

$$x = \frac{F}{k_1} + \frac{F}{k_2} + \frac{F}{k_3}$$

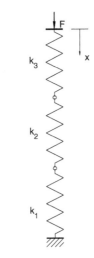

FIGURE 1.8: Three helicoidal springs in series

If we write k_{eq} for the spring constant of the equivalent spring constant, we obtain

$$\frac{F}{k_{eq}} = \frac{F}{k_1} + \frac{F}{k_2} + \frac{F}{k_3} \qquad (1.3)$$

which yields

$$\frac{1}{k_{eq}} = \frac{1}{k_1} + \frac{1}{k_2} + \frac{1}{k_3} \tag{1.4}$$

It is easy to generalize this result for any number of springs and state that *the equivalent spring constant of a series connection of springs is the* **harmonic mean** *of the individual spring constants.* As an example let us assume the following data for Figure 1.8

$$F = 0.5\text{N}, \quad k_1 = 0.05 \text{ N mm}^{-1}, \quad k_2 = 0.08 \text{ N mm}^{-1}, \quad k_3 = 0.04 \text{ N mm}^{-1}$$

In MATLAB we calculate the equivalent spring constant with the commands

```
≫ F = 0.05;
≫ k = [ 0.05 0.08 0.04 ];
≫ Keq = 1/sum(1./k)
Keq =
  0.0174
x = F/Keq
  2.8750
```

It may be interesting to calculate the individual compressions of the three springs

```
≫ xi = F./k
xi =
  1.0000 0.6250 1.2500
```

and check

```
≫ sum(xi) - x
ans =
  0
```

In Example 1.2 and Exercises 1.18 and 1.19 we shall find other applications of the harmonic mean. The Statistics Toolbox contains a function, `harmmean`, which solves the problem. As this function does not exist in the basic MATLAB package, in Exercise 1.17 we ask the reader to program the function.

EXAMPLE 1.2 Series and parallel connections of resistors

Figure 1.9 shows three resistors, R_1, R_2, R_3, connected in series with a dc source, E. The same current, i, passes through all three resistors. We note with E also the voltage of the dc source, and with R_1, R_2, R_3, the resistances

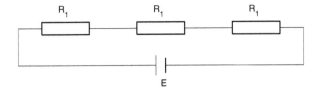

FIGURE 1.9: Three resistors connected in series

of the three resistors. According to Ohm's law, the voltage drops across the individual resistors are

$$R_1 i, \; R_2 i, \; R_3 i$$

The sum of the three voltage drops equals the voltage of the dc source

$$E = R_1 i + R_2 i + R_3 i$$

It appears that the three resistors can be replaced by an equivalent resistor whose resistance is

$$R_{eq} = R_1 + R_2 + R_3 \tag{1.5}$$

so that

$$i = \frac{E}{R_{eq}}$$

It is easy to generalize the above result and state that *the equivalent resistance of a number of resistors connected in series is equal to the sum of the individual resistances.*

As an example let us assume the values, $E = 12/;\mathrm{v}$, $R_1 = 100$ ohms, $R_2 = 200$ ohms, $R_s = 300$ ohms. In MATLAB we calculate the equivalent resistance, the current intensity, and the individual voltage drops with the commands

```
≫ E =12;
≫ R = [ 100 200 300 ];
≫ Req = sum(R)
Req =
  600
≫ i = E/Req
i =
  0.0200
≫ u = i*R
u =
  2 4 6
```

We can check that the sum of the individual voltage drops equals the voltage of the dc source

```
>> sum(u) - E
ans =
    0
```

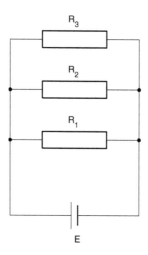

FIGURE 1.10: Three resistors connected in parallel

What happens if the resistors are connected in parallel, as in Figure 1.10? In this case the voltage across any resistor is equal to E, while the current intensities in the three resistors are equal to

$$E/R_1, \ E/R_2, E/R_3$$

According to Kirchoff's current law, the total current, i, through E equals the sum of the three currents and we can write

$$i = E/R_1 + E/R_2 + E/R3$$

If we note by R_{eq} the equivalent resistance that can replace the three given resistors, we can write

$$i = \frac{E}{R_{eq}} = E/R_1 + E/R_2 + E/R3$$

The obvious conclusion is

$$\frac{1}{R_{eq}} = \frac{1}{R_1} + \frac{1}{R_2} + \frac{1}{R_3} \tag{1.6}$$

It is easy to generalize the above result and state that *the equivalent resistance of a number of resistors connected in parallel is equal the the harmonic mean of the individual resistances.* See also Example 1.1.

For a numerical example we use the same values as for the series connection of resistors. Then, we calculate the equivalent resistance and the total current with the MATLAB commands

```
≫ R = [ 100 200 300 ];
≫ E = 12;
≫ Req = 1/sum(1./R)
Req =
  54.5455
≫ it = E/Req
it =
  0.2200
```

We can calculate the currents through the three resistors

```
≫ i = E./R
i =
  0.1200 0.0600 0.0400
```

and check that their sum equals the total current

```
≫ sum(i)
ans =
  0.2200
```

EXAMPLE 1.3 Load matching in dc

Figure 1.11 shows a circuit containing a dc source, E, with internal resistance R_i, and a load of resistance R_L. It is known that the power, P, dissipated in the load, R_L, is maximum when the load resistance equals the internal resistance of the source. To check this result on an example we assume $E = 12$ V, and $R_i = 8$ ohms. Let us calculate the power P for a range of R_L values embracing 8 ohms and plot the power against the load resistance. We do this with the commands

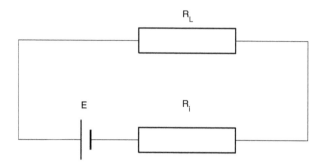

FIGURE 1.11: A dc resistor circuit

```
≫ E = 12;
≫ i = 8;
≫ RL = 0:0.1:20;
≫ i = E./(Ri + RL);
≫ P = RL.*i.^2;
≫ plot(RL, P), grid
≫ xlabel('RL, ohm')
≫ ylabel('P, W')
```

The resulting graph is shown in Figure 1.12 and the maximum appears, indeed, where the load resistance is equal to the internal resistance of the voltage source.

Proceeding as above we have shown, indeed, that the power transmitted to the load has a maximum when the load resistance is equal to the internal resistance of the voltage source. We have done this, however, for a particular case and rely on induction for generalization. There is a better way of illustrating the result. By using **nondimensional** quantities we go directly to the general result.

We begin by finding the expression of the current passing through the circuit

$$i = \frac{E}{R_i + R_L}$$

Next we calculate the power transmitted to the load

$$P = i^2 R_L = (\frac{E}{R_i + R_L})^2 R_L = \frac{E^2}{R_i} \cdot \frac{R_l/R_i}{(1 + R_L/R_i)^2}$$

Letting $r = R_L/R_i$, and dividing both sides of the equation by E^2/R_i, we obtain

$$P\frac{R_i}{E^2} = \frac{r}{(1 + r)^2} \tag{1.7}$$

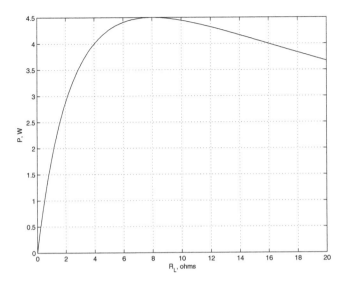

FIGURE 1.12: Power in the dc resistor circuit

The ratio r is *nondimensional* by construction. The quantity E^2/R_i is a power; it is called *the short-circuit power* and it is a characteristic of the voltage source. In other words, E^2/R_i is a constant of the given circuit. Thus, the maximum of the nondimensional quantity on the left side of Equation 1.7 and the maximum of the transmitted power, P, occur at the same value r. The derivative of the right side of Equation 1.7 respective to r is $(1-r)/(1+r)$, yielding one, and only one extremum at $r=1$. Now, we show that this is a maximum by plotting the right side of Equation 1.7 against the ratio r, in a range around $r=1$. The MATLAB commands are

```
≫ r = 0:  0.1:  2;
≫ y = r./((1 + r).^2);
≫ plot(r, y), grid
≫ xlabel('Resistance ratio, r = R_L/R_i')
≫ ylabel('Non-dimensional power, r/(1 + r)^2')
```

In the fourth line the underscore is the TEX command for producing subscripts. For example, typing 'R_L' prints R_L in the label of the x−axis. The resulting graph is shown in Figure 1.13.

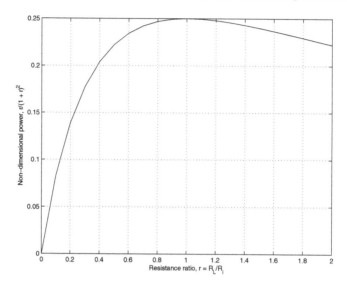

FIGURE 1.13: Nondimensional power in the dc resistor circuit

1.12   More exercises

Exercise 1.10 The base of natural logarithms

The base of natural logarithms is the number $e = 2.71828\ldots$. Try to obtain this number as exponential of 1, in format **long**. The number e is defined as

$$e = \lim_{n\to\infty} \left(1 + \frac{1}{n}\right)^n$$

Use this formula in MATLAB with $n = 50$, 150, 250, 350, and see how you obtain values approaching the exact value. Do not type the formula several times, but retrieve it with the ↑ key. Can you devise a way of writing the above $n-$values in an array and call the formula only once?

Exercise 1.11 Foot-to-meter conversion

A length of one meter (1 m) corresponds to 3.28 feet (3.28 ft) in the old British system of units still used in the USA. Write a function, **Ft2m**, that converts lengths measured in feet into lengths measured in meters.

Exercise 1.12 Meter-to-foot conversion

Write a function, **m2ft**, that converts lengths measured in meters into lengths measured in feet.

Exercise 1.13 Pound-to-kilogram conversion

A mass that measures one pound (1 lb) in the old British system, still in use in the USA, corresponds to 0.454 kg in the metric system. Write a function, **lb2kg** that converts mass values in pounds into mass values in kilograms.

Exercise 1.14 Kilogram-to-pound conversion

Write a function, **Kg2lb**, that converts mass values measured in kilograms to mass values measured in pounds.

Exercise 1.15 Converting minutes and seconds of angle

The angle that subtends a full circumference is usually divided into 360 *degrees*. Thus, this angle equals $360°$, and the right angle, $90°$. Traditionally, the degree is divided into 60 *minutes*, that is $1° = 60'$, and the minute is divided into 60 *seconds*, that is $1' = 60"$. Geographical coordinates are given in these units. On scientific calculators and in MATLAB we can enter angle values in degrees and decimal fractions of degrees. Then, the sine of $30°30'$ is calculated as `sin(30.5)`, and the cosine of $30°30"$ is calculated as `sin(30 + (30/60)/60)`. Write a function, **ms2deg**, that converts angles measured in degrees, minutes and seconds, to angles measured in decimal fractions of degrees. The function should be called as `ms2deg(d, m, s)`, where d is the number of degrees, m, the number of minutes, and s, the number of seconds.

Exercise 1.16 Converting nautical miles

The *nautical mile* is the average length of an arc of meridian of 1 minute. The standard value is 1852 m. Write a function, `nm2km`, that converts distances given in nautical miles, into distances in km.

Exercise 1.17 Harmonic mean

Write a function, `harmmean`, that accepts as input argument an array of m numbers $n = [n_1\ n_2 \dots n_m]$, and yields their harmonic mean

$$m_h = \frac{1}{1/n_1 + 1/n_2 + \dots 1/n_m}$$

Use this function to solve the problems in Examples 1.1 and 1.2.

Exercise 1.18 Parallel and series connections of inductors

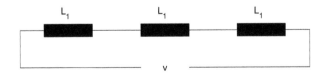

FIGURE 1.14: Three inductors connected in series

Figure 1.14 shows three inductors connected in series, while Figure 1.15 shows three inductors connected in parallel. Let the inductances of the three inductors be $L_1 = 200\ \mu H$, $L_2 = 300\ \mu H$, $L_3 = 400\ \mu H$. For each connection calculate the equivalent inductance and verify that the total voltage and the total current correspond to those predicted from the voltages across and the currents through the individual inductors (as a model see Example 1.2). .

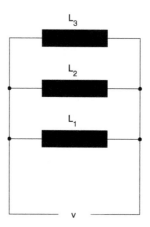

FIGURE 1.15: Three inductors connected in parallel

Hint. In a series connection of n inductances the current, i, passing through each of them is the same, and the voltage drop across each inductance is

$$v_i = L_i \frac{di}{dt}$$

and the sum of voltage drops equals the voltage, v, at the terminals of the connections.

In a parallel connection of inductances the voltage across each inductor is the same and equals the voltage at the terminals of connection, v, while the sum of currents through the individual inductors equals the current at the terminals of the connection.

Exercise 1.19 Parallel and series connections of capacitors
Figure 1.16 shows three capacitors connected in series, while Figure 1.17 shows three capacitors connected in parallel. Let the capacitances of the three capacitors be $C_1 = 2\ \mu F$, $C_2 = 4\ \mu F$, $C_3 = 6\ \mu F$. For each connection calculate the equivalent capacitance and verify that the total voltage and the total charge across each connection correspond to those predicted from the voltages across and charges of the individual capacitors (as a model see Example 1.2).

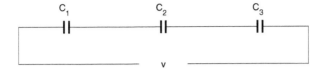

FIGURE 1.16: Three capacitors connected in series

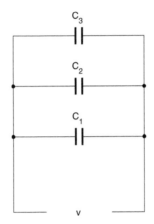

FIGURE 1.17: Three capacitors connected in parallel

Hint. The electric charge, q, of a capacitor with capacitance C can be related to the voltage across the capacitance, u, by

$$q = Cu$$

When several capacitors are connected in series, as in Figure 1.16, the voltage across the connection is equal to the sum of the voltages across the individual capacitors

$$v = u_1 + u_2 + u_3$$

while all capacitors have the same charge

$$q = q_1 + q_2 + q_3$$

We are looking for an equivalent capacity, C_{eq}, such that

$$q = C_{eq}v \tag{1.8}$$

When the capacitors are connected in parallel, as in Figure 1.17, the voltages across each capacitor are equal to the voltage across the whole connection

$$v = u_1 + u_2 + u_3$$

and the total charge of the connection is the sum of the charges of the individual capacitors

$$q = q_1 = q_2 + q_3$$

We are again looking for an equivalent capacitance that satisfies Equation 1.8.

2

Vectors and matrices

In the previous chapter we learned how to store a set of numbers under one name and how to operate on them by using that name. We introduced thus the notion of a one-dimensional *array*. In this chapter we are going to generalize this notion and its uses. First we show that we can store numbers not only as one line, but also as a column. Next we combine one-dimensional arrays into two-dimensional arrays. The array is the basic data structure of MATLAB. We show that the one-dimensional array allows a natural representation of *vectors*, quantities that have a magnitude, a direction and a sense. The operations defined on vectors are the addition, the multiplication by a scalar, the *scalar*, or *dot product* and the *vector*, or *cross product*. This chapter includes a few applications of vectors in Geometry and in Mechanics.

We call two-dimensional arrays **matrices** when perform with them the *matrix product*. Another notion introduced in this chapter is that of *determinant*. We illustrate applications of these notions in Geometry and Mechanics.

In conclusion, this chapter deals with elementary applications of MATLAB in linear algebra. In fact, MATLAB was initially developed for computations in this discipline.

As to programming in MATLAB, in this chapter we are going to learn how to repeat operations in *for loops*, and how to check the correctness of the iterative scheme. We also learn how to choose between sets of operations by using the *If elseif else* construction. To perform all the operations described in this chapter we introduce additional MATLAB commands, including three commands for enhancing graphics.

In this chapter we program three utility functions that we are going to use throughout the book and that may help the reader to develop further script files or functions. As an example of more complex programming, we develop a function that calls other functions, solves a simple problem in statics, and plots the drawing that explains the problem.

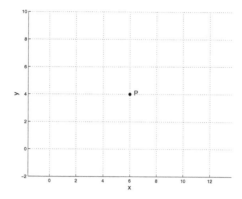

FIGURE 2.1: The point with coordinates $x = 6, y = 4$

2.1   Vectors in geometry

2.1.1   Arrays of point coordinates in the plane

In plane analytical geometry a point is defined by two coordinates; we usually call them the x- and the y-*coordinate*. For example, Figure 2.1 shows a part of the x, y plane and, on it, the point **P** with coordinates $x = 6, y = 4$. As we learned in Chapter 1, we can collect the two coordinates under one name as below

```
≫ P = [ 6; 4 ]
ans =
P =
   6
6  4
```

P is an *array* in the sense defined in Subsection 1.8.1. This time we separated the elements of the array by a semicolon, ';', instead of a comma, ',', or a blank space. Therefore, **B** is displayed as a vertical array. Usually we call such an array a **column vector** and we represent it as

$$B = \begin{bmatrix} 5 \\ 6 \end{bmatrix}$$

Another way of defining a column vector in MATLAB is to type the name of the vector followed by the equality sign, an opening square bracket and the first element:

```
≫ P = [ 6
```

Now press **Enter**. Note that you move to a second line and no prompt appears. Type now the second element and a closing square bracket. The display is

```
4 ]
ans =
P =
6
4
```

We have chosen to define a point as a column vector because in this way we can easily combine points to define a **polygonal line**, and, in later sections, more complex lines. To give an immediate example, let us define three points

```
≫ P1 = [ 1; 2 ];
≫ P2 = [ 2; 4 ];
≫ P3 = [ 5; 3 ];
```

Connecting these points by straight lines we obtain a *polygonal line*. If we close the line by connecting the last point, **P3**, to the first point, **P1**, we obtain a **polygon**. In our case this is the simplest polygon, a *triangle*. Having represented the points by column vectors, we can define the polygon as an object formed by the *juxtaposition*, or *concatenation* of the column vectors

```
P = [ P1 P2 P3 P1 ]
P =
1 2 5 1
2 4 3 2
```

We created thus a *two-dimensional array*, more specifically a *2-by-4 array* called so because it has two rows and four columns. Mind the order: first the number of rows, next that of columns. The numbers *2* and *4* are the *dimensions of the array*. In the usual mathematical notation we write

$$\mathbf{P} = \begin{bmatrix} 1\ 2\ 5\ 1 \\ 2\ 4\ 3\ 2 \end{bmatrix}$$

Any element of a two-dimensional array is identified by two *indices*, the first one indicating the row number, the second, the column number. Thus, for the array defined above,

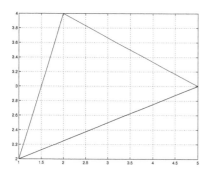

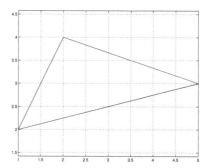

FIGURE 2.2: Plotting a polygonal line

FIGURE 2.3: Plotting a polygonal line at correct scales

```
≫ P(1, 1)
ans =
    1
```

To find the dimensions of an array, we call the function `size` as follows

```
≫ [ m, n ] = size(P)
m =
    2
n =
    4
```

This way of calling a function is new for us. We begin by defining an array
of **output arguments**, m and n. We continue with the equality sign and the
name of the function. The function responds by putting the number of rows
in the element m, and that of columns in n.

To discover the advantage of working with two-dimensional arrays, instead
of separate numbers, let us plot the polygon **P** defined above

```
≫ plot(P(1, :), P(2, :)), grid
```

The notation P(1, :) means *the first row of P, all columns*, that is all
x-coordinates of the given points. Analogously, the notation P(2, :) means
the second row of P, all columns, that is all y-coordinates of the given points.
Thus, we plotted y against x. The resulting plot is shown in Figure 2.2.

A simple look at the figure shows that the x-coordinates are plotted at a different scale than the y-coordinates. By default, MATLAB chooses the scales
so that the plot fills the entire figure window, at the largest possible scales.
Indeed, the corners of the plotted polygon lie on the borders of the figure, but

the polygon is *stretched* in the vertical direction. We can accept such a graph when we plot, for example, a trigonometric function, as in Figure 1.3. We can also accept such a graph when plotting a physical quantity against another physical quantity of a different kind, for example water density against water temperature. This is possible because we do not have in mind any visual representation for such quantities. Things change when we represent geometric objects because we are accustomed to their visual representations. Plotting different axes at different scales produces a distorted image that does not correspond to our experience. Moreover, in engineering we must represent geometric objects so as to be able to appreciate their properties and their relationship to other geometrical objects. For example, how can we appreciate if a given machine element fits into another if their dimensions are not all represented at the same scale? Therefore, in our example we add the command `axis equal` to force the same scale for the x- and y-coordinates

```
≫ plot(P(1, :), P(2, :)), axis equal, grid
```

The resulting plot is shown in Figure 2.3.

The next problem we want to solve with our array representation of points is finding the distance between two points. For instance, let us find the distance between the points $\mathbf{P_1}$ and $\mathbf{P_2}$. Applying Pythagoras' theorem we calculate

$$d = \sqrt{(2-1)^2 + (4-2)^2} = 2.2361$$

To obtain this distance in MATLAB we can calculate the difference of two column vectors in the same way we calculated the difference of two row vectors, and apply Pythagoras' theorem to the elements of the difference vector

```
≫ dP = P2 - P1
dP =
  1
  2
≫ d12 = sqrt(dP(1)^2 + dP(2)^2)
d12 =
  2.2361
```

MATLAB has a built-in function, `norm`, that performs the operation defined in the right side of the command beginning with `d12`. The function `norm` is defined for other, more advanced purposes that we may learn later, but it is also helpful in the simple case treated here

```
>> d12 = norm(dP)
ans =
   2.2361
```

In section 2.1.5 we are going to give an additional meaning to the function *norm*.

2.1.2   The perimeter of a polygon – for loops

In engineering we often have to calculate certain properties of polygons, or of figures that can be approximated by polygons. Such a property is the *perimeter*. The knowledge of the perimeter can be necessary, for example, for calculating the length of a weld that connects a polygonal element to another element. Other examples are the calculation of the length of a fence or the calculation of the area of the developed lateral surface of a prismatic body.

Returning to the triangle shown in Figure 2.3, we can calculate its perimeter by simply adding the distances between the three vertices, that is the lengths of the three sides

```
>> norm(P2 - P1) + norm(P3 - P2) + norm(P3 - P1)
ans =
   9.5215
```

While the above procedure may be acceptable for one case, with few points, it isn't convenient for repeated use, and certainly not for larger numbers of points. To better use the possibilities of a programming language like MATLAB, let us develop a function that performs the operation for any number of points. The solution is simple, but we are going to show how to proceed in a systematic manner that can be used later for more complex problems.

We begin by defining the *input* and the *output*. The input, that is the argument of the function, will be the description of the polygonal line. As shown above, we have chosen for this description a *2-by-n* array, where n is the number of points, in this case the number of vertices. If the object is a polygon, that is a closed polygonal line, the line must be closed by repeating in the array, as the last column, the column vector of the first point (vertex). The output should be the sum of the distances between the points represented in the input array. This is one number.

Next, we choose a strategy for calculating the sum of distances. We shall use for this, as in the preceding subsection, the MATLAB function **norm**. If there are n points, then the function has to call $n-1$ times the function **norm**. For a polygon with m vertices the function must call the function **norm** m times. In other words, the function we are interested in is based on *repetitive* calls of **norm**. We say that each call is an **iteration**. Like other programming languages, MATLAB provides a *repetitive construct* that performs a required number of iterations. This construct is called a **for loop** and its basic form is

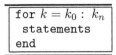

For the moment we assume $k_0 \le k_n$, later we are going to see that this restriction is not necessary. The statements included between the line beginning with `for` and the line `end` are executed for $k = k_0$, $k = k_0 + 1, \ldots k_n$, that is $k_n - k_0 + 1$ times. The statements can include functions of k.

Below is code that calculates the length of the polygonal line using a *for loop*. Write this code in a file with the name `perimeter` and the extension `m`, and note how the MATLAB editor helps in formatting the text.

```
function  s = perimeter(Polyline)

%PERIMETER  Calculates the length of the polygonal line Polyline.
%           For n points (vertices) Polyline is a 2-by-n array.
%           The first line represents the x-coordinates of
%           the n points, the second line, their y-coordinates.
%           If Polyline is a closed polygon, the last point is
%           the repetition of the first point.

s           = 0; % initializes to zero the perimeter length
[ m, n ] = size(Polyline);
for k   = 2: n
          ds = Polyline(:, k) - Polyline(:, k - 1);
          s    = s + norm(ds);
end
```

The first line declares the contents of the file as a function with input argument *Polyline* and output argument *s*. The next six lines, beginning with '%', are *comments* that explain the purpose of the function and how to build its input argument. The first word in the comments is the name of the function, identical to the name of the file. After writing the function, try the command **help perimeter** and see how the function echoes the six comment lines. Comment lines are not executable.

The function **perimeter** creates a variable called *s* and initializes its value to zero. At each iteration of the loop the function will increase the value of the variable *s* by the length of one side of the polygonal line. The MATLAB built-in function **size** is used to determine the number of iterations to be executed. The output argument n, of **size**, is the number of vertices of the polygonal line. The number of segments (sides in the case of a polygon) is $n - 1$, and this is also the number of required iterations. Therefore, the first line of the loop contains the expression k = 2: n. The body of the loop is contained between the line beginning with `for` and the line `end`. As a matter of good style, the body of the loop is *indented*. In this way it is easier to identify the loop and separate it from the rest of the code. As you may have noted while writing the code, the MATLAB editor produces automatically the indentation.

To understand how the loop works, let us *simulate* its behavior when the input is the description of the triangle shown in Figure 2.3. The function is called with the input argument [P1 P2 P3 P1]. The function reads the input and finds that its dimensions are $m = 2$ and $n = 4$. The execution begins with $s = 0$ and $k = 2$. The function calculates

```
ds = Polyline(:, 2) - Polyline(:, 1);
```

afterward the function calculates the norm of *ds*, and adds the result to *s*. At this moment the value of *s* equals the length of the side $\overline{P_1 P_2}$. The value of *k* is incremented to 3. The second iteration of the loop calculates

```
ds = Polyline(:, 3) - Polyline(:, 2);
```

then, the norm of the new *ds* and adds it to the variable *s* making it equal to the length of $\overline{P_1 P_2}$ plus the length of $\overline{P_2 P_3}$. The value of *k* is incremented to 4. The third iteration calculates

```
ds = Polyline(:, 4) - Polyline(:, 3);
```

then, the norm of the new *ds* and adds it to the variable *s* making it equal to the length of $\overline{P_1 P_2}$, plus the length of $\overline{P_2 P_3}$, plus the length of $\overline{P_3 P_1}$. At this moment $k = n$ and the execution of the loop stops. Note that at the end of each iteration, *s* correctly represents the sum of the distances between the points processed up to this point. In computer jargon *s* is called a *loop invariant*.

Now we know that the function works correctly for a polygon with three vertices, that is a triangle. We may see no reason why the function could not work correctly for a polygon with any number of vertices, *n*. However, one may ask for a more convincing, general proof. Such a proof can be obtained by *induction*. To carry on such a proof we must first check that the function works correctly for $n = 1$, that is one vertex. Next, assuming that it works for some *m*, where $1 < m < n$, we must show that it also works for $m + 1$.

When $n = 1$ the polygonal line has only one vertex. The iterations must start with $k = 2$, which is greater than 1. No iteration is executed and *s* remains as initialized, that is equal to zero. This is also the value of the output and it is correct. The distance between a point and itself is, indeed, null.

We assume now that the function worked correctly until $k = m$ and *s* is equal to the sum of distances between the first $m-1$ points. The next iteration increments *k* to $m + 1$ and the function adds to *s* the distance between the points $\mathbf{P}_m$ and $\mathbf{P}_{m+1}$. Thus, the value of *s* equals the length of the line up to the $(m + 1)$th point and this is what we had to prove.

We developed the function `perimeter` starting from the function `norm`. Next, we decided to use it in a loop. This way of designing software, starting from a building block and ending up with the whole function, or program, is called *bottom-up*. Later we shall learn also how to design *top-down*.

2.1.3 Vectorization

A tremendous advantage of MATLAB is that it allows the user to apply an operation directly to a whole array without having to build a loop in which the operation is applied to each element of the array in part. We have seen this in Section 1.5. Readers that have used other programming languages know that the operations described in that section required a loop. Applying operations directly to arrays is called in MATLAB jargon **vectorization**. This way of working not only simplifies programming, but it also shortens execution time. Knowing this we may ask if it would have been possible to avoid the loop in the example developed in the preceding section. To answer this question let us look at the function `perimeter` starting from its end and going backwards towards its beginning. The last line of the loop sums the lengths of the polygon sides. This operation does not require a loop if the said lengths are stored in an array, because the MATLAB function `sum` does this job in one program step. Could we obtain the side lengths without calling `norm` in a loop? The answer is yes, if we know the differences of point coordinates. Then, for two adjacent points, $\mathbf{P}_i$ and $\mathbf{P}_{i+1}$, we can apply Pythagoras' theorem to the differences of the x- and y-coordinates and build the array of side lengths. It remains to see how to obtain the differences of coordinates without a loop. MATLAB provides a function, `diff`, that does the job. Given an array of m numbers

$$A = [n_1 \ n_2 \ \dots \ n_{m-1} \ n_m]$$

the function `diff` yields an array of *m-1* numbers

$$dA = diff(A) = [(n_2 - n_1) \ (n_3 - n_2) \ \dots (n_m - n_{m-1}]$$

We can use the function `diff` to *vectorize* the operation that produces the differences of coordinates. To do this, we decompose the *2-by-n* array *Polyline* into two row vectors, the first of x-coordinates, the second of y-coordinates. Thus, for the points exemplified in Subsection 2.1.1,

```
>> diff(P(1, :))
ans =
   1   3   -4
>> diff(P(2, :))
ans =
   2   -1   -1
```

The function `perim` described below implements the considerations developed in this subsection.

```
function    p = perim(Polyline)
```

```
%PERIM  Calculates the length of the polygonal line Polyline
%       by vectorization. The input is a 2-by-n array where
%       the first line represents the x-coordinates of
%       the n points, the second line, their y-coordinates.
%       If Polyline is a closed polygon, the last point is
%       the repetition of the first point.

dpx = diff(Polyline(1, :));
dpy = diff(Polyline(2, :));
ds     = (dpx.^2 + dpy.^2).^(0.5);  % lengths of sides
p      = sum(ds);
```

2.1.4   Arrays of point coordinates in solid geometry

Up to now we worked in the plane. It is easy to generalize to the three-dimensional space. For example, let us add a third dimension, the z-coordinate, to the three points defined in Subsection 2.1.1:

```
≫ P1 = [ 1; 2; 1 ];
≫ P2 = [ 2; 4; 3 ];
≫ P3 = [ 5; 3; 2 ];
≫ P = [ P1 P2 P3 P1 ];
```

To plot we use this time the command `plot3` with the following arguments: the array of x-coordinates, the array of y-coordinates, the array of z-coordinates, and a string argument that defines the color and the line type. We also define three labels, one for each axis of coordinates.

```
≫ plot3(P(1, :), P(2, :), P(3, :), 'k-')
≫ axis equal, grid
≫ xlabel('x'), ylabel('y'), zlabel('z')
```

Try these commands for yourself. The plot may seem nice, but would appear less nice if we would include it as it is in this book. As we must reduce it at scale, the lines would be too thin and the fonts too small. To prepare the figure for the book we use two additional commands:

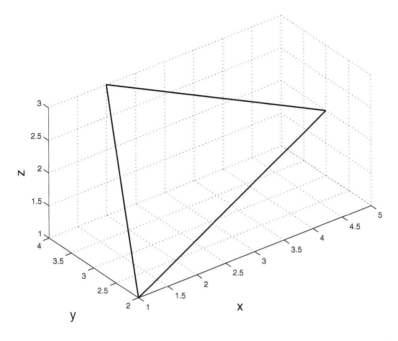

FIGURE 2.4: Plotting a triangle in 3-D space

```
≫ plot3(P(1, :), P(2, :), P(3, :), 'k-', 'LineWidth', 1.5)
≫ axis equal, grid
≫ xlabel('x', FontSize', 16)
≫ ylabel('y, 'FontSize', 16)
≫ zlabel('z','FontSize', 16)
```

The commands LineWidth and FontSize belong to *handle graphics*, a sub-ject that we are going to discuss again in this book. For the moment let us learn that the two commands affect the properties described by their name, that they must be defined as character strings, and that they must be fol-lowed by a numerical value. The default font-size value is 12. This value is measured in *points*, where one point is equal to 1/72 inch. The resulting plot is shown in Figure 2.4. The orientation of the coordinate axes is that usual in American engineering and literature. Later in the book we are going to learn how to change the direction of axes and show them as it is usual under the ISO standards used in Europe.

You can run the function perimeter with the three-dimensional triangle as input argument. The result will be 10.5593. The function perimeter is valid for any number of dimensions. You cannot run the function perim as defined above because it takes into consideration only two dimensions. It is easy to modify it for three dimensions. However, it would not work for two. To let it work with any number of dimensions we either need a loop over the number of dimensions, or a *conditional statement*. We leave the first possibility to an exercise and introduce conditional statements later in this book.

EXAMPLE 2.1 A function for plotting polygonal lines

With the knowledge acquired to this point we can start to develop a few utility functions that we are going to reuse in several examples developed in this book. Moreover, the reader may develop script files or functions and reuse the utility functions in them.

The first function plots a polygonal line through a set of n given points. We assume that each point is defined by an *m-by-1* array whose elements are the point coordinates. Here m is equal to 2, for two-dimensional points, and to 3, for three-dimensional points. The polygonal line is defined by concatenating the arrays of the composing points in an *m-by-n* array. In two-dimensional plots we use the command `plot`, while in three-dimensional plots we use the command `plot3`. To distinguish between these cases, we employ a *conditional construct* of the form

```
if          condition
            statements
elseif      condition
            statements
else
            statements
end
```

The following code does the job.

```
function          pline(P, w, c, s)

%PLINE plots line with given properties, between given points.
%       pline(P, w, c, s) plots a line between the points
%       whose coordinates are described in the array P, the
%       line width being w, the color, c, and the line style s.
%       For m points in an n-dimensional plot, P is an n-by-m
%       array. The input arguments c and s should be supplied
%       as strings.
%       Written by Adrian Biran.

[ m, n ] = size(P);
if m == 2               % two-dimensional plot
    H1 = plot(P(1, :), P(2, :));
    set(H1, 'LineWidth', w, 'Color', c, 'LineStyle', s)
elseif m == 3   % three-dimensional plot
    H2 = plot3(P(1, :), P(2, :), P(3, :));
```

```
    set(H2, 'LineWidth', w, 'Color', c, 'LineStyle', s)
else
    error('Input argument P has inadequate dimensions')
end
```

Consider the line

```
H1 = plot(P(1, :), P(2, :));
```

This line assigns the **handle** H1 to the plot command. A handle is an identifier of a command or function and can be used to define or modify proprieties. In our example we close the line with a semicolon, ';', because otherwise MATLAB will echo the handle. In the next line we use the function set, with the handle H1 as an argument, to define some properties of the plot and not use the default properties. In this case we change the line width, its color and its style. Each property is defined by a conventional name, written as a string, followed by a value.

You may experiment with the three points defined in Subsection 2.1.4. Try, for example, the command pline(P, 1.5, 'r', '-'). Play with the input arguments to see how they influence the plot. See also Example 2.5 for an application.

EXAMPLE 2.2 A function that plots black-filled circles

In this example we develop a second utility function that we are going to reuse in several other examples. In many figures we show 'points' as black circles. The function developed in this example plots such circles in two-dimensional plots. The function is called with two input arguments:

- the center of the circle, defined as a two-dimensional array, C;

- the radius of the circle, r.

The function is based on the parametric equations of the circle; there should be no problem to recognize them in the following listing. The parameter is t, defined as an array of values beginning at 0, increasing with the step $\pi/30$, and ending at 2π, that is a full circle.

```
    function point(C, r)

%POINT plots filled circles marking points
%       in two-dimensional plots.
%       Input arguments:
%       C coordinates of center, a two-dimensional vector
%       r radius of circle.
%       Written by Adrian Biran.
```

```
t = 0: pi/30: 2*pi;      % curve parameter
x = C(1) + r*cos(t);
y = C(2) + r*sin(t);
patch(x, y, [ 0 0 0 ]) % fills in black
```

To fill the circle we call the function `patch` with three arguments:

1. the array of x–coordinates of the polygon to be filled;

2. the array of y–coordinates of the polygon to be filled;

3. the filling color.

We defined here the color as a vector with three elements, the intensities of the *red*, *green* and *blue* components of the color we want to see displayed on the screen. When the values of all elements are equal to zero, no color is projected; this corresponds to black. When the values of all elements are equal to 1 the *additive mixing* of colors results in white.

See Example 2.5 for applications of the function `point`.

EXAMPLE 2.3 A function that plots arrows

In this example we develop a third utility function that we are going to reuse in several other examples. In many drawings we use arrows to point out certain features, or to define dimensions. The toolbar of MATLAB figures includes an icon and a pull down menu that enable us to draw 'manually' three types of arrows. We found this method a bit awkward because placing the arrows in their correct position is sometimes a matter of trial and error. For more precision we developed our own function that draws the arrows 'programmatically', in two-dimensional plots. The function, `arrow`, is called with three input arguments:

- the origin, *o*, defined as a two-dimensional array;

- the end, *e*, defined as a two-dimensional array;

- the width of the support line. This argument also influences the size of the arrow head.

The arrow head is a triangle with an angle of 15°, and is filled in black, as prescribed by ISO-related drawing standards. The function draws first the support line, then calculates the slope of the support line, calculates the vertices of the triangle and, finally, fills the triangle in black. The latter job is done by the function `fill` that, like `patch`, is called with three input arguments, actually the same. To illustrate another possibility, we defined here black by the string `'k'`.

```
      function arrow(o, e, t)
%ARROW draws arrow of thickness t, from origin o to end e
%        in a two-dimensional plot. The input arguments
%        0 and e should be supplied as two-dimensional vectors.
%        The arrows correspond to those recommended in
%        ISO-related drawing standards.
%        Written by Adrian Biran.

% draw support
plot([ o(1) e(1) ], [ o(2) e(2) ], 'k-', 'LineWidth', t)
% hold on
x             = e(1) - o(1);
y             = e(2) - o(2);
alphar = atan2(y, x);
alphad = 180*alphar/pi;
l             = 5*t/cosd(7.5);
xL            = e(1)  - l*cosd(alphad + 7.5);
yL            = e(2)  - l*sind(alphad + 7.5);
xU            = e(1)  - l*cosd(alphad - 7.5);
yU            = e(2)  - l*sind(alphad - 7.5);
fill([ xU e(1) xL ], [ yU e(2) yL ], 'k')
```

In Example 2.5 we call the function **arrow** several times.

2.1.5   Geometrical interpretation of vectors

There are mathematical objects and physical quantities that have only a **magnitude**. Only one number is necessary to define such objects; we call them **scalars**. Examples in geometry include the areas of plane figures and the volumes of solids. An example in physics is the temperature. There are other mathematical objects, or physical quantities for the definition of which one number is not sufficient. In MATLAB we define them by arrays of numbers. In this subsection we are giving a geometrical meaning to certain one-dimensional arrays and define a few operations that can be carried on them. Thus, the affected arrays become **vectors**. It will be immediately evident that the MATLAB basic object, the array, is a natural representation of the concept of vector and allows us to operate on vectors in an easy, intuitive manner. To illustrate these concepts by a concrete example, we show in Figure 2.5 the triangle whose vertices are the points P_1, P_2, P_3 defined in Subsection 2.1.1. We also show, in dot-point lines, the same triangle translated 3 units parallel to the x-axis, and 4 units parallel to the y-axis. The vertices of the translated triangle are marked Q_1, Q_2, Q_3. Figure 2.6 is a repetition of Figure 2.5 with the addition of arrows that connect the vertices of the initial triangle with those of the translated triangle.

Let us consider one of the arrows shown in Figure 2.6, for example, P_1Q_1.

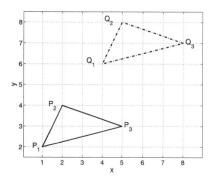 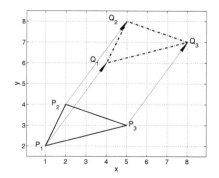

FIGURE 2.5: A triangle and its translation

FIGURE 2.6: Displacement vectors

It has a length, usually called *magnitude*, a *direction*, defined, by example, by the angle it makes with the x-coordinate axis, and a *sense*, that is from $\mathbf{P_1}$ to $\mathbf{Q_1}$. A geometrical object that has a magnitude, a direction and a sense is called **vector**. The arrows $\overrightarrow{P_1Q_1}$, $\overrightarrow{P_2Q_2}$, $\overrightarrow{P_3Q_3}$ are three examples of vectors. They are *equivalent* because all three have the same magnitude, direction, and sense. They have, however, different origins. Each vector represents a *displacement*. We used an arrow above the letters marking the origin and the end of each vector. This way of noting vectors is common in continental Europe. In English-speaking countries it is usual to mark vectors by bold-face letters and this is the convention adopted in this book.

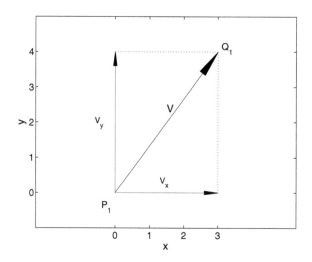

FIGURE 2.7: The components of a two-dimensional vector

Let us consider the vector $\mathbf{V} = \overrightarrow{P_1Q_1}$. We can find it by subtracting the coordinates of the point P_1 from those of the point Q_1

$$\mathbf{V} = Q_1 - P_1 = \begin{bmatrix} 4 \\ 6 \end{bmatrix} - \begin{bmatrix} 1 \\ 2 \end{bmatrix} = \begin{bmatrix} 3 \\ 4 \end{bmatrix}$$

A representation of this vector is shown in Figure 2.7. $V_x = 3$ is the *x-axis component* of $\mathbf{V}$, $V_y = 4$, the *y-axis component*. Thus, the MATLAB array suits very well for the definition of a vector by its components parallel to cartesian axes of coordinates. The magnitude of $\mathbf{V} = \overrightarrow{P_1Q_1}$ is the length of the vector that can calculated with the MATLAB function **norm**:

```
≫ norm(V)
≫ ans =
   5
```

In mathematical texts the magnitude of a vector, $\mathbf{V}$, or its *norm*, is usually noted as $\|\mathbf{V}\|$. The direction of $\mathbf{V}$ is defined, for example, by the angle made by the vector with the $x-$axis. We can find this angle by elementary trigonometry; in Subsection 2.1.8 we are going to learn a method that uses vector operations. As to the sense of $\mathbf{V}$, it is given by the signs of its components.

2.1.6 Operating with vectors

In this subsection we define two operations that can be applied to vectors:

- the addition of two vectors;

- the multiplication of a vector by a scalar.

Given two vectors, V_1 and V_2, their sum is a vector whose components are the sums of the components of the vectors, V_1 and V_2. The following example shows how easily this operation is executed in MATLAB

```
≫ V1 = [ 0.5; 2.5 ], V2 = [ 2.5; 1 ]
V1 =
   0.5000
   2.5000
V2 =
   2.5000
   1.0000
≫ V1 + V2
ans =
   3.0000
   3.5000
```

The geometrical interpretation of this operation is shown in Figure 2.8.

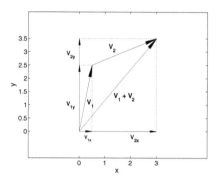

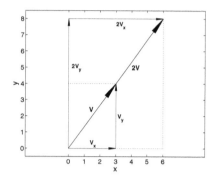

FIGURE 2.8: The addition of two vectors

FIGURE 2.9: Multiplication of a vector by a scalar

Given a vector, $\mathbf{V}$, and a scalar, λ, the multiplication $\lambda\mathbf{V}$ yields a vector whose components are equal to those of $\mathbf{V}$ multiplied by λ. Again, it is easy to carry this operation in MATLAB:

```
≫ V = [ 3; 4 ];
≫ 2*V
ans =
   6
   8
```

The geometrical interpretation of this operation is shown in Figure 2.9. The two operations defined in this subsection have a few proprieties that facilitate calculations with vectors. In Exercise 2.1 we ask the reader to use MATLAB to verify those proprieties. A set of vectors, V, for which the operations of addition of vectors, and multiplication of a vector by a scalar are defined, and which have the proprieties exemplified in Exercise 2.1, is called a **vector space**.

Exercise 2.1 Proprieties of vector spaces

Define in MATLAB the following vectors

$$\mathbf{V_1} = \begin{bmatrix} 1 \\ 3 \\ 4 \end{bmatrix}, \ \mathbf{V_2} = \begin{bmatrix} 2 \\ 4 \\ 3 \end{bmatrix}, \ \mathbf{V_3} = \begin{bmatrix} 5 \\ 0 \\ 1 \end{bmatrix}$$

Check in MATLAB that the following proprieties of vector spaces hold.

Closure under vector addition
$\mathbf{V_1} + \mathbf{V_2}$ is a vector.
Associativity

$$(\mathbf{V_1} + \mathbf{V_2}) + \mathbf{V_3} = \mathbf{V_1} + (\mathbf{V_2} + \mathbf{V_3})$$

Commutativity

$$\mathbf{V_1} + \mathbf{V_2} = \mathbf{V_2} + \mathbf{V_1}$$

Zero
There exists a **zero vector**, say $\mathbf{V_0}$, such that

$$\mathbf{V_1} + \mathbf{V_0} = \mathbf{V_0} + \mathbf{V_1} = \mathbf{V_1}$$

As the exemplified vectors are *3-by-1* arrays, the zero vector of the respective vector space is obtained in MATLAB by the command `zeros(3, 1)`.

Inverse
For each vector, $\mathbf{V_i}$, there exists an *inverse vector*, $-\mathbf{V_i}$, such that adding this inverse vector to the given vector $\mathbf{V_i}$ yields the zero vector. Check this property for all three vectors defined in the beginning of this exercise.

Closure under multiplication with a scalar
Define in MATLAB the scalar $\lambda = 0.5$, and show that $\lambda\mathbf{V_1}$, $\lambda\mathbf{V_2}$, and $\lambda\mathbf{V_3}$, are all vectors and that their components are equal to those of the given vectors multiplied by λ.

Further properties of the multiplication by a scalar
Define $\mu = 1.5$ and check

$$\lambda(\mathbf{V_1} + \mathbf{V_2}) = \lambda\mathbf{V_1} + \lambda\mathbf{V_2}$$
$$(\lambda + \mu)\mathbf{V_1} = \lambda\mathbf{V_1} + \mu\mathbf{V_1}$$
$$(\lambda\mu)\mathbf{V_1} = \lambda(\mu\mathbf{V_1})$$

Note. The elements of the vectors involved in this exercise are all integers. Therefore, the properties described here are always verified. When working with numbers that are not integers, many of the above properties are verified only approximatively. This subject is treated in Chapter 5.

2.1.7   Vector basis

In the preceding section we have seen how a vector can be represented as the sum of other vectors. In general, a vector can be represented as a *linear combination* of other vectors

$$\mathbf{V} = \lambda_1 \mathbf{V_1} + \lambda_2 \mathbf{V_2} + \ldots \lambda_n \mathbf{V_n}$$

where the coefficients λ_1, λ_2, ... λ_n are scalars. If we can find a set of scalars, λ_i, not all zero, such that the sum $\mathbf{V}$ is zero, we say that the vectors $\mathbf{V_1}$, ... $\mathbf{V_n}$ are *linearly dependent*. If all the scalars must be zero to yield a zero sum, we say that the given vectors are *linearly independent*. In linear algebra it is shown that, for a given vector space, V, there is an interest to find a set of linearly independent vectors, $\mathbf{e} = \{\mathbf{e_1}, \mathbf{e_2}, \dots \mathbf{e_n}\}$ such that all the elements of that space can be represented as linear combinations of the vectors $\mathbf{e_i}$. The set $\mathbf{e}$ is called a **basis** of the vector space V. The number of elements of $\mathbf{e}$ is the **dimension** of the space. If all the vectors $\mathbf{e_1}$, $\mathbf{e_2}$, ... $\mathbf{e_n}$ have the magnitude 1 and are perpendicular one to another, the basis is called *orthonormal*. Thus, for the three-dimensional Euclidean space we can define in MATLAB the orthonormal basis:

```
>> e1 = [ 1; 0; 0 ], e2 = [ 0; 1; 0 ], e3 = [ 0; 0; 1 ]
e1 =
    1
    0
    0
. . .
```

Readers familiar with vector algebra may know that the above vectors are often marked as $\mathbf{i}$, $\mathbf{j}$, $\mathbf{k}$, or $\vec{i}$, $\vec{j}$, $\vec{k}$. To show that these vectors have unit magnitudes, try, for example, `norm(e1)`. In the next subsection we are going to check also the perpendicularity. For the moment let us see how we represent, in the given basis, a vector that lies in the $z = 0$ plane:

```
>> V = 3*e1 + 4*e2
V =
    3
    4
    0
```

2.1.8   The scalar product

In this subsection we introduce a new operation on vectors, the **scalar**, or **dot product**. We show that the dot product can be used to check perpendicularity, to find the projection of a vector on an axis, or to find the angle between two vectors.

Given two vectors, $\mathbf{V_1}$, $\mathbf{V_2}$, and knowing that the angle between them is α, their *scalar*, also called *dot product*, is defined as

$$p = \mathbf{V1} \cdot \mathbf{V_2} = \|\mathbf{V_1}\| \cdot \|\mathbf{V_2}\| \cos \alpha \qquad (2.1)$$

The result is a scalar, hence the first name of the product. The dot, '·', used to mark this product explains the other name. The operation is performed in MATLAB by the function dot. As an example let us define two vectors and calculate their dot product:

```
≫ V1 = [ 3.4641; 2 ];
≫ V2 = [ 2.5; 4.3301 ];
≫ p = dot(V1, V2)
p =
   17.3205
```

To verify that the result corresponds to the definition given above, calculate the lengths of the two vectors (using the function norm) and the angle between them.

```
≫ norm(V1)
ans =
   4.0000
≫ norm(V2)
ans =
   5.0000
≫ beta = atand(V1(2)/V1(1));
≫ gamma = atand(V2(2)/V2(1));
≫ alpha = gamma - beta;
≫ p = norm(V1)*norm(V2)* cosd(alpha)
```

We obtain the same result as the one calculated with the command dot.

It is easy to prove that, if the vector $\mathbf{V_1}$ has the components V_{1x}, V_{1y}, and the vector $\mathbf{V_2}$, the components V_{2x}, V_{2y}, then the product $\mathbf{V_1} \cdot \mathbf{V_2}$ equals $V_{1x}V_{2x} + V_{1y}V_{2y}$. Let us check this in MATLAB:

```
≫ p1 = V1(1)*V2(1) + V1(2)*V2(2)
p1 =
   17.3205
≫ p - p1
ans =
   0
```

The generalization to three and more dimensions is straightforward.

The scalar product enables us to 'measure' lengths and angles. The length can be a displacement or the magnitude of a vector. For example consider the vector $\mathbf{V_1}$ with the components defined above; its magnitude can be calculated as

$$\|V_1\| = \sqrt{V_1 \cdot V_1} = \sqrt{V_{1x}^2 + V_{1y}^2}$$

With the values exemplified above,

```
≫ V1 = [ 3.4641; 2 ];
≫ sqrt(dot(V1, V1))
ans =
   4.0000
```

We recovered the same value as that obtained with the MATLAB built-in function **norm**. The use of the latter is simpler; however, it is important to show that the task can be carried on by calling the dot product.

Let us show now how the dot product enables us to 'measure' angles. Above, to calculate the angle between the vectors V_1 and V_2 we used twice the MATLAB function **atan** to find the angles between the given vectors and the x−axis, and then subtracted one angle from the other. This procedure is rather awkward. The elegant solution is based on the definition of the scalar product; it enables us to extract the angle in one step:

```
≫ alpha = acosd(dot(V1, V2)/(norm(V1)*norm(V2)))
alpha =
   29.9998
```

As another application of the scalar product in geometry let us see how it helps us to find the **projection** of a vector on one of the coordinate axes. Remembering how we defined the unit vector along the x−axis, e_1, and the vector **V**, in Subsection 2.1.7, we can find the magnitude of the component of **V** along the x−axis as

```
≫ Vx = dot(V, e1)
Vx =
   3
```

If we want to find the vector V_x we must multiply the scalar result by the unit vector along the corresponding axis:

```
≫ dot(V, e1)*e1
ans =
   3
   0
   0
```

A further application of the scalar product is in checking perpendicularity.

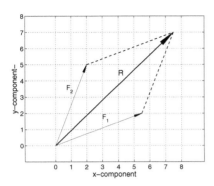

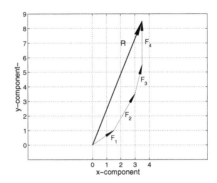

FIGURE 2.10: The parallelogram rule

FIGURE 2.11: Adding several forces by the polygon rule

For example, let us check that the unit vectors $\mathbf{e_1}$, $\mathbf{e_2}$ defined in Subsection 2.1.7 are perpendicular one to another:

```
≫ dot(e1, e2)
ans =
    0
```

We invite the reader to continue this check for the pairs $\mathbf{e_2}$, $\mathbf{e_3}$ and $\mathbf{e_3}$, $\mathbf{e_1}$.

2.2 Vectors in mechanics

2.2.1 Forces. The resultant of two or more forces

Forces can be readily represented as vectors whose elements are the force components parallel to given cartesian coordinate axes. In this representation vectors can be easily added in MATLAB. For example, after defining two forces, $\mathbf{F_1}$, $\mathbf{F_2}$, their resultant is found by *vector addition*:

```
≫ F1 = [ 5.5; 2 ];
≫ F2 = [ 2; 5 ];
≫ R = F1 + F2
R =
    7.5000
    7.0000
```

In the SI system, the forces $\mathbf{F}_1$, $\mathbf{F}_2$, and their resultant, $\mathbf{R}$, can be measured in Newton, noted as N. The graphical representation of the above operation is shown in Figure 2.10. The figure also illustrates the application of the parallelogram rule to the addition of two vectors. When several forces are applied to a body, and they all lie in the same plane, their resultant can be found by building a polygon of forces, for example:

```
≫ F1 = [ 1.5; 1 ];
≫ F2 = [ 1.5; 2.5 ];
≫ F3 = [ 0.5; 2 ];
≫ F4 = [ 0; 3 ];
≫ R = F1 + F2 + F3 + F4
R =
   3.5000
   8.5000
```

The above operation is illustrated in Figure 2.11. To find the magnitude of a force represented by its cartesian components use the function `norm`, and to find the orientation (angle) of such a force use the dot product, as shown in Subsection 2.1.8. Two-dimensional vectors can also be represented in MATLAB by their magnitude and the angle with a given axis of coordinates. This is done using the complex-number representation, a subject treated in Chapter 7.

EXAMPLE 2.4 Weight hanging from two bars

In this example we develop a function that solves a simple problem in statics. In Figure 2.12 two bars are connected to a ceiling by articulations and are connected one to another by a third articulation. A weight, w, is hanging from the latter articulation. We want to find the tension forces, $\mathbf{R}_1$ and $\mathbf{R}_2$, in the two bars. We isolate the common point of the two bars and calculate the forces in the two bars. For equilibrium, their resultant must be a vector opposed to that of the hanging weight, w. We draw this vector together with the parallelogram that shows it as the resultant of the forces $\mathbf{R}_1$ and $\mathbf{R}_2$. By simple trigonometry and geometry we write

$$a = c\tan\alpha, \ b = c\tan\beta, \ \gamma = 180 - (\alpha + \beta)$$

Moreover, the law of sines yields

$$\frac{\sin\alpha}{R_2} = \frac{\sin\beta}{R_1} = \frac{\sin\gamma}{w}$$

These relationships are implemented in the function `statics1` detailed below. Write the function in a file called `statics1.m`.

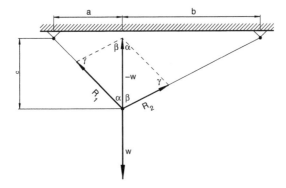

FIGURE 2.12: Calculation of the tensions in the two bars

```
function    [ R1 R2 ] = statics1(w, a, b, c)

STATICS1    Finds tensions in two bars connected to a point
            where a weight w hangs.
            a -  horizontal distance between first
            articulation and vertical of w
            b - horizontal distance between third
            articulation and vertical of w
            c - vertical distance between the plane of
            upper articulations and the point where w hangs.
            Written by Adrian Biran, May 2009
% calculate angles
alpha = atand(a/c);
beta    = atand(b/c);
gamma = 180 - alpha - beta;
% calculate magnitudes of tension vectors
r1          = sind(beta)*w/sind(gamma);
r2          = sind(alpha)*w/sind(gamma);
% define tension vectors
R1          = r1*[ -sind(alpha), cosd(alpha) ];
R2          = r2*[  sind(beta), cosd(beta) ];
```

Let us call our function with the following input arguments: $w = 10$ kN, a = 500 mm, $b = 1000$ mm, $c = 500$ mm.

```
≫ [ R1, R2 ] = statics1(10, 500, 1000, 500)
R1 =
   -6.6667  6.6667
R2 =
    6.6667  3.3333
```

To check the results, define the vector $\mathbf{W} = [0 - w]$ and calculate the sum $\mathbf{R1} + \mathbf{R2} + \mathbf{W}$. In Example 2.5 we turn the function `Statics1` into another function, `Stat1FigFun`, that plots a figure similar to Figure 2.12, but changing as a function of the input length, a.

2.2.2   Work as a scalar product

If a body subjected to a force, $\mathbf{F}$, travels along a path defined by the radius vector r, we define the element of **work** done by the force as

$$dW = F\,dr\cos\varphi$$

where φ is the angle between the force and the element of path. Thus, the element of work is the scalar product of the force by the vector representing the element of path. To calculate the work done by a variable force along a curved path, we must integrate dW. When the force is constant, and the path is a straight line, we can find the work without integration. As a very simple example consider in Figure 2.13 a body of mass $M = 1$ Kg sliding without friction on a slope of 45°. We also assume that the vertical distance traveled by the body is equal to 1 m. To calculate in MATLAB the work done by the weight of the body we first define the weight vector, $\mathbf{W}$, with elements measured in N, and next the path vector, d, with elements measured in m.

```
≫ M = 1;
≫ W = [ 0; 9.81*M ];
≫ d = [ 1 1 ];
≫ Work = dot(W, d)
Work =
    9.8100
```

As expected, the calculated work equals the change in potential energy.

2.2.3   Velocities. Composition of velocities

Another use of vectors in mechanics is in the representation of velocities. As an example, let us consider in Figure 2.14 a ferry that crosses a stream, starting from the point A. Let the velocity of the ferry, relative to water, be $V_f = 10$ knots, and that of the stream, relative to ground, $V_s = 2$ knots. As

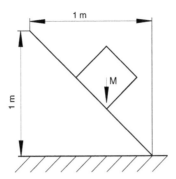

FIGURE 2.13: Body sliding on a 45-degree slope

the ferry has to cross a distance of 1 nautical mile, it will reach the other side at a point B, situated downstream relative to the point A. The velocity of the ferry relative to ground is a vector, V, whose components are V_f and V_s. In MATLAB we define this vector, and calculate its magnitude, as shown below. In this case the magnitude is the *speed over ground*.

```
≫ Vknots = [ 2; 10 ];
≫ norm(Vknots)
ans =
    10.1980
```

Remembering that one knot equals one nautical mile per hour, we invite the reader to calculate the time needed to cross the stream, and the distance traveled down the stream.

2.3 Matrices

2.3.1 Introduction – the matrix product

In Subsection 2.1.5 we have shown that a *vector* is a one-dimensional array for which certain operations have been defined. In this section we are going to introduce the **matrix** as a two-dimensional array for which certain operations are defined.

Let us define a vector, **V**, of length 4, and calculate the dot product of this vector by itself:

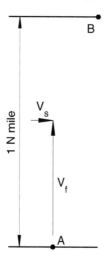

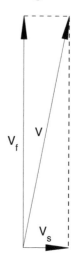

FIGURE 2.14: Ferry traversing a stream

FIGURE 2.15: Composition of velocities

```
≫ V= [ 2 4 1 6 ];
≫ dot(V, V)
ans =
   57
```

We could have obtained the same result by taking the sum of the element-by-element multiplication of the same vector by itself:

```
≫ sum(V.*V)
ans =
   57
```

Trying the simple multiplication of the vector by itself we elicit an error message:

```
≫ V*V
??? Error using ==> mtimes
Inner matrix dimensions must agree.
```

Look now at the slightly modified command:

```
≫ V*V'
ans =
  57
```

This time we succeeded and the difference was produced by the apostrophe added after the second appearance of 'V'. Let us see the effect of this apostrophe:

```
≫ V'
ans =
  2
  4
  1
  6
```

We transformed a row vector into a column vector. The vector V' is the **transpose** of the vector V; in linear algebra it is noted as $\mathbf{V^T}$. With the notion of array dimension defined in Subsection 2.1.1, we can say that the transpose of a *1-by-n* vector is an *n-by-1* vector.

In general, given two vectors of the same dimensions, we can obtain their scalar product by multiplying one vector by the transpose of the other vector:

```
≫ W = [ 5 1 2 3 ];
≫ W*V'
ans =
  34
≫ V*W'
ans =
  34
≫ dot(V, W)
ans =
  34
```

We multiplied a *1-by-n* vector by an *n-by-1* vector and obtained a *1-by-1* vector, that is a scalar. We can write schematically

$$(1 - by - n) \times (n - by - 1) = (1 - by - 1)$$

The *internal dimensions*, n, are equal, and the external dimensions, *1*, yield the dimensions of the result, *1-by-1*. Soon we are going to generalize this rule.

In Subsection 2.1.1 we saw that, given m vectors with dimensions *n-by-1*, we can combine the vectors into an *n-by-m* array. Thus, we obtained the *2-by-4* array P. Let us build again the array P, define a *1-by-2* vector, A, and multiply A by P:

```
≫ P = [ 1 2 5 1; 2 4 3 2 ]
P =
   1   2   5   1
   2   4   3   2
≫ A = [ 5 4 ];
≫ B = A*P
B =
   13   26   37   13
```

As a first generalization of the rule expressed by Equation 2.3.1 we can write

$$(1 - by - 2) \times (2 - by - 4) = (1 - by - 4)$$

The first element of the array B is the scalar product

$$\begin{bmatrix} 5 & 4 \end{bmatrix} \begin{bmatrix} 1 \\ 2 \end{bmatrix}$$

The second element B is the scalar product

$$\begin{bmatrix} 5 & 4 \end{bmatrix} \begin{bmatrix} 2 \\ 4 \end{bmatrix}$$

and so on. Finally, let us add two rows to the array A and multiply it by P:

```
≫ A = [ 5 4; 2 3; 6 7 ];
≫ C = A*P
C =
   13   26   37   13
    8   16   19    8
   20   40   51   20
```

We can write now the general rule

$$(n - by - p) \times (p - by - m) = (n - by - m)$$

C is the **matrix product** of A by P. As we allow this matrix operation on the arrays P, A, B and C, we say that these arrays are **matrices**, the plural of the term **matrix**. The element in place i, j, in the matrix C, is the scalar product of the $i-$th row of the matrix A by the $j-$th column of the matrix P. As previously defined, the index i refers to row, and the index j, to column.

Exercise 2.2 Matrix product compared to array product
Define in MATLAB the **square matrix**

$$\begin{bmatrix} 3 & 4 & 5 \\ 6 & 7 & 8 \\ 9 & 10 & 1 \end{bmatrix}$$

Compare the results obtained with the commands A*A and A.*A.

2.3.2 Determinants

Given a square matrix, **M**, its **determinant**, D, is a number calculated from the elements of **M**. For a *2-by-2* matrix the rule is

$$D\left(\begin{bmatrix} m_{11} & m_{12} \\ m_{21} & m_{22} \end{bmatrix}\right) = m_{11}m_{22} - m_{12}m_{21} \tag{2.2}$$

MATLAB provides a function, det, that does the job.

```
≫ M = [ 5 3; 6 7 ];
≫ det(M)
ans =
17
```

For a *3-by-3* matrix the rules for calculating the determinant are a bit more complex and we refer the reader to textbooks on linear algebra. To give an example, let us refer again to the matrix defined in Example 2.2 above and calculate in MATLAB:

```
≫ M = [ 3 4 5; 6 7 8; 9 10 1 ]
M =
   3    4    5
   6    7    8
   9   10    1
≫ D = det(M)
D =
   30
```

As an exercise we may calculate the same determinant by *Laplace expansion* using the first row:

```
≫ D = 3*det(M(2:3, 2:3)) - 4*det([ M(2:3, 1) M(2:3, 3) ]) +
5*det(M(2:3, 1:2))
D =
   30
```

The function det can be applied to matrices of higher order. One use of determinants is in checking if the rows or columns of a matrix are linearly

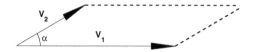

FIGURE 2.16: The area of the parallelogram

independent. If not, the corresponding determinant is zero. For example, let us define a *3-by-3* matrix whose third line is the sum of the first two, and calculate its determinant

```
>> M = [ 1 3 5; 2 6 1; 3 9 6 ];
>> det(M)
ans =
   0
```

Determinants are used in the solution of systems of linear equations, a subject treated in Chapter 3. As shown there, a matrix whose determinant is zero is not *invertible*. In the following sections we are going to show some applications of determinants in geometry and mechanics.

2.4   Matrices in geometry

2.4.1   The vector product. Parallelogram area

In Figure 2.16 the vector $\mathbf{V}_1$, is the base and the vector $\mathbf{V}_2$, one side of a parallelogram. Let α be the angle between the two vectors. The area of the parallelogram is

$$area = \|\mathbf{V}_1\| \times \|\mathbf{V}_2\| \times \sin \alpha$$

To give an example in MATLAB, let the magnitude of $\mathbf{V}_1$ be 10, the magnitude of $\mathbf{V}_2$, 5, and the angle α, 30°

```
>> area = 5*10*sind(30)
area =
   25.0000
```

Given two vectors, $\mathbf{V}_1$, $\mathbf{V}_2$, and the angle between them, α, the **vector**, or **cross product**, $\mathbf{V} = \mathbf{V}_1 \times \mathbf{V}_2$, is defined as a vector perpendicular to the plane defined by the given vectors. The magnitude of $\mathbf{V}$ is given by

$$\mathbf{V} = \|\mathbf{V}_1\| \cdot \|\mathbf{V}_2\| \cdot \sin \alpha \tag{2.3}$$

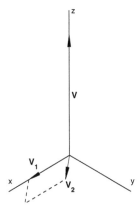

FIGURE 2.17: To the definition of the vector product

and the direction is that in which a right-handed screw would advance when $\mathbf{V_1}$ would be turned towards $\mathbf{V_2}$ (see Figure 2.17).

MATLAB provides a function, **cross**, that calculates the vector product. Let us use it for the example given above:

```
≫ V1 = [ 10; 0; 0 ];
≫ V2 = [ 4.3301; 2.500; 0 ];
≫ Area = cross(V1, V2)
Area =
   0
   0
  25
```

The reader is invited to check that the represented vector, V2, is, indeed, that of the example. To use the function **cross** we have to define at least one of the multiplicands as a three-dimensional vector.

Noting the unit vectors of the cartesian axes of coordinates as **i** for the $x-$axis, **j** for the $y-$axis, and **k** for the $z-$axis we consider the two vectors

$$\mathbf{V} = V_x\mathbf{i} + V_y\mathbf{j} + V_z\mathbf{k}$$
$$\mathbf{W} = W_x\mathbf{i} + W_y\mathbf{j} + W_z\mathbf{k}$$

their vector product can be represented by the determinant

$$\mathbf{V} \times \mathbf{W} = \begin{vmatrix} \mathbf{i} & \mathbf{j} & \mathbf{k} \\ V_x & V_y & V_z \\ W_x & W_y & W_z \end{vmatrix} \tag{2.4}$$

For the vector product we use the symbol $\times$, hence the name *cross product*.

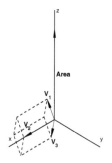

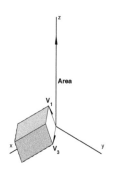

FIGURE 2.18: To the definition of the triple product

FIGURE 2.19: The meaning of the triple product

Returning to the two vectors, $\mathbf{V}_1$, $\mathbf{V}_2$, used in this example, the area of the parallelogram defined by them, that is their vector product, can be calculated also as

$$area = \begin{vmatrix} \mathbf{i} & \mathbf{j} & \mathbf{k} \\ 10 & 0 & 0 \\ 4.3301 & 2.500 & 0 \end{vmatrix} = \begin{vmatrix} 0 & 0 \\ 2.500 & 0 \end{vmatrix} \mathbf{i} - \begin{vmatrix} 10 & 0 \\ 4.3301 & 0 \end{vmatrix} \mathbf{j} + \begin{vmatrix} 10 & 0 \\ 4.3301 & 2.500 \end{vmatrix} \mathbf{k} = 25\mathbf{k}$$

(2.5)

In MATLAB we could have obtained this area simply as

```
≫ det([ 10 0; 4.3301 2.5 ])
ans =
   25
```

This yields immediately a geometrical interpretation of the determinant of a *2-by-2* matrix: the area of the parallelogram defined by the two rows interpreted as vectors. Mind that the MATLAB command **cross** produces a vector, while the command **det** yields a number. A geometrical interpretation of the determinant of a *3-by-3* matrix is given in the next subsection.

2.4.2   The scalar triple product. Parallelepiped volume

Three vectors, $\mathbf{V}_1$, $\mathbf{V}_2$, $\mathbf{V}_3$, can be multiplied in several ways. We are going to deal with one of these ways: first build the vector product $\mathbf{V}_2 \times \mathbf{V}_3$, next calculate the scalar product of $\mathbf{V}_1$ by the vector product, that is

$$w = \mathbf{V}_1 \cdot (\mathbf{V}_2 \times \mathbf{V}_3) \tag{2.6}$$

The scalar w is the **scalar triple product**, or **mixed product** of $\mathbf{V}_1$, $\mathbf{V}_2$, and $\mathbf{V}_3$. As shown in Subsection 2.4.1, the cross product yields a vector whose magnitude equals the area of the parallelogram defined by the vectors

V_2, V_3. Let us call this vector A, an abbreviation of 'Area'. Let the angle between V_1 and A be β. The scalar product of V_1 and A is

$$V_1 \cdot A = \|V_1\| \cdot \|A\| \sin \beta$$

But $\|V_1\| \cdot \sin \beta$ is the height of the end of V_1 above the plane defined by the other two vectors. Multiplying this height by the area of the parallelogram we obtain the volume of the parallelepiped defined by the three vectors. This geometrical interpretation is illustrated in Figures 2.18 and 2.19.

To give an example of calculation in MATLAB, let us reuse, but with a 'shifted' notation, the two vectors, V_1, V_2 defined in Subsection 2.4.1, and define a third vector inclined 45° with respect to the plane defined by the first two vectors (the $xy-$plane).

```
≫ V1 = [ 4.2426; 0; 4.2426 ]
≫ V2 = [ 10; 0; 0 ];
≫ V3 = [ 4.3301; 2.5000; 0 ];
```

We calculate the scalar triple product with the commands

```
≫ volume = dot(V1, cross(V2, V3))
volume =
   106.0660
```

In textbooks on linear algebra we can find the proof that this mixed product equals the determinant

$$\begin{vmatrix} V_{1x} & V_{1y} & V_{1z} \\ V_{2x} & V_{2y} & V_{2z} \\ V_{3x} & V_{3y} & V_{3z} \end{vmatrix}$$

a fact that we can check in MATLAB with the commands

```
≫ Vol = [ V1'; V2'; V3' ]
Vol =
    4.2426    0    4.2426
   10.0000    0    0
    4.3301    2.5000    0
≫ det(Vol)
ans =
   106.0660
```

Thus, the geometric interpretation of a third-order determinant is that it represents the volume of a parallelepiped defined by the rows of the determinant seen as the sides of the solid.

2.5   Transformations

In geometry, in computer graphics and in computer aided design (CAD) it is necessary to move objects from an initial position to another, or to change the shape of objects. Such operations are called `transformations` . Transformations are readily carried on with the help of matrices. In this section we is executed give only a few examples, just to show the reader that there is an entire field of applications in which MATLAB is very helpful. For details on the mathematics of geometric transformations we refer the reader to specialized books, such as Eggerton and Hall (1999) or Marsh (2000). More examples in MATLAB can be found in Biran and Breiner (2002) and in Biran and Breiner (2009).

2.5.1   Translation — Matrix addition and subtraction

In this subsection we give an example of *matrix addition*. Let us return to Subsection 2.1.5 and consider again the triangle defined by the three points, P_1, P_2, P_3, and the `translated` triangle, Q_1, Q_2, , Q_3. We can translate, or *shift*, each point separately by adding to its vector of coordinates the corresponding vector of translation, T, for example:

```
≫ P1 = [ 1; 2 ];
≫ T = [ 3; 4 ];
≫ Q1 = P1 + T
Q1 =
   4
   6
```

As we defined the triangle by a single matrix (obtained by the concatenation of the vectors of coordinates), it would be convenient to perform the operation in one step only, and not for each point in part. We cannot add directly T to P because the dimensions of the former are *2-by-4*, while the dimensions of the latter are *2-by-1*. To obtain a translation matrix of the same size as P we have to repeat T four times. Therefore, we first obtain the dimensions of the given matrix of point coordinates and next build a matrix of ones having the same number of columns as the matrix of coordinates.

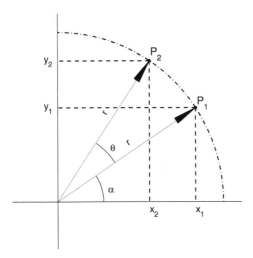

FIGURE 2.20: Rotation around the origin

```
≫ P = [ 1 2 5 1; 2 4 3 2 ];
≫ [ m, n ] = size(P);
≫ Q = P + T*ones(1, n)
Q =
    4  5  8  4
    6  8  7  6
```

We invite the reader to experiment with the command **ones**.

2.5.2   Rotation

In Figure 2.20 we consider the point $\mathbf{P_1}$ whose position vector makes an angle α with the $x-$axis. We want to rotate counterclockwise by the angle θ, to the position marked $\mathbf{P_2}$. The coordinates of the two points are

$$\begin{bmatrix} x_1 \\ y_1 \end{bmatrix} = r \begin{bmatrix} \cos\alpha \\ \sin\alpha \end{bmatrix}, \quad \begin{bmatrix} x_2 \\ y_2 \end{bmatrix} = r \begin{bmatrix} \cos(\alpha+\theta) \\ \sin(\alpha+\theta) \end{bmatrix}$$

Expanding and rearranging the equation that defines the coordinates of the rotated point, x_2, y_2, we obtain

$$\begin{bmatrix} x_2 \\ y_2 \end{bmatrix} = r \begin{bmatrix} \cos\alpha\cos\theta - \sin\alpha\sin\theta \\ \sin\alpha\cos\theta + \sin\theta\cos\alpha \end{bmatrix}$$
$$= r \begin{bmatrix} \cos\alpha \\ \sin\alpha \end{bmatrix} \begin{bmatrix} \cos\theta & -sin\theta \\ \sin\theta & \cos\theta \end{bmatrix} \tag{2.7}$$

In conclusion, to rotate the point $\mathbf{P_1}$ by the angle θ we multiply its vector of coordinates by the *rotation* matrix

$$T_r = \begin{bmatrix} \cos\theta & -sin\theta \\ \sin\theta & \cos\theta \end{bmatrix} \tag{2.8}$$

For reasons explained in Subsection 2.5.3, as we represent points by the column vector of their coordinates, we prefer to multiply the latter vector at left by the transformation matrix, that is

$$\mathbf{T}_{rotation}\mathbf{P}_{given} = \mathbf{P}_{rotated}$$

As an example let us consider the point $\mathbf{P_1}$ in Figure 2.20 and assume the coordinates $x_1 = 3$, $y_1 = 2$. Let us rotate this point by 22.6198 degrees. We want to find the coordinates, x_2, y_2, of the rotated point, $\mathbf{P_2}$. In MAT-LAB we define the initial point, build the rotation matrix, and carry on the transformation as follows

```
≫ P1 = [ 3; 2 ];
≫ theta = 22.6198;
≫ Tr = [ cosd(theta) -sind(theta)
sind(theta) cosd(theta) ];
≫ P2 = Tr*P1
P2 =
   2.0000
   3.0000
```

Figure 2.20 shows the initial and the rotated point at the correct angles.

2.5.3   Homogeneous coordinates

We carried on translation by matrix addition, and rotation by matrix multiplication. Isn't it possible to unify procedures and perform all transformations by multiplication? Moreover, to perform the translation we had to build an *ad-hoc* transformation matrix that takes into account the dimensions of the shifted object. We are interested in a transformation matrix that works for all object dimensions. These aims can be achieved by using **homogeneous coordinates**. While the position vectors used up to now have a number of elements (or components) equal to that of the space they are defined in, in homogeneous coordinates they have an additional component. Thus, to a vector $\mathbf{P}$, defined in the plane, that is in a two-dimensional space, corresponds in homogeneous coordinates a vector $\mathbf{P_h}$, defined by the rule

$$\mathbf{P} = \begin{bmatrix} P_x \\ P_y \end{bmatrix}, ; \mathbf{P}_h = \begin{bmatrix} wP_x \\ wP_y \\ w \end{bmatrix} \tag{2.9}$$

In other words, given the homogeneous coordinates x_h, y_h, w, we recover the usual coordinates as follows

$$x = \frac{x_h}{w}, \ y = \frac{y_h}{w}$$

In most cases we assume $w = 1$, and this greatly simplifies calculations. The geometrical interpretation of homogeneous coordinates is very interesting, but it goes beyond the scope of this book. Let us point out, however, that letting $w = 0$ we can represent points at infinity. Adding these points to the *Euclidean plane* we obtain the *projective plane* and we do not have to distinguish between ordinary points and points at infinity. Neither have we to distinguish between intersecting and parallel lines, and we can say that parallels also intersect, but at infinity. What interests us in this section is that

- in homogeneous coordinates all transformations are performed by **multiplying** the vector of coordinates by a transformation matrix;

- to concatenate two transformations, T_1, T_2, in this order, it is sufficient to use a matrix obtained by multiplying the matrices of the transformations, that is $\mathbf{T} = \mathbf{T_1 T_2}$.

In homogeneous coordinates, the transformation matrix that shifts an object by the horizontal distance a, and the vertical distance b is

$$\mathbf{T}_{tr} = \begin{bmatrix} 1 & 0 & a \\ 0 & 1 & b \\ 0 & 0 & 1 \end{bmatrix} \tag{2.10}$$

For an illustration we return to the triangle and the translation exemplified in Subsection 2.5.1. We first have to convert the matrix of coordinates, $\mathbf{P}$, into a matrix of homogeneous coordinates. Next we build the matrix, $\mathbf{T}$, which translates the triangle by 3 units horizontally, and 4 units vertically, and left multiply $\mathbf{P}$ by $\mathbf{T}$.

```
≫ P = [ P; 1 1 1 1 ]
P =
   1   2   5   1
   2   4   3   2
   1   1   1   1
≫ T = [ 1 0 3; 0 1 4; 1 1 1 ];
≫ Q = T*P
Q =
   4   5   8   4
   6   8   7   6
   4   7   9   4
```

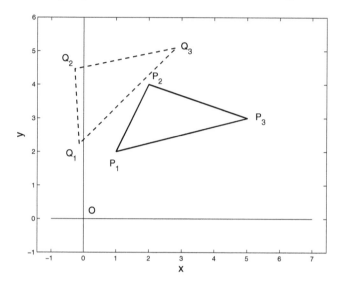

FIGURE 2.21: Rotation around the origin, in homogeneous coordinates

And now, let us convert the transformation matrix 2.8 into a matrix for homogeneous coordinates

$$\mathbf{T}_{RotHom} = \begin{bmatrix} \cos\theta & -\sin\theta & 0 \\ \sin\theta & \cos\theta & 0 \\ 0 & 0 & 1 \end{bmatrix} \qquad (2.11)$$

For example, let us consider again the triangle defined by the matrix of co-ordinates, $\mathbf{P}$, and let us rotate it by 30 degrees counterclockwise. In MATLAB we

1. define the triangle in homogeneous coordinates;

2. define the rotation matrix in homogeneous coordinates;

3. perform the transformation;

4. plot the initial and the rotated triangles.

The calculations are

```
≫ P = [ 1 2 5 1; 2 4 3 2 ; 1 1 1 1];
≫ theta = 30;
≫ Tr = [ cosd(theta) -sind(theta) 0; sind(theta) cosd(theta)
0; 0 0 1];
≫ Q = Tr*P
Q =
  -0.1340  -0.2679  2.8301  -0.1340
   2.2321   4.4641  5.0981   2.2321
   1.0000   1.0000  1.0000   1.0000
≫ plot([ -1 7 ], [ 0 0 ], 'k-', [ 0 0 ], [ -1 6 ], 'k-');
≫ axis equal
≫ ht = xlabel('x');
≫ set(ht, 'FontSize', 16)
≫ ht = ylabel('y');
≫ set(ht, 'FontSize', 16)
≫ ht = text(0.15, 0.25, '0');
≫ set(ht, 'FontSize', 14)
≫ hold on
≫ hp = plot(P(1,:), P(2,:), 'k-');
≫ set(hp, 'LineWidth', 1.5)
≫ ht = text(0.8*P(1,1), 0.8*P(2, 1), 'P_1');
≫ set(ht, 'FontSize', 14)
≫ ht = text(1.05*P(1, 2), 1.05*P(2, 2), 'P_2');
≫ set(ht, 'FontSize', 14)
≫ ht = text(1.05*P(1, 3), P(2, 3), 'P_3');
≫ set(ht, 'FontSize', 14)
≫ hp = plot(Q(1,:), Q(2,:), 'k--');
≫ set(hp, 'LineWidth', 1.3)
≫ ht = text(4*Q(1,1), 0.85*Q(2, 1), 'Q_1');
≫ set(ht, 'FontSize', 14)
≫ ht = text(2.5*Q(1, 2), 1.06*Q(2, 2), 'Q_2');
≫ set(ht, 'FontSize', 14)
≫ ht = text(1.05*Q(1, 3), Q(2, 3), 'Q_3');
≫ set(ht, 'FontSize', 14)
```

Above we used larger fonts to make the plot readable at the reducing scale of Figure 2.21. We have found the positions of the notations by trial and error.

The distance between points is an invariant of translation, rotation, and other transformations. Therefore, such transformations are included under the general name *isometries*, a term that should not be confounded with the *isometric projection* taught in courses on drawing. In Example 2.8 we propose the reader to check the invariance of distance for the rotation shown above.

Often we have to carry on several transformations. The order in which we

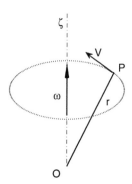

FIGURE 2.22: The vector of rotational velocity

do this defines the final result. In general, changing the order changes the result. This is shown in Exercise 2.9 in which the reader is asked to perform first a translation, next a rotation. In continuation, the reader is asked to apply the same transformation, but first the rotation, next the translation.

2.6   Matrices in Mechanics

2.6.1   Angular velocity

Figure 2.22 shows a point, P, rotating around an axis $O\zeta$ with an *angular velocity* ω. Let the position of the point be identified by the radius $\mathbf{r}$, drawn from an origin that lies on the axis of rotation. The *linear velocity of the point* is a vector, $\mathbf{V}$, which lies in in the plane that contains the point, is perpendicular to the axis of rotation, and has the magnitude $\|\mathbf{V}\| = wr\sin\alpha$, where α is the angle between $\mathbf{r}$ and the axis of rotation. The vector $\mathbf{V}$ is tangent to the trajectory of the point P. These considerations explain why it is convenient to represent the angular velocity by a vector, ω, which is perpendicular to the plane of rotation of the point P. The linear velocity is a vector perpendicular to the plane defined by ω and $\mathbf{r}$ and is calculated as the cross product

$$\mathbf{V} = \omega \times \mathbf{r} \tag{2.12}$$

To give a simple example in MATLAB, let the axis $O\zeta$ coincide with the

$z-$axis, $\|\omega\| = 0.5$ rad s^{-1} and $\mathbf{r} = [2\ 0\ 4]$ m:

```
≫ omega = [ 0 0 0.5 ];
≫ r = [ 2 0 4 ];
≫ V = cross(omega, r)
V =
    0   1   0
```

Using SI units, the angular velocity, ω, is measured in rad $\cdot$ s^{-1}, and r in m. Then V is measured in m s^{-1}; it is parallel to the $xOy-$ plane.

2.6.2 Center of mass

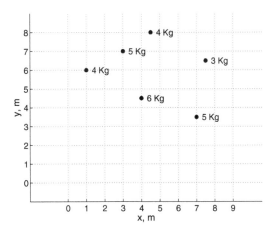

FIGURE 2.23: A system of point masses

Figure 2.23 shows a system of six point masses lying in the same plane. The mass values are shown near the points, and the coordinates of the points can be read in the figure. The **center of mass** of the system is, by definition, the point whose moments about the given axes of coordinates is equal to the sum of the moments of the six points. To calculate the total mass and the center of mass in MATLAB we must organize adequately the data. One possibility is to store the mass values in a vector, noted here as M, and the coordinates in an array noted here as XY. In our case we say that the array XY is a *matrix* because we involve it in a matrix operation. The following screen shows the calculations in MATLAB. We define M as a six-by-one column vector, and XY as a six-by two matrix. The total mass is obtained by calling the function sum with M as argument. To calculate the sums of the moments of the six points about the $x-$ and $y-$axes, we multiply the transpose of the vector M by the matrix XY. The law of corresponding dimensions is respected

$$(1 - by - 6) \times (6 - by - 2) = (1 - by - 2)$$

The result is the row vector of the sum of moments about the two coordinate axes. It remains to divide the result by the total mass to obtain the coordinates of the center of mass about the given coordinate axes.

```
≫ M = [ 4; 5; 6; 4; 5; 3 ];
≫ XY = [ 1 6; 3 7; 4 4.5; 4.5 8; 7 3.5; 7.5 6.5 ]
XY =
   1.0000   6.0000
   3.0000   7.0000
   4.0000   4.5000
   4.5000   8.0000
   7.0000   3.5000
   7.5000   6.5000
≫ Mass = sum(M)
Mass =
   27
≫ Moments = M'*XY
Moments =
   118.5000   155.0000
≫ CG = Moments/Mass
CG =
    4.3889   5.7407
```

Can't we store all the mass data in a single matrix and still carry on the same operations? The answer is yes and one possibility is shown in the following screen.

```
≫ MXY = [ M XY ]
MXY =
   4.0000   1.0000   6.0000
   5.0000   3.0000   7.0000
   6.0000   4.0000   4.5000
   4.0000   4.5000   8.0000
   5.0000   7.0000   3.5000
   3.0000   7.5000   6.5000
≫ CG = MXY(:, 1)'*MXY(:, 2:3)/sum(MXY(:, 1))
CG =
    4.3889   5.7407
```

2.6.3 Moments as vector products

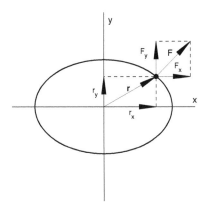

FIGURE 2.24: The moment of a force in the $xy-$plane

In Figure 2.24 we see a force, **F**, acting on a body in a point defined by the radius **r**. Both vectors, **F** and **r**, are defined by their components parallel to the given axes of coordinates. Considering that a moment is positive when it tends to rotate the body *counterclockwise*, the moment of the force about the origin of coordinates is given by

$$M = -r_y F_x + r_x F_y = \begin{vmatrix} r_x & r_y \\ F_x & F_y \end{vmatrix}$$

To give an example consider the point shown in Figure 2.24 and a force whose components parallel to the coordinate axes are both equal to 1.5 N. If the radius **r** is measured in m, the moment in Nm is

```
>> r = [ 3*cosd(40) 2*sind(40) ]
r =
   2.2981   1.2856
>> F = [ 1.5 1.5 ];
det([ r; F ])
ans =
   1.5188
```

The body tends to turn counterclockwise. The fact that we could calculate the moment as a determinant suggests that it is equal to the vector product of **r** by **F**. The moment about the origin, of a force, **F**, acting in a point identified by the vector **r**, is defined, indeed, as

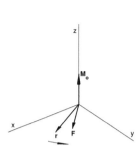

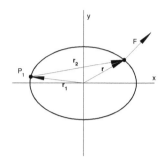

FIGURE 2.25:   The   moment
about a point

FIGURE 2.26:   Calculating the
moment about a shifted point

$$\mathbf{M}_o = \mathbf{r} \times \mathbf{F}$$

Thus, the moment is a vector perpendicular to the plane defined by $\mathbf{r}$ and $\mathbf{F}$ and pointing in the direction in which a screw would advance when $\mathbf{r}$ will turn towards $\mathbf{F}$. To continue the example in MATLAB we must first convert the vectors to three-dimensional ones and afterward call the command **cross**:

```
≫ r = [ r 0 ]
≫ F = [ F 0 ];
≫ cross(r, F)
ans =
   0  0 1.5188
```

Figure 2.25 shows at correct scale the three vectors involved in these calculations and the positive sense of rotation of $\mathbf{r}$ towards $\mathbf{F}$.

As an exercise let us calculate the moment of the force $\mathbf{F}$ about another point, P_1 in Figure 2.26. The new point is defined by the radius $\mathbf{r_1}$. The lever of the force about this point is $\mathbf{r_2} = \mathbf{r} - \mathbf{r_1}$. Then, the moment about P_1 is equal to

$$\mathbf{M}_1 = \mathbf{r_2} \times \mathbf{F} = (\mathbf{r} - \mathbf{r_1}) \times \mathbf{F}$$

As $\mathbf{r} \times \mathbf{F} = \mathbf{M}$, we conclude that

$$\mathbf{M}_1 = \mathbf{M} - \mathbf{r_1} \times \mathbf{F}$$

and this is the rule for calculating the moment of a force about a shifted point. With the values used previously, and defining the position vector of the shifted point, r_1, we calculate in MATLAB:

```
≫ r1 = [ -2.9544 0.3473 0 ];
≫ r2 = r - r1
r2 =
   5.2526   0.9383   0
≫ M1 = cross(r2, F)
M1 =
    0   0   6.4714
≫ M - cross(r1, F)
ans =
    0   0   6.4714
```

2.7   Summary

In this chapter we generalize the notion of *array* and its uses. First, we show that we can store numbers not only as a line, but also as a column. Next, we combine one-dimensional arrays into two-dimensional arrays. The array is the basic data structure of MATLAB. We show that the one-dimensional array allows a natural representation of *vectors*, quantities that have a magnitude, a direction and a sense. The operations defined on vectors are the addition, the multiplication by a scalar, the *scalar*, or *dot product*, and the *vector*, or *cross product*. This chapter includes a few applications of vectors in geometry and in mechanics.

We turn two-dimensional arrays into *matrices* by defining the *matrix product*. Another notion introduced in this chapter is that of *determinant*. We illustrate applications of these notions in geometry and mechanics.

In this chapter we learn to program in MATLAB:

For loops - constructions that perform **iterations** of operations or groups of operations;

If Elseif Else constructions - these are schemes that perform **conditional branching**.

When programming, mind the following recommendations.

Comment lines - The first line of a script file, or the first line following a function declaration should be a comment line beginning with the name of the file, preferably written in upper case letters. More comment lines can follow and they may include a short description of the purpose of the program, the input and output arguments, and the date of writing or last revision. These comments can be retrieved with the command

help followed by the name of the file, which is also the first word of the comments. This helps the writer of the program who wants to use it after some time and has forgotten the details. It also helps another user to call correctly a function. Ending the comments with a date helps the user in choosing the last version.

Indentation - The body of a repetitive construction, such as a **For loop**, or of a conditional construction, such as *if elseif else*, should be indented. Do this to enhance readability.

Check correctness of repetitive constructions - Tools for doing this are *simulation*, a loop *invariant*, or proof by *induction*.

The MATLAB commands and functions introduced in this chapter are:

axis([a b c d)] - defines the size of the figure.

axis equal - forces the same scale on the horizontal and the vertical axes;

axis off - suppresses the plotting of axes.

cross - calculates the vector, or cross product.

det - calculates the determinant of a square matrix.

diff - a function that calculates differences between adjacent elements of an array.

dot - a function that yields the scalar product of two vectors of the same length.

fill - a command that fills a figure with a required color.

hold on - a command that allows us to add plots over an existing graph. It is recommended to add the closing command **hold off** after completing the sequence of plotting commands. These two commands are introduced below, in Example 2.5.

norm - a function defined for more advanced purposes, but that we use in this chapter to calculate the length, or magnitude of a vector.

patch - another command command that fills a figure with a required color;

plot3 - the command for plotting lines in three-dimensional graphs.

size - a command that yields the dimensions of an array, that is the number or rows and the number of columns.

sum - the command that yields the sum of the elements in a given array.

zeros(m, n) - creates the *m-by-n* zero matrix.

To enhance graphs we introduce five elements belonging to *handle graphics*:

'color' - a command to define the color of a line, for example `'color'`, `'r'`.

'FontSize' - a command that allows us to define the size of fonts. The common unit is the *point* equal to 1/72 in.

'LineStyle' - a command that defines the line style, which can be, for example, '-', or ':'.

'LineWidth' - a command that allows us to define the thickness of lines.

set - a command of general use in *Handle graphics* and employed with a handle as first argument, for example `set(h, 'LineWidth', 1.5)`.

In this chapter we also build three utility functions that we are going to reuse throughout this book in creating figures:

arrow - in 2-D plots draws an arrow between two points defined by their coordinates;

pline - in 2- and 3-D drawings connects with straight-line segments of a set of points defined by their coordinates;

point - in 2-D plots draws a black circle of given radius, with center defined by its coordinates.

EXAMPLE 2.5 The graphics of the hanging weight

In this example we return to the function `Statics1`, developed in Example 2.4, and turn it into another function that plots the scheme of the problem adapted to the input variable a. In Chapter 10 we develop a graphical user interface, shortly GUI, that calls the function described in this example. The function developed here, `Stat1FigFun`, is called with one input argument, the horizontal distance a between the first and the second articulation, and an array of four output arguments

- the magnitude, $r1$, of the force **R1**;

- the magnitude, $r2$, of the force **R2**;

- the two-dimensional vector **R1**;

- the two-dimensional vector **R2**.

The figure is drawn in accordance with the argument a.

```
function    [ r1 r2 R1 R2 ] = Stat1FigFun(a)

%STAT1FIGFUN  plots the figure of the GUI TwoBarStatics
%             The input argument a is given in mm, as usual
%             in Mechanical-Engineering drawings.
%             The output arguments are measured in N.
%             This function calls the utility functions arrow,
%             pline and point.
%             Written by Adrian Biran for Taylor &
%             Francis, June 2009.

% Define parameters for drawing
ad = a/10;
b    = 150 - ad;
c    =    50;
ha   = 110;        % higher articulations z
w    =    50;      % length of W vector, N
% define articulations;
A    = [ 30; ha ];
B    = [ (30+ ad); (ha -c) ];
C    = [ (30 + 150); ha ];
% plot something
pline([ A B C ], 1, 'k', '-')
axis([ 0 190 0 135 ])
axis equal, axis off
hold on
% plot ceiling
hp = plot([ 20 190 ], [ 115 115 ], 'k-');
set(hp, 'LineWidth', 1.5)
plot([ 20 22.5 ], [ 117.5 120 ], 'k-')

% plot articulations
point(A, 1)      % first articulation from left
plot([ 25 30 35 25 30 ], [ 115 110.5 115 115 110.5 ], 'k-')
point(B, 1)      % second articulation from left
x0 = 20;         % initialize hatching
for k = 1:67
    plot([ x0 (x0 + 5) ], [ 115 120 ], 'k-')
    x0 = x0 + 2.5;
end
plot([ 187.5 190 ], [ 115 117.5 ], 'k-')    % end hatching
point(C, 1)      % third articulation from left
plot([ 175 180 185 175 180 ], [ 115 110.5 115 115 110.5 ], 'k-')
% plot hanging weight
arrow(B, [ B(1); (B(2) - w) ], 2.1)
```

```
ht = text((B(1) + 2), 30, 'W = 50 N');
set(ht, 'FontSize', 14)

% Plot inverse of W vector
arrow(B, [ B(1); (B(2) + w) ], 1.3)
text((B(1) + 3), (B(2) + w/2), '-W')
% calculate angles
alpha = atand(ad/c);
beta    = atand(b/c);
gamma   = 180 - alpha - beta;
% calculate magnitudes of tension vectors
r1          = sind(beta)*w/sind(gamma);
r2          = sind(alpha)*w/sind(gamma);
% define tension vectors   % two-dimensional
R1          = r1*[ -sind(alpha); cosd(alpha) ];
R2          = r2*[  sind(beta); cosd(beta) ];
% coordinates of vector extremities
L1          = B + R1;
arrow(B, L1, 1.6)
ht          = text((L1(1) - 8), (L1(2) - 8), 'R_1');
set(ht, 'FontSize', 12, 'FontWeight', 'bold')
L2          = B + R2;
ht          = text((L2(1) + 5), (L2(2) - 4), 'R_2');
set(ht, 'FontSize', 12, 'FontWeight', 'bold')
arrow(B, L2, 1.6)
% dimensions
plot([ 30 30 ], [ 115 126 ], 'k-')
plot([ B(1) B(1) ], [ 115 126 ], 'k-')
plot([ 180 180 ], [ 115 126 ], 'k-')
arrow([ (A(1) + B(1))/2 ; 125 ], [ 30; 125 ], 0.8)
arrow([ (A(1) + B(1))/2 ; 125 ], [ B(1); 125 ], 0.8)
ht = text((30 + ad/2), 129, 'a');
set(ht, 'FontSize', 10)
arrow([ (B(1) + C(1))/2 ; 125 ], [ B(1); 125 ], 0.8)
arrow([ (B(1) + C(1))/2; 125 ], [ C(1); 125 ], 0.8)
ht = text((B(1) + C(1))/2, 129, 'b');
set(ht, 'FontSize', 10)
plot([ 4 30 ], [ 110 110 ], 'k-')
plot([ 4 B(1) ], [ 60 60 ], 'k-')
arrow([ 5; 85 ], [ 5; 110 ], 0.8)
arrow([ 5; 85 ], [ 5; 60 ], 0.8)
ht = text(0, 65, 'c = 500', 'rotation', 90);
set(ht, 'FontSize', 10)
% plot parallelogram of forces
pline([ L1 [ B(1); (B(2) + w) ] ], 1, 'r', '--')
```

```
pline([ L2 [ B(1); (B(2) + w) ] ], 1, 'r', '--')
hold off
```

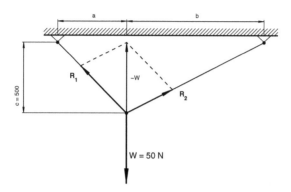

FIGURE 2.27: Figure generated with the command `Stat1FigFun(500)`

The first plotted elements are the straight-line segments that connect the three articulations. MATLAB sizes the figure according to the first plotted elements. We need a slightly larger figure for adding more elements. To avoid the resizing of the figure, we define its size from the beginning by using the command `axis` with an array argument. The four elements of the array are the x−coordinate of the left corners of the plotting area, the x−coordinate of the right corners, the y−coordinate of the lower corners, and the y−coordinate of the upper corners. Further, we use the command `axis equal` to force the same scales along the x− and y−axes, and the command `axis off` to suppress the plotting of axes.

2.8   More exercises

Exercise 2.3 Plotting a square

a) Given the three points

$$\mathbf{P_1} = \begin{bmatrix} 1 \\ 1 \end{bmatrix}, \ \mathbf{P_2} = \begin{bmatrix} 1 \\ 3 \end{bmatrix}, \ \mathbf{P_3} = \begin{bmatrix} 3 \\ 3 \end{bmatrix}$$

define a fourth point, $\mathbf{P_4}$, such that $\mathbf{P_1}$, $\mathbf{P_2}$, $\mathbf{P_3}$, $\mathbf{P_4}$ are the vertices of a square.

b) Plot the square.

c) Calculate the perimeter of this square and verify that it corresponds to what you expect.

Exercise 2.4 A function for calculating the perimeter

Convert the function `perim`, developed in Subsection 2.1.3, into a function, `perim1`, valid for n−dimensional vectors. To do this, use the function `size` to determine the dimension, n, of the vectors and call the function `diff` in a loop that is executed n times, each time for another coordinate.

Exercise 2.5 The angle between two vectors

Write a function that finds the angle between two vectors according to the specification

Input arguments: two vectors V1, V2, defined by their components parallel to cartesian axes of coordinates;

Output argument: the angle between the two vectors, in degrees.

Exercise 2.6 Elementary operations on matrices

Define in MATLAB the matrices

$$A = \begin{bmatrix} 6 & -4 & 10 \\ 2 & 0 & -6 \end{bmatrix}, \quad B = \begin{bmatrix} -2 & 10 & 0 \\ 4 & -6 & 14 \end{bmatrix}, \quad C = \begin{bmatrix} 1 & 3 & 2 \\ 3 & 1 & 2 \end{bmatrix},$$

1) Check that the addition of matrices is *commutative*, that is

$$A + B = B + A$$

2) Check that

$$A - B = -(B - A)$$

3) Check that the addition of matrices is *associative*, for example

$$A + (B + C) = (A + B) + C$$

4) Define in MATLAB

$$D = \begin{bmatrix} 5 & 10 \\ 1 & 2 \end{bmatrix}$$

Try the operations

$$A + D, \ A - D$$

and explain what happens.

5) Define in MATLAB the *2-by-3 zero matrix*

```
O = zeros(size(A))
```

and check that

$$A + O = O + A, \ B + O = O + B$$

7) Define in MATLAB the scalar $k = 2$. See what happens if you multiply it by a matrix, for example $k * A$, with A defined as above.

8) Check that the multiplication of matrices by scalars is *associative*, for example

$$k * A + k * B = k * (A + B)$$

Also, define $h = 3$ and check that

$$h * A + k * A = (k + h) * A$$

Note. The elements of the matrices involved in this exercise are all integers. Therefore, the proprieties described here are always verified. When working with numbers that are not integers, many of the above properties are verified only approximatively. This subject is treated in Chapter 5.

Exercise 2.7 The cross products of base vectors
For the vector basis defined in Subsection 2.1.7 use the cross product to show that

$$e_1 \times e_2 = e_3, \ e_2 \times e_3 = e_1, \ e_3 \times e_1 = e_2$$

Mind the circular permutation of indices in the above equations.

Exercise 2.8 Properties of rotation
For the rotation shown in Figure 2.21 you are asked to:

1. use the function norm to check that the distances between points are not changed by rotation. In other words check that $\overline{P_1 P2} = \overline{Q_1 Q_2}$, a.s.o;

2. use the dot product to check that the points are rotated, indeed, by 30 degrees. In other words check that $\widehat{P_1 O Q_1} = 30°$, a.s.o.

Exercise 2.9 Combining translation and rotation
Suppose that we have to perform several transformations. In general, the order in which we carry them defines the final result. Consider in Subsection 2.1.5 the triangle defined by the three points, $\mathbf{P_1}$, $\mathbf{P_2}$, $\mathbf{P_3}$. Your assignment incudes the following tasks:

1. convert the matrix of coordinates, $\mathbf{P}$, into a matrix of homogeneous coordinates;

2. plot the initial triangle defined by **P**. The axes of the plot should allow the addition of the two plots required below;

3. define a translation matrix, in homogeneous coordinates, that shifts objects 3 units horizontally, and 4 units vertically;

4. define a rotation matrix, in homogeneous coordinates, that rotates objects 30 degrees counterclockwise;

5. using the matrices defined above, first translate, next rotate the triangle. Call the resulting points Q_1, Q_2, Q_3. Plot the transformed triangle in the same figure that shows the initial triangle;

6. using the matrices defined above, first rotate, next translate the triangle. Call the resulting points R_1, R_2, R_3. Plot the transformed triangle in the same figure that shows the initial triangle.

Exercise 2.10 Weight data of a containership

FIGURE 2.28: Coordinates used for the containership

Table 2.1 shows the weight data of a containership. These data are taken from an example found in Taggart (1980) and are slightly simplified for the purpose of the exercise. The coordinates of the centers of gravity of the various weight groups are:

VCG (*Vertical Center of Gravity*), is the height coordinate measured parallel to the $z-$axis, from the *baseline*, as shown in Figure 2.28;

LCG (*Longitudinal Center of Gravity*), is the length coordinate measured parallel to the $x-$axis, from the line marked AP (Aft Perpendicular), as shown in Figure 2.28. The Aft Perpendicular coincides here with the $z-$axis.

The $y-$ coordinate of the center of gravity of the lightship and that of the fully-loaded ship are expected to be zero; therefore, the table contains only two coordinates.

The ship without any load consists of items 1 to 3. The sum of the masses of the three first items is known as *lightship*. Items 4 to 10 constitute the *deadweight*. The sum of the Lightship and Deadweight masses is the *full load* mass. In this exercise you are required to calculate

Table 2.1: Weight data of containership

	Weight group	Mass, t	VCG, m	LCG, m
1	Hull	8624	9.8	99.5
2	Outfit	2383	14.2	94.6
3	Machinery	1158	8.4	32.1
4	Miscellaneous deadweight	152	11.6	16.5
5	Passengers, crew and stores	85	15.3	26.5
6	Fuel oil	6727	8.0	113.3
7	Fresh water	588	4.5	32.9
8	Salt water ballast	2470	1.0	108.4
9	Stabilizer tank	163	9.3	80.2
10	Container cargo	15485	14.1	102.6

1. the total mass and the VCG and LCG of the lightship;

2. the total mass and the VCG and LCG of the deadweight;

3. the total mass and the VCG and LCG of the full load.

Solve this exercise with the techniques described in Subsection 2.6.2.

3

Equations

3.1 Introduction

The first part of this chapter deals with systems of linear equations. To solve them in MATLAB they must be written in the matrix form $\mathbf{AX} = \mathbf{B}$. Then, the preferred MATLAB solution uses the **backslash operator**, '\', called as $\mathbf{X} = \mathbf{A}\backslash\mathbf{B}$. The backslash operator always yields a solution; it can be a unique solution, a particular solution, a solution in the *least-squares* sense, or no solution at all. To explain these cases we distinguish between fully determined, underdetermined, overdetermined and homogeneous systems. The Kronecker-Capelli theorem gives a criterion for knowing whether a given system of linear equations has a solution or not. The criterion is based on the notion of *rank*, a number that can be calculated in MATLAB.

The second part of the chapter treats to some extent algebraic equations. To explain how to solve them in MATLAB we show first how to represent polynomials and how to deal with them in MATLAB. Equating a polynomial to zero yields an equation that can be solved in MATLAB with the command *roots*. This command, however, has its limitations, a subject discussed later in another chapter.

Equations that are not polynomials can be solved by iterative methods. We show how to implement in MATLAB a popular iterative method, that is the *Newton-Raphson method*.

3.2 Linear equations in geometry

3.2.1 The intersection of two lines

Let us consider the following system of two linear equations in two unknowns

$$0.5x - y + 2.5 = 0$$
$$3x - y - 2.5 = 0$$

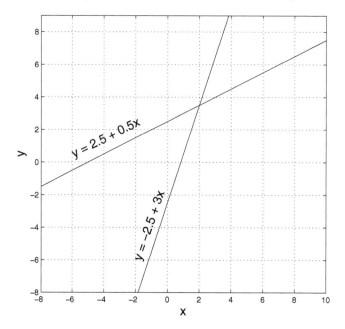

FIGURE 3.1: The intersection of two straight lines

If we rewrite them as

$$y = 2.5 + 0.5x$$
$$y = -2.5 + 3x \tag{3.1}$$

we can interpret these equations as representing the two straight lines plotted in Figure 3.1. The solution of the system is the pair of numbers, x, y, that simultaneously satisfies both equations. This is the intersection point of the two lines. To find the solution we can, for example, retrieve x from the first solution and substitute it in the second. We obtain

$$x = 5/2.5 = 2$$

Substituting the above value into the second equation yields $y = 3.5$.

3.2.2 Cramer's rule

The system examined in the previous subsection was simple and it was easy to solve it in a particular way. It is interesting, however, to look for more general solutions that would suit more complex systems. A first step in this direction is to use Cramer's rule. To do so we must first define the *coefficient matrix*, A, and the free vector, B, so that the vector X is the solution of the matrix equation

$$AX = B \qquad (3.2)$$

For a system of two equations in two unknowns we can detail Equation 3.2 as

$$\begin{bmatrix} a_{11} & a_{12} \\ a_{21} & a_{22} \end{bmatrix} \begin{bmatrix} x_1 \\ x_2 \end{bmatrix} = \begin{bmatrix} b_1 \\ b_2 \end{bmatrix}$$

The solution by Cramer's rule is

$$x_1 = \frac{\begin{vmatrix} b_1 & a_{12} \\ b_2 & a_{22} \end{vmatrix}}{\begin{vmatrix} a_{11} & a_{12} \\ a_{21} & a_{22} \end{vmatrix}}, \ x_2 = \frac{\begin{vmatrix} a_{11} & b_1 \\ a_{21} & b_2 \end{vmatrix}}{\begin{vmatrix} a_{11} & a_{12} \\ a_{21} & a_{22} \end{vmatrix}},$$

where we used the notation '$|\ |$' for determinants. Let us use this method to solve Equations 3.1 in MATLAB. We first define the matrices involved and solve for x :

```
≫ A = [ 0.5 -1; 3 -1 ];
≫ B = [ -2.5; 2.5 ];
≫ D1 = [ B A(:, 2) ];
≫ x = det(D1)/det(A)
x =
    2
```

Next, we solve for y :

```
≫ D2 = [ A(:, 1) B ];
≫ y = det(D2)/det(A)
y =
    3.5000
```

3.2.3 MATLAB's solution of linear equations

Cramer's method is easy to explain, but computationally inefficient; it is acceptable for 'small' systems. We are looking for methods that suit also large systems of equations. One idea is be to multiply the free vector $\mathbf{B}$ by the inverse of the matrix $\mathbf{A}$. We say that the matrix $\mathbf{A}^{-1}$ is the **inverse** of the matrix $\mathbf{A}$ if $\mathbf{A}\mathbf{A}^{-1} = \mathbf{A}^{-1}\mathbf{A} = \mathbf{I}$, where $\mathbf{I}$ is the *identity*, or *unit* matrix with the same dimensions as $\mathbf{A}$. In MATLAB the inverse of a matrix is calculated with the command inv, for example, with the same data as above,

```
≫ Ainv = inv(A)
Ainv =
   -0.4000  0.4000
   -1.2000 0.2000
≫ A*Ainv
ans =
   1  0
   0  1
≫ Ainv*A
ans =
   1  0
   0  1
```

The matrix

$$\begin{bmatrix} 1 & 0 \\ 0 & 1 \end{bmatrix}$$

is the identity or unit $2-by-2$ matrix; it can be obtained in MATLAB with the command eye

```
≫ I2 = eye(2)
I2 =
   1  0
   0  1
```

Using the command inv, the solution of the given system is found by

```
≫ X = inv(A)*B
X =
   2.0000
   3.5000
```

It worked, but this method is applicable only to square matrices that are not singular. In this chapter we give also examples that do not answer to these conditions. Therefore, MATLAB provides an efficient method that uses *Gaussian elimination* for the general case, and special methods for particular cases. These efficient solutions are obtained with the *backslash* operator, '\,

```
≫ X = [ x; y ];
≫ X = A\B
X =
    2.0000
    3.5000
```

To verify the solution we calculate the *residual*:

```
≫ X = [ x; y ];
≫ A*X - B
ans =
    0
    0
```

In this simple case the residual is zero. In other cases, however, due to the way in which numbers are stored in the computer, there can be a very small residual. See Chapter 5 for more details.

3.2.4 An example of an ill-conditioned system

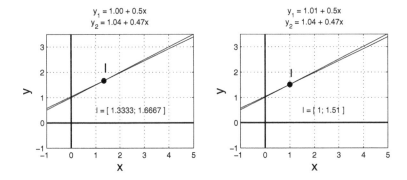

FIGURE 3.2: An ill-conditioned system

Whoever has tried to find the intersection of two lines that have nearly the same slope knows that the solution is not accurately defined. Such an example is shown in Figure 3.2. Therefore, textbooks of navigation teach the reader that when a point must be found as the intersection of two *position lines*, it is recommended to look for lines that are as nearly as possible perpendicular one to another. In linear algebra, finding the intersection of two lines that have nearly the same slope leads to an *ill-conditioned* system whose study has important implications in engineering. Let us see an example treated in MATLAB. Below, we first enter the matrices that describe the system shown in the left-hand side of Figure 3.2 and solve them obtaining the solution vector X_1. Next, we consider the right-hand side of the same figure. The first equation (the one in x_1, y_1) is slightly changed, while the second (the one in x_2, y_2) remains as in the left-hand side plot. We enter these equations too, and solve them obtaining the solution vector X_2. Finally, we calculate the relative difference, in percentage, between the two solutions.

```
≫ A1 = [ 0.5 -1; 0.47 -1 ];
≫ B1 = [ -1; -1.04 ];
≫ X1 = A1 \ B1
X1 =
   1.3333
   1.6667
≫ A2 = A1;
≫ B2 = [ -1.01; -1.04 ];
≫ X2 = A2 \ B2
X2 =
   1.0000
   1.5100
≫ e1 = 100*(X2 - X1)/X1
e1 =
   0   -20.0000
   0   -9.4000
```

In the second system of equations b_1 changed by 1%. The change of the $x-$coordinate of the intersection point is 20%, and that of the $y-$coordinate, 9%. In other words, a small change in one parameter of the system resulted in a much larger change in the solution of the system.

One measure of the sensitivity of a system to changes in its parameters is the *condition number* of the coefficient matrix. The definition of the condition number in MATLAB can be found by typing **doc cond**. For our purposes it is sufficient to mention here that the change in the solution of a system of linear equations is smaller than the product of the condition number and the change in the parameters of the system. The condition number is calculated in MATLAB with the command **cond**

```
≫ cond(A1)
ans =
   82.3512
```

We say that the system considered in this subsection is *ill conditioned*. Such systems should be avoided in engineering and the reason is simple. Real-life measurements in engineering systems are always affected by errors. If such measurements are used as input to an ill-conditioned system, the results of calculations may be affected by large errors and be unreliable.

3.2.5   The intersection of three planes

Let us consider the system of three linear equations

$$\frac{x}{10} + \frac{y}{3} + \frac{z}{3} = 1$$
$$\frac{x}{2} + \frac{y}{20} + \frac{z}{2} = 1$$
$$\frac{x}{5} + \frac{y}{10} + \frac{z}{0.5} = 1 \tag{3.3}$$

Their solution is given by

```
≫ A = [ 1/10 1/3 1/3; 1/2 1/20 1/2; 1/5 1/10 1/0.5 ];
≫ B = [ 1; 1; 1 ];
≫ A\B
ans =
   1.5385
   2.3077
   0.2308
```

Equations 3.3 can be interpreted as representing three planes in the x, y, $z-$ space. The denominator 10 in the coefficient that multiplies x in the first equation is the coordinate of the point in which the first plane intercepts the x axis. Similarly, the denominator 3 in the coefficient that multiplies y in the first equation is the coordinate of the point in which the first plane intercepts the y axis, a.s.o. In Figure 3.3 we detail this interpretation. The upper, left-hand plot shows the plane defined by the first equation. The upper, right-hand plot shows the planes defined by the first two equations and their intersection. The lower, left-hand plot shows the intersection line of the first two planes, and the third plane. The lower, right-hand figure shows the plane defined by the third equation and, on it, the solution point.

3.3   Linear equations in statics

3.3.1   A simple beam

Figure 3.4 shows a beam with two supports. The beam is acted upon by a force, F_1, perpendicular to the beam, and an inclined force, F_2. Let the angle between the latter force and the beam be α. The reaction, R_1, in the

The plane x/10 + y/3 + z/3 = 1

Plus the plane x/2 + y/20 + z/2 = 1

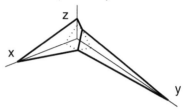

Intersection plus plane x/5 + y/10 + z/0.5 = 1

Third plane and solution point

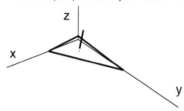

FIGURE 3.3: The intersection of three planes

left-hand support is perpendicular to the beam. The reaction in the right-hand support has a component, R_{2y}, perpendicular to the beam, and an *axial* component, R_{2x}, that opposes the axial component of F_1. We have three unknowns, R_1, R_{2x}, R_{2y}. Writing that the sum of vertical forces is zero, the sum of horizontal forces is zero, and the sum of moments about the left-hand support is zero, we obtain the three equations

$$R_1 + R_{2y} = F_1 + F_2 \sin \alpha$$
$$R_{2x} = F_2 \cos \alpha$$
$$l R_{2y} = l_1 F_1 + l_2 F_2 \sin \alpha$$

To give an example in MATLAB we assume the values that are shown below

```
≫ l = 5;
≫ l1 = 1;
≫ l2 = 3;
≫ alpha = 60;
≫F1 = 800;
≫F2 = 500;
```

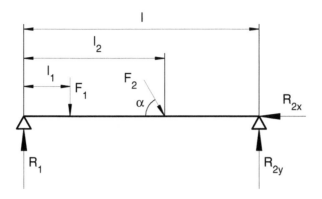

FIGURE 3.4: A simply-supported beam

Next, we build the coefficient matrix and the free vector, and solve the system:

```
≫ A = [ 1 0 1; 0 1 0 ; 0 0 1 ];
≫ B = [ F1 + F2*sind(alpha);
F2*cosd(alpha);
l1*F1 + l2*F2*sind(alpha) ];
≫ X = A \ B
X =
   813.2051
   250.0000
   419.8076
```

We separate the elements of the solution vector and check the correctness of the solution by calculating the moments about the right-hand support:

```
≫ R1 = X(1)
R1 =
   813.2051
≫ R2x = X(2)
R2x =
   250.0000
≫ R2y = X(3)
R2y =
   419.8076
≫ (1- l1)*F1 + (1 - l2)*F2*sind(alpha) - 1*R1
ans =
   0
```

3.4   Linear equations in electricity

3.4.1   A DC circuit

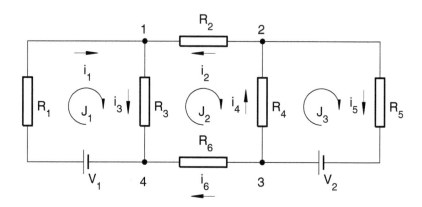

FIGURE 3.5: A DC circuit

Figure 3.5 shows a DC circuit that includes two voltage sources, V_1 and V_2, and six resistors marked R_1 to R_6. Applying Kirchoff's first law (the law of currents) to nodes 1, 2, and 3 we obtain the equations

$$-i_1 - i_2 + i_3 = 0$$
$$i_2 - i_4 + i_5 = 0$$
$$i_4 - i_5 + i_6 = 0 \qquad (3.4)$$

Above we considered as positive the currents that exit a node. Next we apply Kirchoff's second law (the law of voltages) to the left, the right and the middle loops

$$i_1 R_1 + i_3 R_3 = V_1$$
$$i_4 R_4 + i_5 R_5 = V_2$$
$$i_2 R_2 + i_3 R_3 + i_4 R_4 - i_6 R_6 = 0 \qquad (3.5)$$

Equations 3.4 and 3.5 constitute a system of six linear equations in six unknowns that can be rewritten in matrix form as

$$
\begin{vmatrix}
-1 & -1 & 1 & 0 & 0 & 0 \\
0 & 1 & 0 & -1 & 1 & 0 \\
0 & 0 & 0 & 1 & -1 & 1 \\
R_1 & 0 & R_3 & 0 & 0 & 0 \\
0 & 0 & 0 & R_4 & R_5 & 0 \\
0 & R_2 & R_3 & R_4 & 0 & -R_6
\end{vmatrix}
\begin{vmatrix} i_1 \\ i_2 \\ i_3 \\ i_4 \\ i_5 \\ i_6 \end{vmatrix}
=
\begin{vmatrix} 0 \\ 0 \\ 0 \\ V_1 \\ V_2 \\ 0 \end{vmatrix}
\tag{3.6}
$$

To calculate an example in MATLAB, let us assume the following values:

Resistors, in Ω : $R_1 = 20$, $R_2 = 40$, $R_3 = 20$, $R_4 = 40$, $R_5 = 40$, $R_6 = 10$.

Voltage sources, in V : $V_1 = 9$, $V_2 = 12$.

We enter these values in MATLAB and define the matrix of resistances, A, and the vector of voltage sources, S:

```
≫ R = [ 20 40 20 40 40 10 ];
≫ V1 = 9;
≫ V2 = 12;
A = [ -1 -1 1 0 0 0
0 1 0 -1 1 0
0 0 0 1 -1 1
R(1) 0 R(3) 0 0 0
0 0 0 R(4) R(5) 0
0 R(2) R(3) R(4) 0 -R(6) ];
≫ S = [ 0; 0; 0; V1; V2; 0 ];
```

We use the backslash operator to solve the system and obtain the branch currents measured in *amperes*:

```
≫ I1 = A\S
I1 =
0.2906
-0.1313
0.1594
0.0844
0.2156
0.1312
```

To verify the correctness of our results we can calculate the sums of currents at nodes 0 to 4:

```
>> I1(1) - I1(3) - I1(6)
   1.1102e-016
>> -I1(1) - I1(2) + I1(3)
ans =
   0
>> I1(2) - I1(4) + I1(5)
ans =
   -5.5511e-017
>> I1(4) - I1(5) + I1(6)
ans =
   -5.5511e-017
```

For reasons explained in Chapter 5 we obtained very small numbers instead of the expected zeros. We can also apply Kirchoff's second law to the outer loop:

```
>> R(1)*I1(1) - R(2)*I1(2) + R(5)*I1(5) + R(6)*I1(6) - V1 -
V2
ans =
   0
```

3.4.2   The method of loop currents

In the preceding subsection we considered the circuit shown in Figure 3.5 and we analyzed it by applying Kirchoff's laws. In doing this we had to write six equations because the unknowns are six branch currents. There are other methods that require fewer equations and reveal a helpful pattern in the resulting system. In this subsection we are going to analyze the same system, but employ the *method of loop, or mesh currents*. We consider again Figure 3.5 and identify three loops. The first loop, for example, includes the voltage source V_1 and the resistors R_1 and R_3. We assume that the current through it is J_1 and it runs in a clockwise sense. The second loop includes only resistors, R_3, R_2, R_4, and R_6. We assume again a clockwise current, J_2. The third loop includes the voltage source V_2 and the resistors R_4, R_5. The clockwise current through this loop is J_3. Kirchoff's second law applied to the three loops yields the equations

$$R_1 J_1 + R_3(J_1 - J_2) = V_1$$
$$R_3(J_2 - J_1) + R_2 J_2 + R_4(J_2 - J_3) + R_6 J_2 = 0$$
$$R_4(J_3 - J_2) + R_5 J_5 = V_2 \qquad (3.7)$$

The matrix form of the system represented by Equations 3.7 is

$$\begin{vmatrix} (R_1 + R_3) & -R_3 & 0 \\ -R_3 & R_2 + R_3 + R_4 + R_6 & -R_4 \\ 0 & -R_4 & R_4 + R_5 \end{vmatrix} \begin{vmatrix} J_1 \\ J_2 \\ J_3 \end{vmatrix} = \begin{vmatrix} V_1 \\ 0 \\ V_2 \end{vmatrix} \qquad (3.8)$$

Visual inspection of the matrix of resistances reveals an important pattern. The elements on the diagonal are the sums of the resistances in a loop. For example, the element in position 1,1 is the sum of the resistances R_1 and R_3 included in the first loop. Similarly, the element in position 2,2 is the sum of the four resistors that belong to the second loop. The element in position 1,2 is the resistor common to the the first and the second loop. The matrix is symmetric. Thus, the element in position 2,1 is equal to the element in position 1,2 because the same resistor that connects the first loop to the second also connects the second loop to the first. The elements in positions 1,3 and 3,1 are zero because no resistor connects the first loop to the third. The elements that are off diagonal appear with a negative sign. Knowing this pattern one can directly write the matrix of resistors.

Assuming the same values as in the preceding subsection, we calculate in MATLAB:

```
≫ RM = [ (R(1) + R(3)) -R(3) 0
 -R(3) (R(2) + R(3) + R(4) + R(6)) -R(4)
 0 -R(4) (R(4) + R(5)) ];
≫ VV = [ V1; 0; V2 ];
≫ J = RM \ VV
≫ J =
   0.2906
   0.1313
   0.2156
```

The branch currents we are looking for are given by

$$i_1 = J_1$$
$$i_2 = -J_2$$
$$i_3 = J_1 - J_4$$
$$i_4 = J_3 - J_2$$
$$i_5 = J_3$$
$$i_6 = J_2$$

We calculate them in MATLAB by

```
≫ I2 = zeros(5, 1);
≫ I2(1) = J(1);
≫ I2(2) = -J(2);
≫ I2(3) = J(1) - J(2);
≫ I2(4) = J(3) - J(2);
≫ I2(5) = J(3)
≫ I2(6) = J(2)
I2 =
   0.2906
  -0.1313
   0.1594
   0.0844
   0.2156
   0.1313
```

These are exactly the results obtained in the previous subsection. If we still have in the workspace the results of the first subsection we can easily check:

```
≫ I1 - I2
ans =
   1.0e-016 *
   0.5551
  -0.2776
   0.2776
   0
  -0.2776
  -0.8327
```

The differences are either zero or very small numbers due to numerical errors.

3.5   On the solution of linear equations

3.5.1   Homogeneous linear equations

If the free vector, **B**, is zero we say that the system is **homogeneous**

$$\mathbf{AX} = \mathbf{0}$$

Such a system has non-trivial solutions – that is solutions other than $\mathbf{X} = \mathbf{0}$ – only if the determinant of the matrice of coefficients is zero. The case of a system of two homogeneous equations in two unknowns lends itself to a simple, but interesting geometrical interpretation (see Banchoff and Wermer, 1983). Let us consider the system

$$a_1 x + b_1 y = 0$$
$$a_2 x + b_2 y = 0 \qquad (3.9)$$

and define

$$\mathbf{V_1} = \begin{bmatrix} a1 \\ b1 \end{bmatrix}, \ \mathbf{V_2} = \begin{bmatrix} a2 \\ b2 \end{bmatrix}, \ \mathbf{X} = \begin{bmatrix} x \\ y \end{bmatrix},$$

Equations 3.9 show that the scalar products $\mathbf{V_1} \cdot \mathbf{X}$ and $\mathbf{V_2} \cdot \mathbf{X}$ are zero. This means that the solution vector $\mathbf{X}$ must be perpendicular to both $\mathbf{V_1}$ and $\mathbf{V_2}$. The condition can be fulfilled if and only if the two vectors $\mathbf{V_1}$ and $\mathbf{V_2}$ lie on the same line that passes through the origin. In other words, there must exist some scalar k such that

$$\mathbf{V_2} = k\mathbf{V_1}$$

Then we can rewrite Equations 3.9 in matrix form as

$$\begin{bmatrix} a_1 & b_1 \\ ta_1 & tb_1 \end{bmatrix} \begin{bmatrix} x \\ y \end{bmatrix} = 0$$

The determinant of the coefficient matrix is obviously zero, as required. As the vector $\mathbf{X}$ is perpendicular to the vector $\mathbf{V_1}$, its general form is

$$\mathbf{X} = \begin{bmatrix} -tb_1 \\ ta_1 \end{bmatrix} = t \begin{bmatrix} -b_1 \\ a_1 \end{bmatrix}$$

As an example, enter in MATLAB the coefficient matrix

$$\mathbf{A} = \begin{bmatrix} 2 & -7 \\ 10 & -35 \end{bmatrix}$$

and show that the system

$$\mathbf{AX} = \mathbf{0} \qquad (3.10)$$

has solutions of the form

$$\mathbf{X} = c \begin{bmatrix} 7 \\ 2 \end{bmatrix}$$

where c is a scalar of your choice. MATLAB cannot solve this case with the backslash operator, that is by the command X = A\B, where all the components of B are zero. Try it for yourself.

And what about a system of three homogeneous equations in three unknowns? Let us consider the system

$$a_1 x + b_1 y + c_1 z = 0 \tag{3.11}$$
$$a_2 x + b_2 y + c_2 z = 0$$
$$a_3 x + b_3 y + c_3 z = 0$$

and define

$$\mathbf{V_1} = \begin{bmatrix} a_1 \\ b_1 \\ c_1 \end{bmatrix}, \ \mathbf{V_2} = \begin{bmatrix} a_2 \\ b_2 \\ c_2 \end{bmatrix}, \ \mathbf{V_3} = \begin{bmatrix} a_3 \\ b_3 \\ c_3 \end{bmatrix}, \ \mathbf{X} = \begin{bmatrix} x \\ y \\ z \end{bmatrix}$$

Equations 3.11 show that if a solution vector exists, it must be perpendicular to the vectors $\mathbf{V_1}$, $\mathbf{V_2}$, $\mathbf{V_3}$. All three vectors pass through the origin. Let us assume that none of the vectors $\mathbf{V_1}$ and $\mathbf{V_2}$ is a multiple of the other. Then, these vectors define a plane; let us call it Δ. The solution vector, $\mathbf{X}$, can be perpendicular also to the vector $\mathbf{V_3}$ only if the latter lies in the plane Δ. This happens if $\mathbf{V_3}$ is a linear combination of the two vectors $\mathbf{V_1}$ and $\mathbf{V_2}$ and the system of Equations 3.11 can be rewritten as

$$\begin{bmatrix} a_1 & b_1 & c_1 \\ a_2 & b_2 & c_2 \\ ma_1 + na_2 & mb_1 + nb_2 & mc_1 + nc_2 \end{bmatrix} \begin{bmatrix} x \\ y \\ z \end{bmatrix} = \begin{bmatrix} 0 \\ 0 \\ 0 \end{bmatrix} \tag{3.12}$$

The determinant of the coefficient matrix is obviously zero satisfying thus the condition for a non-trivial solution. As the solution vector, $\mathbf{X}$, must be perpendicular to the plane Δ defined above; the solution vector is any multiple of the vector product of $\mathbf{V_1}$, $\mathbf{V_2}$. To give an example in MATLAB let us consider a homogeneous system whose coefficient matrix is

$$\mathbf{A} = \begin{bmatrix} 3 & 5 & 7 \\ 2 & 9 & 3 \\ 5 & 14 & 10 \end{bmatrix}$$

First, check that the third row is, indeed, a linear combination of the first two rows. Next, check in MATLAB that the determinant of $\mathbf{A}$ is zero. Now calculate in MATLAB:

```
>> A = [ 3 5 7; 2 9 3; 5 14 10 ];
>> V1 = A(1, :);
>> V2 = A(3, :);
>> X = cross(V1, V2)
X =
   -48 5 17
>> A*X'
ans =
   0
   0
   0
```

In conclusion, the general solution vector is

$$\mathbf{X} = c \begin{bmatrix} -48 \\ 5 \\ 17 \end{bmatrix}$$

where c is any scalar, in the trivial case 0.

To see what is the solution yielded by the MATLAB backslash operator try the following:

```
>> B = [ 0; 0; 0 ];
>> A\ B
Warning:  Matrix is close to singular or badly scaled.
   Results may be inaccurate.  RCOND = 1.348128e-017.
ans =
   0
   0
   0
```

This is only the trivial solution.

3.5.2 Overdetermined systems — least-squares solution

A system of equations is said to be **overdetermined** if the number of equations is larger than the number of unknowns. We consider here only the case in which the given equations are linearly independent, that is no equation can be obtained as a linear combination of other equations of the system, a case that can yield a unique solution. One example of an overdetermined system that has no solution is

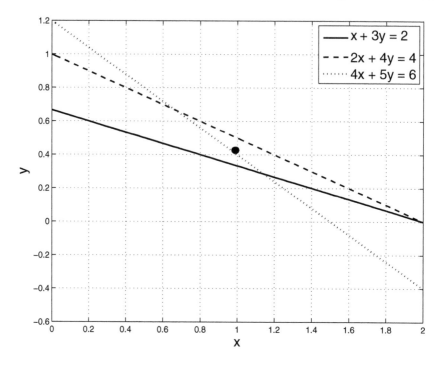

FIGURE 3.6: The geometrical interpretation of an overdetermined system

$$x + 3y = 2$$
$$2x + 4y = 4$$
$$4x + 5y = 6 \qquad (3.13)$$

Equations 3.13 can be interpreted as the three lines shown in Figure 3.6. Each pair of equations has an intersection, but there is no point common to all three lines. The MATLAB backslash operator yields, however, a solution:

```
≫ A = [ 1 3; 2 4; 4 5 ];
≫ B = [ 2; 4; 6 ];
≫ X = A\B
X =
   0.9888
   0.4270
```

The point having these coordinates is shown as a black circle in Figure 3.6. Obviously this is not a solution as it can also be seen by calculating the residual:

```
≫ R = A*X - B
R =
    0.2697
   -0.3146
    0.0899
```

It can be shown that what we obtained is a solution in the **least-squares** sense, that is one that minimizes $\|AX - B\|$. In other words, the solution point is the one for which the sum of its distances to the given lines is a minimum. This solution is that given by

$$\mathbf{X} = (\mathbf{A^T A})^{-1} \mathbf{A^T B}.$$

or, in MATLAB:

```
≫ X2 = inv(A'*A)*A'*B
X2 =
    0.9888
    0.4270
```

as obtained with the backslash operator.

EXAMPLE 3.1 Overdetermined system resulting from measurements

Overdetermined systems frequently result from experiments. For instance, let us assume an experiment that involves an input variable, x, and an output variable, y. The experiment was repeated six times, for six different values of the input variable, and the measured values of the output variable were recorded as shown in the following table.

x	0	1	2	3	4	5
y	2.2246	5.3826	8.3534	10.6516	14.1781	16.2095

Figure 3.7 shows a plot of the measured points. Measured values are always affected by errors; therefore, they may not lie exactly on a line or 'smooth' curve. Visual inspection, however, hints that the measured points correspond to a linear relationship of the form

$$y = ax + b \tag{3.14}$$

where a and b are two unknowns. To determine them we can write the six equations

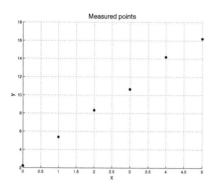

FIGURE 3.7: Measured points

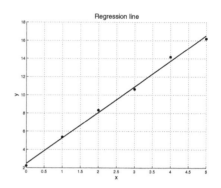

FIGURE 3.8: Plotting the regression line

$$2.2246 = 0 + b$$
$$5.3826 = x + b$$
$$8.3534 = 2x + b$$
$$10.6516 = 3x + b$$
$$14.1781 = 4x + b$$
$$16.2095 = 5x + b \tag{3.15}$$

Rewriting these equations in matrix form we obtain

$$\begin{bmatrix} 0 & 1 \\ 1 & 1 \\ 2 & 1 \\ 3 & 1 \\ 4 & 1 \\ 5 & 1 \end{bmatrix} \begin{bmatrix} a \\ b \end{bmatrix} = \begin{bmatrix} 2.2246 \\ 5.3826 \\ 8.3534 \\ 10.6516 \\ 14.1781 \\ 16.2095 \end{bmatrix}$$

Solving in MATLAB yields:

```
≫ x = 0:  5;
≫ A = [ x' ones(6, 1) ];
≫ B = [ 2.2246 5.3826 8.3534 10.6516 14.1781 16.2095 ];
≫ X = A\ B
X =
   2.8174
   2.4565
```

The solution minimizes the sum of squares of the residuals. In other words it is the one for which

$$\sum_{i=1}^{6}(ax_i + b - y_i)^2$$

is a minimum. The equation we looked for is

$$y = 2.8174x + 2.4565$$

The line corresponding to this equation is plotted in Figure 3.8 together with the measured points. As we went back from measurements to the required relationship, we say that we performed a **regression**, in this case a *linear regression*. More about such calculations can be found in Chapter 4 under *curve fitting*. For the moment, try for yourself the plot in Figure 3.8.

3.5.3 Underdetermined system

When the number of equations is smaller than the number of unknowns, the system is *underdetermined*. For example, let us return to the system described in Subsection 3.2.5, and suppose that we are given only the first two Equations 3.3

$$\frac{x}{10} + \frac{y}{3} + \frac{z}{3} = 1$$
$$\frac{x}{2} + \frac{y}{20} + \frac{z}{2} = 1 \tag{3.16}$$

There is a test that lets us know whether a linear system has a solution or not. It is called the *Kronecker-Capelli theorem*. This theorem is based on the notion of **rank** of matrix for which there are several equivalent definitions. For example, it is the number of linearly independent rows or columns of a matrix. Another definition says that the rank of a matrix is the order of the largest non-zero determinant that can be extracted from the given matrix. The rank can be calculated in MATLAB with the command with the same name, for example,

```
≫ A = [ 1 2 3; 2 3 4; 4 5 6 ];
≫ rank(A)
ans =
sg  2
```

We invite the reader to check that the determinant of the given equation is, indeed, zero. We also define the **augmented matrix** of the system $\mathbf{AX} = \mathbf{B}$ as the matrix $[\mathbf{A}\ \mathbf{B}]$.

The Kronecker-Capelli theorem states that if the rank of the matrix of coefficients is equal to the rank of the augmented matrix, the system admits a solution, if the rank of the matrix of coefficients is less than the rank of augmented matrix, the system has no solution. Another statement, given in Abate (1996) as the *Rouché-Capelli theorem*, adds that the solution is unique if and only if the rank of the augmented matrix is equal to that of the unknowns.

Applying the Kronecker-Capelli theorem we see that the system has solutions:

```
≫ A = [ 1/10 1/3 1/3; 1/2 1/20 1/2 ];
≫ B = [ 1; 1 ];
≫ rank(A)
ans =
   3
≫ rank([ A B ])
ans =
   3
```

Actually, as predicted by the Rouché-Capelli theorem, the given system admits an infinity of solutions, among them certainly the solution found in Subsection 3.2.5. Let us call that solution X0. Solving the undetermined system with the backslash operator we obtain

```
≫ X1 = A \ B
X1 =
     0
   1.1111
   1.8889
```

As the $2 - by - 3$ matrix **A** is not square, it has no inverse. Without elaborating on this subject, it can be mentioned that one can define a $3 - by - 3$ matrix **P**, called the **pseudoinverse** of **A**, such that it fulfils certain conditions, especially

$$A * P * A = A$$
$$P * A * P = P$$

Other names of the pseudoinverse are *Moore-Penrose pseudoinverse*, or *generalized inverse*; it can be calculated in MATLAB by calling the **pinv** command:

```
>> pinv(A)
ans =
   -1.0670   1.4905
    2.5035  -1.0630
    0.8166   0.6158
```

We invite the reader to check that the resulting pseudoinverse fulfills the conditions shown above. We are using it to obtain another solution:

```
>> X2 = pinv(A)*B
X2 =
    0.4235
    1.4405
    1.4325
```

To obtain the general solution we express two unknowns as functions of the third, for example

$$\frac{x}{10} + \frac{y}{3} = 1 - \frac{z}{3}$$
$$\frac{x}{2} + \frac{y}{20} = 1 - \frac{z}{2} \tag{3.17}$$

The solution of Equations 3.17 is

$$x = \frac{10}{97} \cdot (9z - 17)$$
$$y = \frac{10}{97} \cdot (7z - 24) \tag{3.18}$$

To verify that the general solution includes the three particular solutions, $\mathbf{X0, X1, X2}$, we write Equations 3.18 as two **anonymous functions** with the *handles* x and y, and call the functions with the $z-$values obtained in the three solutions $\mathbf{X0, X1, X2}$. *Anonymous functions* are defined at the command line or in one single line in a script file or another function, without the need of writing a special M-file. The syntax is

        handle = @(independent variables) expression

To evaluate the values of x and y at the three solution points we call the function **feval** with two arguments in this order

1. the handle of the anonymous function;

2. the point at which we want to valuate the function.

The calculations in MATLAB are

```
≫ D = -10/97;
≫ x = @(z) (9*z - 17)*D;
≫ y = @(z) (7*z - 24)*D;
≫ x0 = feval(x, X0(3))
x0 =
   1.5385
≫ y0 = feval(y, X0(3))
y0 =
   2.3077
≫ x1 = feval(x, X1(3))
x1 =
   0
≫ y1 = feval(y, X1(3))
y1 =
   1.1111
≫ x2 = feval(x, X2(3))
x2 =
   0.4235
≫ y2 = feval(y, X2(3))
y2 =
   1.4405
```

We recovered the values of the three particular solutions. The solution X2 obtained with the pseudoinverse has the smallest norm among all possible solutions. We can check this for three solutions:

```
≫ N0 = norm(X0
N0 =
   2.7831
≫N1 = norm(X1)
N1 =
   2.1915
≫N2 = norm(X2)
N2 =
   2.0752
```

3.5.4   A singular system

Let us consider the system

$$\begin{vmatrix} 2 & 3 \\ 4 & 6 \end{vmatrix} \begin{vmatrix} x \\ y \end{vmatrix} = \begin{vmatrix} 5 \\ 10 \end{vmatrix}$$

As it can be seen at a first glance, the second equation equals the first multiplied by two. Writing this equation in MATLAB and applying the Kronecker-Capelli theorem we conclude that it has a solution:

```
≫ A = [ 2 3; 4 6 ];
≫ B =[ 5; 10 ];
≫ rank(A)
ans =
  1
≫ rank([ A B ])
ans =
  1
```

However, the backslash operator does not yield any solution:

```
≫ X = A\B
Warning:  Matrix is singular to working precision.
X =
   NaN
   NaN
```

We cannot use Cramer's rule because the determinant of the system is zero. With the pseudoinverse of the coefficient matrix we obtain a solution, and we verify that it is a solution within the precision of the computer:

```
≫ X = pinv(A)*B
X =
   0.7692
   1.1538
≫ A*X - B
ans =
   1.0e-014 *
  -0.0888
  -0.1776
```

The two equations considered here represent the same straight line. Solving the system for y we obtain the equation of this line. We enter this equation in MATLAB as an anonymous function and, calling the command **feval**, we check that the solution found with the aid of the pseudoinverse lies, indeed, on the line:

```
≫ y = @(x) -2*x/3 + 5/3;
≫ feval(y, X(1))
ans =
   1.1538
```

3.5.5   Another singular system

Let us consider the system

$$x_1 + 2x_2 + 3x_3 = 12$$
$$3x_1 + 2x_2 + x_3 = 15$$
$$4x_1 + 4x_2 + 4x_3 = 27$$

As one can immediately see, the last equation is the sum of the first two. The system is singular, but the Kronecker-Capelli theorem indicates that there is a solution. We cannot find it with the backslash operator. The calculations in MATLAB are

```
≫ A = [ 1 2 3; 3 2 1; 4 4 4 ];
≫ B = [ 12; 15; 27 ];
≫ rank(A), rank([ A B ])
ans =
   2
ans =
   2
≫ X1 = A\B
Warning:  Matrix is singular to working precision.
X1 =
   NaN
   NaN
   NaN
```

Using the pseudoinverse we obtain a solution:

```
≫ X2 = pinv(A)*B
X2 =
   3.0000
   2.2500
   1.5000
```

The residual is practically zero:

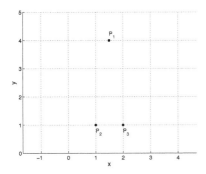

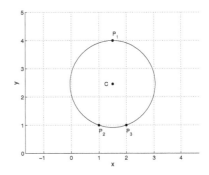

FIGURE 3.9: Three non-collinear points

FIGURE 3.10: Circle through three non-collinear points

```
>> A*X2 - B
ans =
 1.0e-014 *
 0.3553
 0
 0.3553
```

EXAMPLE 3.2 Circle through three points

Let us assume that given three non-collinear points we are asked to draw the circle passing through them. An example is shown in Figure 3.9 where the points are defined by their coordinates:

$$P_1 = \begin{bmatrix} x_1 \\ y_1 \end{bmatrix} =, \quad P_2 = \begin{bmatrix} x_2 \\ y_2 \end{bmatrix}, \quad P_3 = \begin{bmatrix} x_3 \\ y_3 \end{bmatrix}$$

We give a first solution following Anton and Rorres (2005); it uses determinants. Next, we show a solution that uses the backslash operator. The equation of the circle with the center $C = [C_x, C_y]$, and the radius r is

$$(x - C_x)^2 + (y - C_y)^2 = r^2 \tag{3.19}$$

This equation can be rewritten in the form used by the above mentioned authors

$$c_1(x^2 + y^2) + c_2 x + c_3 y + c_4 = 0 \tag{3.20}$$

Writing that Equation 3.20 is fulfilled in the three given points we obtain three additional equations and we combine them with the general equation obtaining the system

$$c_1(x^2 + y^2) + c_2 x + c_3 y + c_4 = 0$$
$$c_1(x_1^2 + y_1^2) + c_2 x_1 + c_3 y_1 + c_4 = 0$$
$$c_1(x_2^2 + y_2^2) + c_2 x_2 + c_3 y_2 + c_4 = 0$$
$$c_1(x_3^2 + y_3^2) + c_2 x_3 + c_3 y_3 + c_4 = 0$$

As this system is homogeneous, it has a non-trivial solution only if its determinant is zero. To obtain the coefficients $c_1, ... c_4$ we can expand the determinant using the first line. This is done in the following function.

```
function    [ c, C, r ] = Circle3PA(P1, P2, P3)

%CIRCLE3PA Given three non-collinear points, Circle3PA
%          Finds the equation, the radius and the
%          center of the circle passing through them.
%          The solution follows Anton and Rorres, 2005
%          Input arguments: coordinates of three points
%          given as three column vectors, e.g
%          P1 = [ x1; y1 ];
%          Output arguments:
%           coefficients of the equation, c
%           coordinates of center
%           given as C = [ Cx; Cy ];
%           circle radius, R

% check that the points are not collinear
if cross([ (P2 - P1); 0 ], [ (P3 - P1); 0 ]) == 0
     errordlg( 'Given points are collinear', 'Input error')
end
% Separate variables
x1 = P1(1); y1 = P1(2);
x2 = P2(1); y2 = P2(2);
x3 = P3(1); y3 = P3(2);
% Matrix of coefficients
A = [ (x1^2 +y1^2)   x1  y1  1
      (x2^2 +y2^2)   x2  y2  1
      (x3^2 +y3^2)   x3  y3  1 ] ;
% Calculate the coefficients of the equation of the circle
c      = zeros(1, 4); % allocate space
c(1) =   det(A(1:3, 2:4));
c(2) = -det([ A(1:3, 1) A(1:3, 3:4) ]);
c(3) =   det([ A(1:3, 1:2) A(1:3, 4) ]);
c(4) =   det([ A(1:3, 1:3) ]);
% calculate coordinates of center of circle
```

```
C = [ -c(2)/(2*c(1)); -c(3)/(2*c(1)) ];
r = norm(C - P1);
```

Below we call the function with the coordinates of the three points shown in Figure 3.9.

```
≫ P1 = [ 1.5; 4 ];
≫ P2 = [ 1; 1 ];
≫ P3 = [ 2; 1 ];
≫ c =
    3.0000   -9.0000   -14.7500   -17.7500
C =
    1.5000
    2.4583
r =
    1.5417
```

The elements of the array c are the coefficients of Equation 3.20. The other data defines the circle shown in Figure 3.10. We now show another solution based on the system obtained by expanding and rearranging Equation 3.19

$$2C_x x + 2C_y y - (C_x^2 + C_y^2 - r^2) = x^2 + y^2 \qquad (3.21)$$

We define

$$2C_x = Z_1; \ 2C_y = Z_2, \ C_x^2 + C_y^2 - r^2 = Z_3$$

and applying Equation 3.21 to the three given points we obtain the matrix equation

$$\begin{vmatrix} x_1 & -y_1 & -1 \\ x_2 & -y_2 & -1 \\ x_3 & -y_3 & -1 \end{vmatrix} \begin{vmatrix} Z_1 \\ Z_2 \\ Z_3 \end{vmatrix} = \begin{vmatrix} x_1^2 + y_1^2 \\ x_2^2 + y_2^2 \\ x_3^2 + y_3^2 \end{vmatrix}$$

This is the basis of the function displayed below. We invite the reader to write it on a file Circle3P.m and run it for the points already used in this example. The results will obviously be the same as those found with the function Circle3PA.

```
function    [ C, R ] = Circle3P(P1, P2, P3);

%CIRCLE3P    Given three points in the xOy plane, this function
%            finds the center and the  radius of the circle
%            passing through them.
%            Input arguments: coordinates of three points
%            given as three column vectors, e.g
%            P1 = [ x1; y1 ];
```

```
%                   Output arguments: coordinates of center
%                   given as C = [ Cx; Cy ];
%                   circle radius, R

% check that the points are not collinear
if cross([ (P2 - P1); 0 ], [ (P3 - P1); 0 ]) == 0
    errordlg( 'Given points are collinear', 'Input error')
end
% build matrix of coefficients
A = [ P1(1), P1(2), -1;
      P2(1), P2(2), -1;
      P3(1), P3(2), -1 ]
% build free vector
B = [ P1(1)^2 + P1(2)^2; P2(1)^2 + P2(2)^2; P3(1)^2 + P3(2)^2 ]
Z = A\B
C = [ Z(1)/2; Z(2)/2 ]
R = sqrt(C(1)^2 + C(2)^2 - Z(3))
```

We have used **errordlg**, a command for calling a built-in *dialog box*. This command is called with two string arguments, the first is the error message, the second, the title of the box. The user has to acknowledge the message by clicking on OK.

3.6   Summary 1

We consider systems of linear equations expressed in the matrix form

$$\mathbf{AX} = \mathbf{B}$$

where $\mathbf{A}$ is the *matrix of coefficients*, $\mathbf{X}$, the solution vector, and $\mathbf{B}$, the *free vector*. The preferred solution in MATLAB is by Gauss elimination, a procedure carried on with the *backslash operator*:

```
≫ X = A\B
```

The backslash operator always yields an answer that can be the general solution, a particular solution, a solution in the *least-squares* sense, or no solution at all. Therefore, the user must analyze the nature of the system under consideration before deciding whether it is possible to accept the solution or not. One test we use is the *Kronecker-Capelli theorem*. To check if a system has a solution we can calculate the rank of the matrix $\mathbf{A}$ and the rank of the augmented matrix $[\mathbf{A}\ \mathbf{B}]$. In MATLAB we calculate the rank of a matrix by calling the command `rank` with the name of the matrix as argument. If the two ranks are equal, the system admits a solution, and if the rank of the matrix of coefficients is less than the rank of augmented matrix, the system has no solution. According to the *Rouché-Capelli theorem*, if and only if the rank of the augmented matrix is equal to the rank of the number of unknowns, the solution is unique.

To analyze the nature of the solutions of linear systems we distinguish between several possible situations, as shown below.

general systems of form $\mathbf{AX} = \mathbf{B}$ of n equations in n unknowns with $det(\mathbf{A} \neq 0/$

ill-conditioned systems that are sensible to small changes of data.

homogeneous systems of the form $\mathbf{AX} = \mathbf{0}$. They have non-trivial solutions only if $det(\mathbf{A}) = 0$. Then, there are infinite solutions.

overdetermined solutions that is, systems with more equations than the number of unknowns. If all the equations are linearly independent, there are no solutions and the backslash operator yields a solution in the *least-squares* sense.

underdetermined systems that is, systems with fewer equations than the number of unknowns. The number of solutions is infinite.

singular systems whose matrix of coefficients is singular; they may have a solution.

In this part of the chapter we used *anonymous functions*, that is functions defined at the command line or in one single line in a script file or another function, without the need of writing a special M-file. The syntax is

```
handle = @(independent variables) expression
```

The commands introduced in this part of the chapter are

\ the *backslash operator*. Given a system of linear equations, $AX = B$, the command `A\B` yields the solution vector X.

cond the command `cond(A)` yields the *condition number* of the matrix $\mathbf{A}$. This number is an indication of the sensibility of the solution of a system of linear equations to changes in its data. If the condition number is large the results may be unreliable.

errordlg a built-in dialog box for displaying error messages.

eye the command `eye(n)` yields the *identity*, or *unit matrix* of size $n-by-n$. For other possibilities see the MATLAB help.

Dimensions

Forces

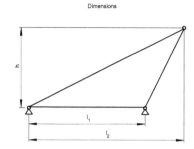

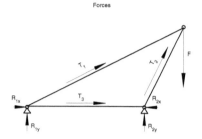

FIGURE 3.11: Crane dimensions   FIGURE 3.12: Crane forces

feval the command [y1, y2, ...] = feval(funhandle, x1, ..., xn) causes the evaluation of the function whose handle is funhandle, at the points x1 to xn.

inv given a square, not singular matrix, $\mathbf{M}$, inv(M) yields $\mathbf{M}^{-1}$, the inverse matrix of $\mathbf{M}$, such that $\mathbf{MM}^{-1} = \mathbf{M}^{-1}\mathbf{M} = \mathbf{I}$, where $\mathbf{I}$ is the *unit*, or *identity* matrix of the same size as $\mathbf{M}$.

pinv the command pinv(A) yields the pseudoinverse of the non-invertible matrix $\mathbf{A}$. If the matrix is invertible, pinv(A) is identical to inv(A). The pseudoinverse can be used to find least-squares solutions of under-determined systems.

rank the command rank(A) yields the rank of the matrix $\mathbf{A}$.

3.7   More exercises

Exercise 3.1 Overdetermined system

Solve the overdetermined system described in Subsection 3.5.2 with the help of the Moore-Penrose pseudoinverse

```
X3 = pinv(A)*B
```

Check that X3 equals the solutions found before.

Exercise 3.2 Forces in crane

Figure 3.11 shows a simply supported crane. Let its dimensions be

$$l_1 = 3 \text{ m}, \ l_2 = 4 \text{ m}, \ h = 2 \text{ m}$$

and assume a force $F = 10$ kN. Write and solve the matrix equations that yield the reactions in the supports and the tensions in the bars, as noted in Figure 3.12.

3.8 Polynomial equations

3.8.1 MATLAB representation of polynomials

Let P be a polynomial in one variable, x. P can be represented in MATLAB by the coefficients of its terms. These coefficients are stored in an array, in the descending order of the variable x. For example, the two polynomials

$$P_1 = 3x^4 + 2x^3 + 4x^2 + x + 6$$
$$P_2 = 2x^5 + 3x^3 + x$$

are defined in MATLAB by

```
≫ c1 = [ 3 2 4 1 6 ];
≫ c2 = [ 2 0 3 0 1 0 ];
```

Suppose that we want to calculate the value of the first polynomial for $x = 5$. The trivial way of doing it is

```
≫ x = 5;
≫ P_1 = c1(1)*x^4 + c1(2)*x^3 +c1(3)*x^2 + c1(4)*x +c1(5)
P_1 =
    2236
```

This is rather tedious and, as shown in Subsection 5.11, computationally inefficient. MATLAB provides a function, `polyval`, which called with the arguments $c1$ and x yields the desired result:

```
≫ P_2 = polyval(c1, x)
P_2 =
    2236
```

3.8.2 The MATLAB `root` function

Equating a polynomial to zero results in an algebraic equation. For example, corresponding to the polynomial P_1 defined in the preceding section, we can write the equation

$$3x^4 + 2x^3 + 4x^2 + x + 6 = 0 \tag{3.22}$$

An algebraic equation of degree n has n *roots* that can be real or complex. If the coefficients of all terms are real, the complex roots can come only in *conjugate* pairs. There are formulae for solving equations up to the degree 4, but for the degrees 3 and 4 they are rather complicated. The theory shows that there are no formulae for solving equations of degree 5 and higher. For the degree 2 we have a popular, simple formula. For higher degrees it is usual to solve the equation by iterative procedures. MATLAB provides a general command, **roots**, which called with the array of coefficients as argument yields the polynomial roots. For example, we solve Equation 3.22 by calculating

```
≫ X = roots(c1)
X =
   -0.8085 + 1.0120i
   -0.8085 - 1.0120i
   0.4751 + 0.9830i
   0.4751 - 0.9830i
```

Above, the letter 'i' stands for the **imaginary unit**, $\sqrt{-1}$. As seen, Equation 3.22 has four **complex** roots; they come in pairs of *conjugate numbers*. The second root is the conjugate of the first, that is, its **imaginary part**, 1.0120, is numerically equal, but has the opposite sign of this part in the first root. Similarly, the first root is the conjugate of the second, the fourth the conjugate of the third, and the third the conjugate of the fourth. Working in MATLAB with complex numbers is as easy as working with real numbers. These matters are the subject of Chapter 7.

MATLAB provides also the 'inverse' function of **roots**; it is **poly**, which called with the array of roots as argument, yields the array of polynomial coefficients, possibly *scaled*. Continuing our example we calculate

```
≫ Pc = poly(X)
ans =
    1.0000   0.6667   1.3333   0.3333   2.0000
```

The scaling factor is evidently the coefficient of the highest-degree term, 3:

```
≫ 3*Pc
ans =
    3.0000   2.0000   4.0000   1.0000   6.0000
```

The function `roots` has its limitations; they are discussed in Subsection 5.2.4. For the moment we can show that the roots may be approximations only. Let us check if the elements of the array X are, indeed solutions of the given equations:

```
>> polyval(c1, X)
ans =
1.0e-014 *
  -0.0888 - 0.1776i
  -0.0888 + 0.1776i
  -0.1776 - 0.6661i
  -0.1776 + 0.6661i
```

Instead of the expected zeroes we obtained very small numbers. In the following subsection we show another way of checking our results.

3.8.3   The MATLAB function `conv`

Once the roots of an n-th degree polynomial are known, the polynomial can be represented as the product of n monomials. For example, the polynomial exemplified in the preceding two subsections can be *factored* as

$$P_1 = (x - X(1))(x - X(2))(x - X(3))(x - X(4))$$

Knowing the roots, the equation shown above gives us another possibility of recovering the polynomial coefficients. MATLAB provides a command, `conv`, that enables us to do the job. Given the arrays of coefficients, `c1`, `c2`, of two polynomials p_1, p_2, the command `conv(c1, c2)` yields the array of coefficients of the polynomial product $p_1 \times p_2$. In specialized terms we say that `conv` performs the **convolution** of the argument arrays. Returning to the data of the preceding two subsections, we first perform the multiplication of the first two monomials in the above equation and obtain a 2nd-degree trinomial, `P1`, with real coefficients. Next, we multiply the last two monomials in the above equation and obtain another 2nd-degree trinomial, `P2`, also with real coefficients. Now we multiply `P1` by `P2` and recover the scaled coefficients of the left-hand side of Equation 3.22. Finally we multiply by the scale factor, that is the coefficient of the highest-degree term of the given equation

```
≫ P1 = conv([ 1 -X(1) ], [ 1 -X(2) ])
P1 =
 1.0000  1.6169  1.6777
≫ P2 = conv([ 1 -X(3) ], [ 1 -X(4) ])
P2 =
 1.0000  -0.9502  1.1921
≫ P3= conv(P1, P2)
P3 =
 1.0000  0.6667  1.3333  0.3333  2.0000
≫ 3*P3
ans =
 3.0000  2.0000  4.0000  1.0000  6.0000
```

EXAMPLE 3.3 Vertical throw

Let us assume that an object is thrown upwards with the vertical velocity $V_0 = 5$ m · s^{-1}. After how many seconds will it reach the height of 5 m? To answer this question we start by observing that during the time t the object travels upward a distance equal to $V_0 t$, and falls a distance equal to $gt^2/2$. In total, the height reached by the body during the time t is

$$h = V_0 t - \frac{1}{2}gt^2$$

We have to solve the resulting second-degree equation for which there is a well-known formula. To practice the techniques learned in this chapter, we try, however, the command **roots**. To do this we first define the coefficients of the governing equation:

```
≫ g = 9.18;
≫ V0 = 10;
≫ c = zeros(3, 1);
≫ c(1) = g/2;
≫ c(2) = -V0;
≫ c(3) = 5;
```

Using the function **roots** we solve the equation and obtain, as expected, two solutions, $t_1 = 1.4013$, $t_2 = 0.7774$. The calculations in MATLAB are

```
≫ roots(c)
ans =
    1.4013
    0.7774
≫ t_end = 2*V0/g;
≫ t = 0:  t_end/50:  t_end;
≫ z = V0*t - g*t.^2/2;
≫ plot(t, z, '-', t, 5*ones(size(t)), 'k-',...
'LineWidth', 1.5), grid
≫ ht = xlabel('Time, s');
≫ set(ht, 'FontSize', 14)
≫ht = ylabel('Height, m')
≫set(ht, 'FontSize', 14)
```

The thrown object crosses twice the height 5 m, first while going up, second, while falling down. MATLAB delivered us the solutions in reverse order! To see this we plot the graph of height, h, against time, t, as shown in Figure 3.13.

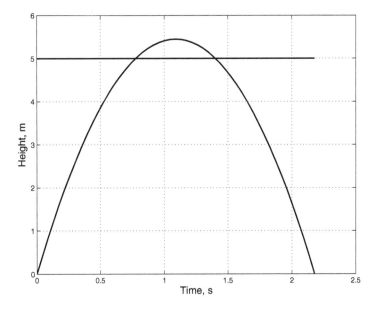

FIGURE 3.13: Vertical throw time history

To get the numerical values of the two instants in which the thrown object crosses the given height, click the **Data cursor** icon on the figure toolbar and then click the point you are interested in. For details see Figure 3.14. If

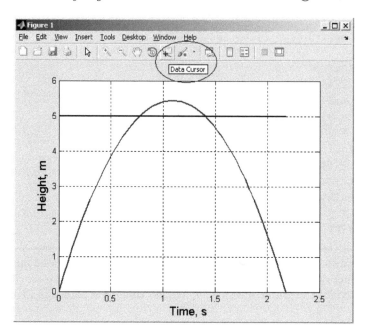

FIGURE 3.14: Reading data on the time-history plot

you want to use the numerical data in further calculations, use instead the command `ginput`. Below we call this command with the output arguments t, h, and the input argument 2, which means that we want to use the command twice. A cross-hair will appear over the plot; drag it upon the first intersection point and click the mouse. The cross-hair will appear a second time. Repeat the operation for the second intersection point. The time and height values are stored in the variables t and h.

```
≫ [ t, h ] = ginput(2)
t =
    0.7805
    1.4084
h =
    4.9912
    5.0088
```

As seen above, depending on the precision with which we pick up the intersection point, the results may differ slightly from those obtained by solving the equation.

EXAMPLE 3.4 A simple barge – Example of 3d degree equation

Figure 3.15 shows the longitudinal section, the top view and the side ele-

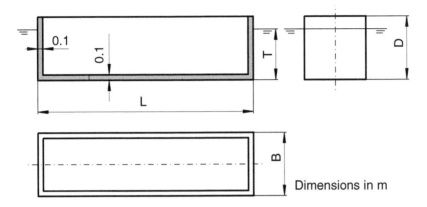

FIGURE 3.15: A simple barge

vation of a simple barge built of wood. The drawing is arranged according to first angle views. The barge carries a load, M. It is required to find the dimensions of the barge, given the data

length-to-breadth ratio	$k = L/B = 3.5$	
depth	$D = B$	
draft	$T = 0.8D$	
sea-water density	$\rho_{SW} = 1.025$	$\text{t} \cdot \text{m}^{-3}$
wood density	$\rho_{wood} = 0.8$	$\text{t} \cdot \text{m}^{-3}$
load	$M = 1$	t

The design equation, based on Archimedes' principle, states that the mass of the displaced volume equals the sum of the hull mass and the load mass.

$$\rho_{SW} LBT = \rho_{wood}\{LBD - (L - 0.2)(B - 0.2)(D - 0.1)\} + M \qquad (3.23)$$

Substituting part of the given data we obtain

$$\frac{0.8}{k^2}\frac{\rho_{SW}}{\rho_{wood}}L^3 = LBD - (L - 0.2)(BD - 0.1B - 0.2D + 0.02) + \frac{M}{\rho_{wood}}$$
$$= LBD - LBD + 0.1LB + 0.2LD - 0.02L$$
$$+ 0.2BD - 0.02B - 0.04D + 0.004 + \frac{M}{\rho_{wood}}$$

Substituting the given values for D and T and rearranging the equation yields

$$\frac{0.8}{k^2}\frac{\rho_{SW}}{\rho_{wood}}L^3 - \left(\frac{0.1}{k} + \frac{0.2}{k} + \frac{0.2}{k^2}\right)L^2$$
$$+ \left(0.02 + \frac{0.02}{k} + \frac{0.02}{k} + \frac{0.04}{k}\right)L - \left(0.04 + \frac{M}{\rho_{wood}}\right) = 0 \quad (3.24)$$

To work in MATLAB we define the left-hand of this equation as the polynomial

$$c(1)L^3 + c(2)L^2 + c(3)L + c(4)$$

calculate the coefficients, and call the command **roots**:

```
≫ k = 3.5;
≫ wood = 0.8;
≫ M = 1;
≫ rho = 1.025;
≫ c = zeros(1, 4);
≫ c(1) = 0.8*rho/(wood*(k^2));
≫ c(2) = -(0.3/k + 0.2/(k^2));
≫ c(3) = 0.02 + 0.06/k;
≫ c(4) = -(0.004 + M/wood);
≫ L = roots(c)
L =
   2.8765
  -0.8285 + 2.1269i
  -0.8285 - 2.1269i
```

We obtain an array of three roots, as expected for a third-degree equation. Only one root is real, that is, technically possible. We choose this root and calculate the dimensions and the *displacement*, that is, the mass of the volume of water displaced by the barge:

```
≫ L = L(1);
≫ B = L/k
B =
   0.82185
≫ D = B;
≫ T =0.8*D
T =
   0.65748
≫ Displacement = rho*L*B*T
Displacement =
   1.5932
```

The result is a mass measured in tons. We calculate now the hull volume, in m^3, and the hull mass:

```
>> HullVolume = (L*B*D - (L - 0.2)*(B- 0.2)*(D - 0.1));
>> HullMass = wood*hullV
HullMass =
   0.59317
```

Adding to this mass the value of the load, 1 t, we obtain, indeed, the displacement, 1.593 t. The results were displayed with an excessive number of digits. For example, in naval-architectural practice it makes no sense to detail lengths beyond millimeters. If the reader intends to write a script file, or a function for a similar task, it will be reasonable to display the results with commands like

```
disp([ 'B = ' num2str(B, 3) ' m' ])
```

The argument of the above command is an array whose elements are a string, the command num2str(B, 3) that converts the number B to a string in which the number is displayed with 3 decimal digits, and a string that displays the units of the result. Continuing our example:

```
>> disp([ 'B = ' num2str(B, 3) ' m' ])
B = 0.822 m
```

3.9 Iterative solution of equations

3.9.1 The Newton-Raphson method

In engineering we can encounter equations that cannot be reduced to polynomials. To solve them we can start with a guess solution, say x_0, that we substitute in the equation. In general the result will be not zero, but a number; let it be δ_0. We can assume a rule for correcting the initial guess as a function of δ and substitute the corrected value, x_1, in the equation. Again the result may be different from zero, say δ_1. We apply again the correction rule and continue so until some *stopping rule* is fulfilled. For example, we can stop the procedure when δ_i is smaller than some given error limit, or when two successive solutions, x_i, x_{i+1}, differ by a number that is less than a given *tolerance*, ε. We call such a procedure **iterative** and say that each repetition of the calculation is an **iteration**, a term that comes from the Latin *iterare*. Iterative procedures can be easily programmed in MATLAB by using the scheme

while *some condition*
                        statements
end

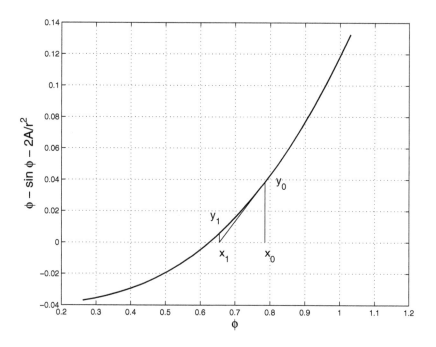

FIGURE 3.16: Solution of Equation 3.25 by the Newton-Raphson procedure

In this subsection we describe the *Newton-Raphson* procedure, probably the most popular iterative method. To explain it we refer to an example given in Biran and Breiner (2002), Section 7.5. There we calculate the angle, ϕ, of a segment of a circle that has a given area, A. The equation of the area as a function of the angle is

$$A = \frac{r^2}{2}(\phi - \sin \phi) \tag{3.25}$$

We want to find the angle x for which

$$y = x - \sin x - \frac{2A}{r^2} = 0 \tag{3.26}$$

Let us assume the values $A = 0.02$, $r = 1$. In Figure 3.16 we plot Equation 3.26 in a guess interval. Assuming a first guess $x_0 = \pi/4$, we calculate the corresponding value, y_0, of the function given by Equation 3.26. It is not zero. Therefore, we seek a better value by descending (in other cases ascending) along the tangent in $y-$) to the given curve. The new value is

$$x_1 = x_0 - \frac{y_0}{\dot{y}(x_0)} \qquad (3.27)$$

Equation 3.26 yields for this a new value, y_1. As this value is not zero, we repeat the procedure. We end the iterations when we receive an approximation consistent with our needs. The function `Newton` shown below implements this procedure. It receives four arguments:

1. the handle of the function whose zero we seek;

2. the handle of the derivative of the above function;

3. a guess solution,

4. the *tolerance* admitted for the stopping rule. When the difference between two successive iterations is less than the given tolerance, the iterations stop. This argument is optional; when it is not supplied, the function assumes a built-in tolerance, `eps`.

```
function  x = newton(h, hdot, x0, tol)

%NEWTON  Newton-Raphson procedure
%   Input arguments:
%       h, handle of given equation defined as
%           anonymous function
%       hdot, handle of derivative of h, defined as
%            anonymous function
%       x0, initial guess
%       tol, difference between two successive solutions,
%            optional
%            Output argument: x, solution within approximation tol
if nargin == 3
        tol = 10^(-6);
end
xi    = x0;
eps  = 1;
while  eps > tol
        y           = feval(h, xi);
        ydot = feval(hdot, xi);
        xi1    = xi - y/ydot;
        eps    = abs(xi1 - xi);
        xi        = xi1;
end
x = xi;
```

In the above function we used `nargin`, a command that returns the number of input arguments. In our case, if the argument *tol* is given, the calculations end when the difference between the results of two successive iterations is smaller than *tol*. If the fourth input argument is not given, the number of input arguments equals 3 and the conditional statement is executed setting *tol* to 10^{-6}.

As an example, let us assume that the two coordinates of a point, P_1, are x_1, y_1. It is required to find the angle of rotation around the origin, α, that will bring the point to the x-coordinate x_2. From Subsection 2.5.2 we know that the relationship between the initial and the rotated x-coordinate is

$$x_2 = x_1 \cos\alpha + y_1 \sin\alpha \qquad (3.28)$$

The unknown in our problem is the angle α. Equation 3.28 can be rewritten as an equation in either $\sin\alpha$ or $\cos\alpha$ and solved as a second-degree equation in the chosen trigonometric function. This requires some algebraic manipulation. To avoid this treatment, we can directly solve the equation in α by some iterative procedure, for example the Newton-Raphson procedure. We begin by entering the data of the problem and by plotting the given point and the circle on which the given point will move during rotation:

```
≫ x1 = 3;
≫ y1 = 2;
≫ r = norm([ x1; y1 ])
≫ x2 = 2;
≫ t = 0:  2.5:  90;
≫ xc = r*cosd(t);
≫ yc = r*sind(t);
≫ hp = plot(xc, yc, 'k--'), grid;
≫ set(hp, 'LineWidth', 1.3)
≫ axis equal
≫ ht = xlabel('x');
≫ set(ht, 'FontSize', 16)
≫ ht = ylabel('y');
≫ set(ht, 'FontSize', 16)
≫ hold on
≫ point([ x1; y1 ], 0.03)
≫ ht = text(1.01*x1, 1.05*y1, 'P_1');
≫ set(ht, 'FontSize', 16)
```

Next, we define our equation as an *anonymous function* with the *handle* `h1`, and the derivative of the given equation as another *anonymous function* with the handle `h2`. Now we call our function `newton` with the output argument `alpha` and use the result to plot the new position of the rotated point:

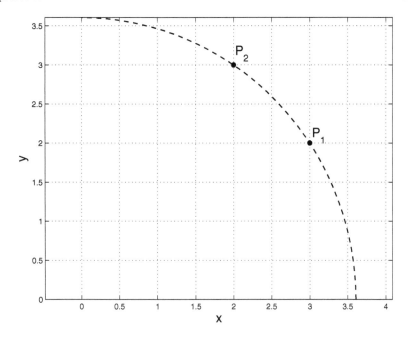

FIGURE 3.17: Rotating the point P_1 around origin

```
≫ h1 = @(alpha) x1*cosd(alpha) - y1*sind(alpha) - x2;
≫ h2 = @(alpha) -x1*sind(alpha) - y1*cosd(alpha);
≫ alpha = newton(h1, h2, 1)
alpha =
   22.6198
≫ y2 = -x1*sind(alpha) + y1*cosd(alpha);
≫ point([ x2; y2 ], 0.03)
≫ ht = text(1.01*x2, 1.05*y2, 'P_2');
≫ set(ht, 'FontSize', 16)
≫ hold off
```

As shown in textbooks on numerical methods, the Newton-Raphson procedure can fail for certain initial guesses. An example can be found in Exercise 3.4.

3.9.2 Solving an equation with the command `fzero`

MATLAB provides the command `fzero` for finding one zero of a continuous function of one variable. This command can be called with two arguments, the handle of the function, and an initial guess. Let us return to the example in the previous section and try to solve it with `fzero`:

```
≫ x1 = 3;
y1 = 2;
x2 = 2;
h = @(alpha) x1*cosd(alpha) - y1*sind(alpha) - x2;
alpha1 = fzero(h, -10)
alpha1 =
    22.6199
```

Alternatively we can call `fzero` taking as the second argument a guess interval:

```
≫ alpha2 = fzero(h, [ -10 -30 ])
alpha2 =
    -22.6199
```

If the given function does not change sign in the guess interval, `fzero` fails and issues an error message. Try for yourself `fzero(h, [ 10 30 ])`. While the command `roots` shows how many zeros a given equation has, the command `fzero` finds one zero at most. This command can fail or give no results even in apparently simple cases. An example can be possibly found in Exercise 3.4.

3.10   Summary 2

In Chapter 2 we introduce a first *conditional structure*, the `for loop` scheme. In the second part of this chapter we explain a second conditional structure, the *while* loop. The scheme is

> `while` *some condition*
> > *statements*
>
> `end`

where *some condition* can be an equality, or an inequality condition expressed by means of one *relational operator*, or more relational operators combined by some *logical operator* (see the MATLAB help for details). The *statements* are the calculations to be executed. They will be *iterated* until the *condition* is fulfilled.

The commands introduced in the second part of this chapter are

conv given the arrays of coefficients, u, v, of two polynomials, P_1, P_2, P3 = conv(u, v) yields the array of coefficients of the product $P_3 = P_1 P_3$.

disp displays text or array.

fsolve finds one zero of a continuous function. This command should be called with two arguments, the handle of the function and a guess value or guess interval.

ginput the command [x. y] = ginput(n) enables the user to choose with the mouse n points and store their coordinates in the arrays x, y.

nargin this command returns the number of input arguments.

num2str converts a number into a string that can be displayed or printed. The command s = num2str(x, n) converts the number x into a string s, while the number is displayed with the precision n. The default option is $n = 4$.

pinv calculates the *pseudoinverse* of a matrix that is not square.

poly the command poly(r) generates the polynomial whose roots are stored in the array r.

polyval given the array of polynomial coefficients, p, using the command polyval(p ,x) we obtain the value of the polynomial evaluated at x. The argument x can be one number or an array of numbers.

roots given the polynomial equation $c(1)x^n + c(2)x^{n-1} + \ldots c(n+1) = 0$, the command roots(c) yields the roots of the equation.

3.11   More exercises

Exercise 3.3 Binomial formulae by convolution
Use the command conv to calculate

$$(x + 1)^2, \ (x + 1)^3, \ (x - 1)(x + 1)$$

Exercise 3.4 Solving a 7th degree polynomial equation
The equation

$$(x + 1)^7 = 0$$

has, obviously, seven zeros, all equal to -1. In other words, the equation has a root, -1, of multiplicity 7. You are required

1. to expand the polynomial using convolution;

2. to solve the equation with the command `roots`;

3. to solve the equation with the function `newton` developed in this chapter;

4. to solve the equation with the command `solve`.

The command `roots` yields solutions that are not zeroes of the given equation. The other methods may yield wrong results, or no results, depending on your initial guess.

4

Processing and publishing the results

Students carry on calculations for their homework, for the interpretation of laboratory experiments, for projects. Engineers perform calculations for design projects. All this work must be presented in reports. Scientists make calculations during their research; they want to publish them. More and more students, engineers and scientists want to show their work also in *presentations*, such as by Microsoft® PowerPoint. Once it was necessary to write again the equations used and the results obtained. This means additional work and a potential source of errors. Today MATLAB provides several possibilities for preparing reports and articles without the need of writing again the work done while calculating. In this chapter we are going to show only the simplest ways of exporting to reports, articles and presentations; they are sufficient for the level and extent of work covered by this book.

In most cases students, engineers and scientists do not limit themselves to presenting or publishing calculations, but want to complete them with graphs and illustrations. In the preceding chapters we have described the basic techniques for plotting. In this chapter we introduce more possibilities and show how the newest versions of MATLAB enable also interactive plotting. We also show how to store the graphs in files that can be inserted in reports, articles and presentations. A fruitful technique is to *fit curves*, in fact equations, to the graphs of measured data. This chapter contains also an introduction to this subject.

4.1 Copy and paste

The simplest way of exporting the work done on the computer screen is to use the *copy* and *paste* facilities. Point to the first line you want to copy, scroll down the screen keeping pressed the `Shift` key and 'paint' all the lines you are interested in. Open the `Edit` menu on the toolbar and click on `Copy`. Alternatively, press the right-hand button of the mouse. In the menu that opens click on `Copy`. Open the file on which you want to write. Point to the line in which you want to insert the copied calculations, open the `Edit` menu and click `Copy`. Alternatively you can do the same thing with the right-hand button of the mouse. If you are preparing your report or article in Microsoft®

Word, the file you opened should have the extension doc. If you are using LATEX, the file should have the extension tex. For PowerPoint you can copy your calculations directly on a slide. Now you can delete the lines that don't interest you, change fonts, and insert titles and explanations.

A small detail. In the default, *loose* format, MATLAB displays empty lines. Most of them must be canceled when processing the text. To avoid this, before beginning your calculations enter the command format loose.

4.2 Diary

If you want to store the screens of your calculations use the *diary* facility. Before beginning your calculations enter the command diary followed by the name of the file on which you want to save your work. At the end of the work you want to store, enter the command diary off. For example,

```
≫ diary Mywork.dia
≫ x = 0:  0.1:  5;
≫ y = x.^2 + 3*x + 5;
≫ plot(x, y, 'k-')
≫ diary off
```

We have used the extension dia to remember that it is a diary. You may use another extension, for example txt. To open the file in MATLAB click on the Open file icon – the second from left on the toolbar–. A dialog box opens. In its Files of type window open the pull-down menu and click on All Files(*.*). You may also open the file in your word processor or typesetter. You may use Copy and Paste to take lines in the file you are writing, or insert the whole diary file if your software allows this.

The latest versions of MATLAB provide extended facilities for producing a report in HTML or LATEX. These facilities go beyond the scope of this book. The reader interested in them may use the following Web address:

http://www.mathworks.com/products/matlab/description6.html

4.3 Exporting and processing figures

The plot command produces a figure on the screen. To print the figure click on File on the toolbar and, in the menu that opens click on Print. If you want to store the figure in a file you can use the command line. Thus, if you want to obtain a JPEG file to insert in a Word document, type

> print -djpeg Myfile.jpg

In Word click on the place you want to insert the figure. In the toolbar click **Insert**, then **Picture**, **From File**, and in the dialog box that opens look for the folder you worked in MATLAB and then choose the file.

If you want to produce a black-and-white Encapsulated Postscript file for inserting in a LATEXdocument, type

> print -deps Myfile.eps

If you want to obtain a color Encapsulated Postscript file for inclusion in LATEXdocument, type

> print -depsc Myfile.eps

To insert the picture in your **tex** file use

```
\begin{figure}[hbt]
 \centering
   \includegraphics[scale=0.80]{Myfile.eps}
     \caption{Example of graphics}
       \label{f:Myfile}
 \end{figure}
```

If you want to produce a graphic file interactively click in the toolbar on the third icon from left. This icon looks like 3.5" diskette and has the name **save Figure**. A dialog box opens; its name is **Save As**. In the upper window look for the folder in which you want to save the file. In the window **File name** type the name, and below this open the menu **Save as type** and make your choice.

4.4 Interpolation

4.4.1 Interactive plotting and curve fitting

Table 4.1 shows values of the kinematic viscosity of fresh water in the range 0 to 28 degrees Celsius. Write this data on a file **kvisc.m** following the model

```
%KVISC  kinematic viscosity of fresh water
% Meaning of columns:
%     Temperature       kinematic viscosity
```

Table 4.1: Kinematic viscosity of fresh water

Temperature,^{0}C	ν, m^2s^{-1}	Temperature,^{0}C	ν, m^2s^{-1}
0	1.79	15	1.140
1	1.73	16	1.110
2	1.67	17	1.080
3	1.62	18	1.060
4	1.57	19	1.030
5	1.52	20	1.010
6	1.47	21	0.983
7	1.43	22	0.960
8	1.39	23	0.938
9	1.33	24	0.917
10	1.31	25	0.896
11	1.27	26	0.876
12	1.24	27	0.857
13	1.20	28	0.839
14	1.17		

```
%   degrees Celsius           m^2/s
nu = [  0    1.79
        . . .
       28   0.839 ];
```

To plot the viscosity against temperature use these commands:

```
≫ kvisc
≫ plot(nu(:, 1), nu(:, 2), 'ro')
≫ xlabel('Temperature, degrees Celsius')
≫ ylabel('Kinematic viscosity of fresh water, m^2 s^-1')
```

The commands shown above are simple and fast. There is also a possibility of plotting interactively without using commands. To do this enter kvisc to load the data shown in Table 4.1. The icon of the variable *nu* appears in the *Workspace Window* (1 in Figure 4.1). Clicking twice on this icon opens a new window called *Variable Editor - nu*; it displays the two columns of the array *nu*. They are marked by 2 and 4 in Figure 4.1. Click on the heading 1 of the first column (2 in the figure) and on **Copy** in the **Edit** menu of the main toolbar. Next write 'x = ' in the *Command Window* and click on **Paste** in the **Edit** menu. A new array will be displayed, as shown under 3 in Figure 4.1. Enter the semicolon ';' after it and press **Enter**. The icon of the new variable, *x*, will appear in the *Workspace Window*. Repeat this action, this time for the second column of the array *nu* (marked by 4 in the figure) and writing 'y

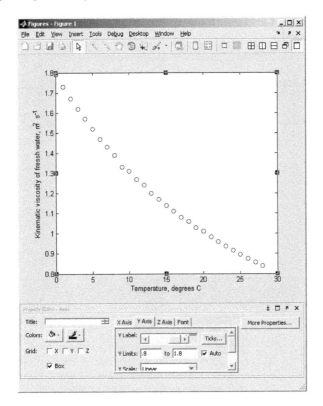

FIGURE 4.1: Interactive plotting and curve fitting

=' in the command window. Enter the semicolon ';' after it and press Enter. The resulting array is marked 5 in Figure 4.1 and its icon appears in the *Workspace Window*.

Close the *Variable Editor*. Holding down Ctrl click the icons of x and y in the *Workspace Window*. While these icons are 'painted', open the *Plot* menu (marked 6 in the figure) in the toolbar of the *Workspace Window*, and click on the Plot as two series icon. From the opening menu choose plot(x, y). The plot appears in a new window with the title Figure 1. We are not interested in the plotted curve, but want to fit a curve for the given points. Therefore, in the toolbar of the figure click on the arrow, Edit Plot, and after this on the curve. Open the Edit menu of the figure and click on Axes properties. Click again on the curve. Below the figure, in the Marker box, open the pull-down menu and click on o, and in the Line box open the menu and click on no line. Open again the Edit menu and click again on Axes properties. Below the figure, navigate through X Axis → X Label and write Temperature, degrees C. Next, navigate through Y Axis → Y Label and write Viscosity of fresh water, m^2 s^{-1}. At this stage we see what appears in Figure 4.2. To exit the editing mode click on the `Exit` icon, `X`, of the editing window.

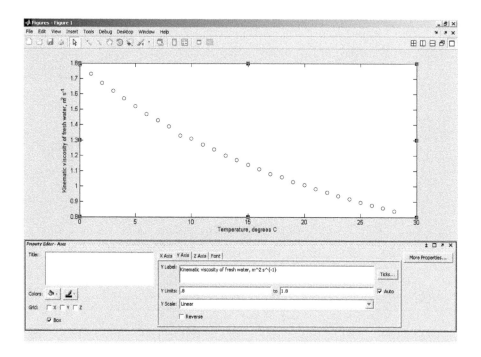

FIGURE 4.2: Interactive plot editing

In the toolbar open the *Tools* menu and choose *Basic Fitting*. A dialog opens as shown in Figure 4.3. Check *cubic, Show equations* and *Plot residuals*, as marked by arrows in the above-mentioned figure. The plot is immediately updated and it looks as in Figure 4.4. If you click on the arrow in the bottom, right-hand side of the dialog box (Figure 4.3), the graphical interface expands and shows the equation of the fitted curve, its coefficients, and the norm of the residuals.

If you did not plot interactively, but worked with the *Command Line*, it is necessary to add the commands that produce the title and label the axes. To save the figure, under the toolbar open the menu *Save Figure* (the third icon from left) and choose the file format. To generate a file that produces the same result that you obtained above, open the menu *File* and click on *Generate M-file*. Run the resulting function with the x and y input arguments used until now. You may run the file also with another pair of arrays. Moreover, you may customize the file for other regressions.

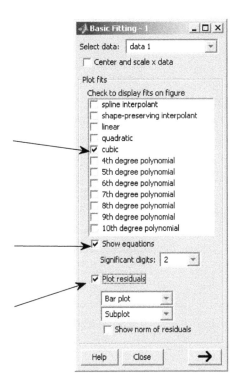

FIGURE 4.3: Dialog box for curve fitting

4.5 The MATLAB®spline function

In this section we describe a simple way of fitting a curve over a number of
points, we show why it should be preferred to regular polynomial interpolation, and we explain how it functions. To give an interesting example, we show
in Figure 4.5 a series of points situated along a transversal section, near the
stern of a passenger ship. As this section is symmetric, it is sufficient to show
only half of it. For those familiar with the terminology of naval architecture,
we may say that the points belong to a *ship station*. We are looking for a way
of fitting a *fair* curve over the given points. The naval architectural term *fair
curve* says much more than 'smooth curve'; it covers not only continuity, but
also continuities of the first and the second derivatives and some aesthetical
features that are difficult to define. To compare different possibilities without
having to input several times the same point coordinates, let us write a script
file that we call `SplineTrial`. We first try a polynomial of the highest degree

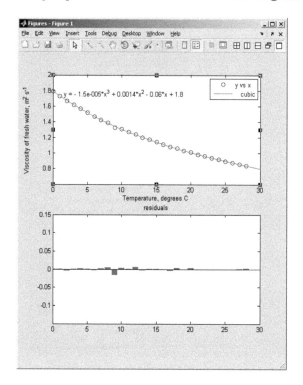

FIGURE 4.4: Curve fitted over fresh-water, kinematic-viscosity points

compatible with the given number of points. The first part of the script file is

```
%SPLINETRIAL    Experiments with the MATLAB spline
% enter station points
x    = [ -9.1361 -7.7103 -6.1394 -3.9366 -2.5037 -1.5628 -0.9030 ...
 -0.5933 -0.4555  ];
y    = [ 11.0276  8.6891  7.0619  5.3533  4.1221  2.8927  1.6643 ...
  0.6349  0.1994   ];
% plot given points
plot(x, y, 'ro'), grid
axis equal
xlabel('Half-breadth')
ylabel('Depth')
pause
%%%%%%%%%%%%%  polynomial interpolation %%%%%%%%%%%%%%%%%
n    = length(x);
m    = n - 1;
% coefficients of polynomial interpolation, degree m
c    = polyfit(x, y, m)
xi = x(1): (x(n) - x(1))/30: x(n);  % interpolation scale
```

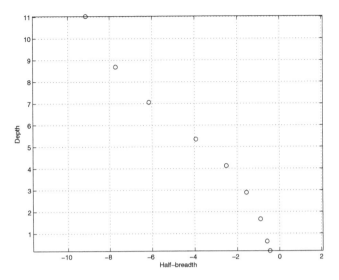

FIGURE 4.5: Points on a ship station

```
yi = polyval(c, xi);          % polynomial interpolation
plot(x, y, 'ro', xi, yi, 'k-'), grid
axis equal
title([ 'Polynomial fit, degree  '  num2str(m) ])
xlabel('Half-breadth')
ylabel('Depth')
```

The statements before the **pause** command plot the given points as small circles. The statements that follow

- use the function **polyfit** to calculate the array, c, of the coefficients of the polynomial that passes through all given points;

- use the function **polyval** to interpolate points between the given ones;

- over the given points plot the curve defined by the interpolated points.

The result of the polynomial interpolation is shown in Figure 4.6; it is far from being satisfactory. Experience shows that the higher the degree of the interpolating polynomial, the larger are its oscillations. The phenomenon is known as *polynomial inflexibility*. The quest for an acceptable solution leads to the idea of *piecewise interpolation*, in which the interval is divided into several arcs. Lower order polynomials are fitted for each arc, while conditions of continuity are imposed at the points where two arcs meet. The degree of the polynomial is usually chosen as three, the lowest degree for which the second derivative exists. This allows the representation of points of inflexion. The basic MATLAB package includes a function called **spline** that implements

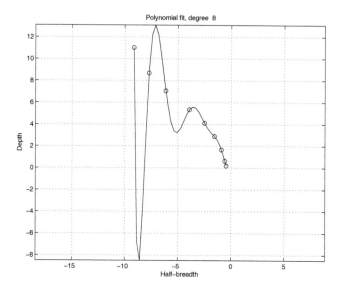

FIGURE 4.6: Ship station - Polynomial interpolation

this idea. The term *spline* was taken from naval architecture where it refers to a flexible, metallic or wooden stripe used to draw curves. The first mathematician to use the term for new mathematical objects was Isaac Schoenberg (Galatz-Romania 1903, 1990). Let us first try the MATLAB `spline` function and explain its functioning afterwards. To continue our script file we first add the command `pause` and after it the lines that call the MATLAB function `spline`:

```
pause
%%%%%%%%%%%%%%%% spline interpolation % %%%%%%%%%%%%%%%%%%%
yi = spline(x, y, xi);
% plot MATLAB spline
plot(x, y, 'ro', xi, yi, 'k-')
axis equal
title('Cubic spline interpolation')
xlabel('Half-breadth')
ylabel('Depth')
```

The function `spline` is called with three arguments:

- the array of $x-$ coordinates of the given points;

- the array of $y-$coordinates of the given points;

- the array, xi, of $x-$coordinates of the points to be interpolated.

The output is the array of y−coordinates of the interpolated points. In Figure 4.7 we plot the given points as circles, and the interpolated curve as a solid line. In this case the result is satisfactory. The MATLAB `spline` function has its limitations and, in some cases, it may yield unsatisfactory results. Then, other splines may be used. Being satisfied, for the moment, by the spline function, let us explain how it is calculated. Given n points with the coordinates x_i, y_i, $i = 1 \ldots n$, we want to fit in each interval $[x_i, x_{i+1}]$, a cubic polynomial of the form

$$S_i = c_{i1}(x - x_i)^3 + c_{i2}(x - x_i)^2 + c_{i3}(x - x_i) + c_{i4} \qquad (4.1)$$

subject to the conditions of continuity to be described immediately. We need $n - 1$ splines, each one having four coefficients, in total $4(n - 1)$ coefficients. Each polynomial must pass through the two points that define the interval of the spline. As we have $n - 1$ intervals, we obtain $2(n - 1)$ conditions of the form $S_i(x_i) = y_i$, $S_i(x_{i+1}) = y_{i+1}$. We ask also for the continuity of tangents at the common point of two intervals. This yields $n - 2$ conditions of the form $\dot{S}_i(x_{i+1}) = \dot{S}_{i+1}(x_{i+1})$. Finally, we ask for the continuity of curvature at the common point of two intervals. This yields $n - 2$ conditions of the form $\ddot{S}_i(x_{i+1}) = \ddot{S}_{i+1}(x_{i+1})$. Up to now we have defined $4n - 6$ conditions; two more are needed. The usual way to complete the set of conditions is to impose zero curvature at the first and last points, that is $\ddot{S}_1(x_1) = 0$, $\ddot{S}_{n-1}(x_n) = 0$. This condition is called *natural*, and the resulting curve, *natural spline*. It is possible to find out the details of the spline fitted by MATLAB. To exemplify this let us first verify that we still have in our workspace the arrays of coordinates, x and y used in Figure 4.5. Next we call the `spline` function with two arguments only, and the output argument **pp**, an acronym for *piecewise polynomial*. The resulting display shows that **pp** is a *structure* with five *fields*:

breaks a *1-by-9* array of *double-precision* numbers;

coefs an *8-by-4* array of *double-precision* numbers;

pieces containing the value 8;

order containing the value 4;

dim containing the value 1.

```
≫ x
x =
Columns 1 through 6
 -9.1361   -7.7103   -6.1394   -3.9366   -2.5037   -1.5628
Columns 7 through 9
 -0.9030   -0.5933   -0.4555
≫ y
y =
Columns 1 through 6
  11.0276   8.6891   7.0619   5.3533   4.1221   2.8927
Columns 7 through 9
   1.6643   0.6349   0.1994
≫ pp = spline(x, y)
pp =
   form:   'pp'
 breaks:   [1x9 double]
  coefs:   [8x4 double]
 pieces:   8
  order:   4
    dim:   1
```

We defer the discussion of the term *double-precision* until Chapter 5, and that of the term *structure* until Chapter 6. For the moment we are interested only in the meaning of the information contained in **pp**. To access this information we use the function **unmkpp** called with **pp** as input argument, and the five fields of **pp** as output arguments:

```
≫ [ breaks, coefs, l, k, d ] = unmkpp(pp)
```

From the results shown below it appears that the array **breaks** is identical to the array x. As to the array **coefs**, it contains the coefficients c_{ij} of the splines S_i. There are eight rows, the i-th row corresponding to the spline S_i fitted between the breaks i and $i + 1$. In each row the coefficients appear in the descending order of the powers of $(x - x_i)$. The variable **l** contains the number of splines; as expected, it is equal to the number of breaks minus one. The variable **order** is the *order* of the spline, in spline terminology the highest power of the piecewise polynomials plus one. The variable **dim** shows that the spline is one-dimensional.

```
breaks =
Columns 1 through 6
 -9.1361  -7.7103  -6.1394  -3.9366  -2.5037  -1.5628
Columns 7 through 9
 -0.9030  -0.5933  -0.4555
coefs =
 -0.0313   0.3402  -2.0615  11.0276
 -0.0313   0.2062  -1.2824   8.6891
 -0.0078   0.0586  -0.8665   7.0619
 -0.0711   0.0067  -0.7228   5.3533
  0.1314  -0.2990  -1.1416   4.1221
 -1.2725   0.0719  -1.3553   2.8927
  3.7140  -2.4469  -2.9223   1.6643
  3.7140   1.0038  -3.3692   0.6349
l =
  8
k =
  4
d =
  1
```

To show that the MATLAB spline is that calculated by Equation 4.1, we continue our file SplineTrial by adding the lines listed below. After calling the function unmkpp, we use the recovered information to implement Equation 4.1 in a loop that is repeated l times. The output yielded by unmkpp allows us to do even more. In some applications we may be interested in the area under the fitted curve. For our spline, and with the notation used above, this area is given by

$$A = \sum_{i=1}^{l} \int_{x_i}^{x_{i+1}} S_i dx$$

$$= \sum_{i=1}^{l} \left[\frac{c_{i1}}{4}(x - x_i)^4 + \frac{c_{i2}}{3}(x - x_i)^3 + \frac{c_{i3}}{2}(x - x_i)^2 + c_{i4}(x - x_i) \right]_{x_i}^{x_{i+1}}$$

$$= \sum_{i=1}^{l} \frac{c_{i1}}{4}(x_{i+1} - x_i)^4 + \frac{c_{i2}}{3}(x_{i+1} - x_i)^3 +$$

$$\frac{c_{i3}}{2}(x_{i+1} - x_i)^2 + c_{i4}(x_{i+1} - x_i)$$

$$(4.2)$$

We include the corresponding MATLAB statements in the lines to be added to our file. The lines for integration are called within the loop over the number of splines.

```
pause
%%%%%%%%%%% Write own spline and calculate area under it %%%%%%%%%%
hold on
% find details of MATLAB spline
pp = spline(x, y)
[breaks,coefs,l,k,d] = unmkpp(pp)
I   = 0                              % initialize integral
% loop over spline segments
for n = 1:l
    % create interpolating scale
    xi = breaks(n): 0.1: breaks(n+1);
    % calculate interpolated points
    yi = coefs(n, 1)*(xi - breaks(n)).^3 + ...
         coefs(n, 2)*(xi - breaks(n)).^2 + ...
         coefs(n, 3)*(xi- breaks(n)) + coefs(n, 4);
    % calculate area under the spline curve
    dx = breaks(n+1) - breaks(n)
    I0  = coefs(n, 1)*dx^4/4 + coefs(n, 2)*dx^3/3 + ...
          coefs(n, 3)*dx^2/2 + coefs(n, 4)*dx
    % plot interpolated spline in red, over the MATLAB spline in black
    plot(xi, yi, 'r-')
    I   = I + I0
end
% end loop over spline segments
t = [ 'Cubic spline and its integral, \int ydx =  ' num2str(I, 2) ]
title(t)
legend('Given points', 'MATLAB spline', 'My spline')
hold off
```

Run the completed file. In the last figure the spline drawn in red by the last program lines cannot be distinguished from the curve plotted by the MATLAB built-in function **spline**; the two curves are identical. The title of the figure contains the value of the area under the curve. Note that in the string printed as title we used the TEX command '\int' to show the symbol of integration.

Exercise 4.1 Checking the spline on the cosine function

Return to the file **SplineTrial** developed in Section 4.5 and modify it to interpolate the function $\cos \alpha$ and calculate its integral in the interval $0 \ldots \pi/2$. In continuation, calculate analytically the area under the curve, in the given interval, and compare with the results yielded by integrating the spline function.

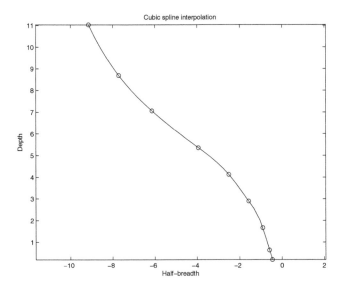

FIGURE 4.7: Ship station – Interpolation by the MATLAB spline function

4.6 Importing data from Excel® – histograms

EXAMPLE 4.1 Building a histogram of data contained in an Excel file

Excel® is an excellent environment for preparing lists of students and their grades. The software also allows for the immediate calculation of average grades and for identifying the minimum and maximum values. MATLAB, on the other hand, can plot the histogram of the grades. In this example we show how to combine the advantages of the two software packets. To do so we build a function, ghistogram, that can be used for the various examinations of a course, let's say in engineering drawing. The function is called with three string arguments:

1. the name of the Excel file of grades;

2. the range of the cells that contain the relevant data;

3. the details that will appear in the title of the figure, in continuation of the general title *Engineering Drawing*.

Our listing is

```
%GHISTOGRAM  Builds histogram of grades
%     ghistogram(filename, grange, term)    plots the histogram
%     of the grades contained in Excel file  'filename', in the
```

```
%       range specified by 'grange', and   completes the title with
%       the data specified in 'term'. All input arguments should be
%       entered as strings.

function        ghistogram(filename, grange, term)

% extract data from the Excel file
A = xlsread(filename, grange);
B = isfinite(A);  % identifies cells that contain numbers
I = find(B == 1); % finds indexes of cells containing numbers
C = A(I);         % builds vector of numerical data

average = mean(C);
minimum = min(C);
maximum = max(C);
failed  = find(C < 55); % identifies grades under 55
lF      = length(failed)% number of those who scored under 55
lS      = length(C);    % sample size
f       = 100*lF/lS;
% plot now histogram
[ n, x ] = hist(C);
bar(x, n)
xlabel('Grades')
% build title string
t = ['Engineering drawing, ' term ', sample size = ',...
     num2str(lS)];
title(t)
N   = max(n);
for i = 1:10                       % label bins
text(x(i), (n(i) + N/50), int2str(n(i)));
end
% print main statistics in left, upper corner
t2 = [ 'Percentage that failed ' num2str(f, 2) ]
text(13, 0.85*N, t2);
t3 = [ 'Average grade ' num2str(average, 2) ]
text(13, 0.90*N, t3)
t4 = [ 'Minimum grade ' num2str(minimum, 2) ];
text(13, 0.95*N, t4)
t5 = [ 'Maximum grade ' num2str(maximum, 2) ];
text(13, N, t5)
```

Figure 4.8 shows a histogram plotted with the following command:

```
>> ghistogram('Grades.xls', 'C6:C123', 'Spring semester 2008,
Term A examination')
```

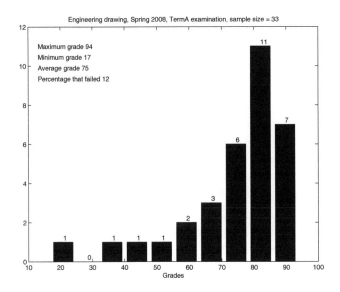

FIGURE 4.8: Histogram of grades contained in an Excel file

Obviously, teachers may adapt the file for their own courses.

4.7 Summary

The simplest way of exporting calculations performed in MATLAB is to use the `Copy` facility in the Workspace, open the file on which you want to include the calculations, and to employ there the `Paste` facility. If you want to keep a record of the work session, type at the beginning `diary` followed by the name of the file. All your commands and the replies of the software will be written in the diary file. When you want to stop recording type `diary off`. You can edit the diary file, insert it in a document, or copy parts of it and insert them in the document you want to produce.

If you produce a graph and want to put it on a file, for example `myfile`, which can be inserted in a Word document, enter the command `print -djpeg myfile.jpg`. If you want to have a file that can be inserted in a `tex` file (to be processed in LaTeX) use `print -deps myfile.eps`. For a colored, Encapsulated Postscript file write instead `print -depsc myfile.eps`.

Data, such as that obtained from experiments, can be plotted also interactively using the buttons, the pull-down menus and the icons appearing in the main toolbar and in the toolbar of the workspace window. Interpolations can be carried on interactively opening the *Tools* menu and choosing *Basic Fitting*.

An easy way of performing a cubic interpolation is to use the MATLAB built-in `spline` function. Given n data points, this function fits between them $n-1$ cubic curves with common ordinates, first and second derivatives at the junction points. The function `unmkpp` extracts the parameters of the fitted curves. In this chapter we have shown how to use these parameters to calculate the area under the spline.

It is possible to import in MATLAB data from Microsoft® Excel sheets. Use for this the command `xlsread`. Use the command `hist` to calculate the histogram, and `bar` to plot it.

The commands introduced in this chapter are

bar the command `bar(x, n)` plots the histogram produced by the command shown in the following item.

diary `diary myfile.dia` opens a file with the given name and stores in it the commands and the answers that appear on the screen.

diary off stops writing to the diary file and closes it.

hist the command `[n, x] = hist(C)` distributes the elements of C into ten equally spaced bins and returns in `n` the frequency count at the bin locations `x`.

print `print -dformat filename` prints the current figure in the specified format and gives the file the specified name. Examples of graphic formats: `-djpeg`, `-deps`, `-depsc`.

spline yi = spline(x,y,xi), cubic spline interpolation over the points $y = f(x)$. The output yi consists of the values interpolated at the points xi.

unmkpp if we call pp = spline(x, y), the command `[breaks,coefs,l,k,d] = unmkpp(pp)` yields the details of interpolated spline.

xlsread the command `xlsread(filename, grange)` extracts from the Excel sheet `filename` the data stored in the range `grange`.

4.8 Exercises

Exercise 4.2 Checking the spline on a simple polynomial

Adapt the file `SplineTrial` developed in Section 4.5 for the function $x^2 + x + 1$ in the interval $0 \ldots 1$. Calculate analytically the area under the curve, in the given interval, and compare with the results yielded by the spline integral.

Part II

Programming in MATLAB®

5

Some facts about numerical computing

5.1 Introduction

As computers have entered all fields of science and technology, even invading our private lives, a new mythology has arisen around them. There are people who think that the computer can solve any problem. There are people, and not only laymen among them, who regard the computer as an infallible tool. On the other hand, there are also those who find the computer a convenient scapegoat. Did you make a mistake and someone else found it? Put the blame on the computer; the machine will not defend itself against the accusation.

In this chapter we show that the computer has its limitations. The computer can deal only with a finite set of numbers. Therefore, computer arithmetic differs from the arithmetic we learned in schools. Approximating a real number by the closest computer number is a source of errors. In chained or iterated calculations the errors can accumulate and become catastrophic. The technical literature cites several examples of disasters caused by defective software, among them the failure of a *Mariner 1* space probe, an error in a radiation therapy system that caused at least five deaths and some serious injuries, the explosion of the *Ariane 5* rocket, the failure of a *Patriot* missile to intercept a *Scud* missile. We study several examples of errors due to the computer representation of numbers and show how to modify calculations in order to minimize or avoid some errors.

In continuation, we introduce **computational complexity**, a method of estimating the amount of computational resources required for the execution of a given algorithm. In this chapter we refer only to time resources considered proportional to the number of execution steps. If a given problem can be solved by several algorithms, one should prefer the algorithm with the lowest complexity.

A part of this chapter reuses material from Biran nd Breiner (2002).

5.2   Computer-aided mistakes

Computations can be affected by several types of errors. First, except in simple cases, we usually work with **models** that are a simplification of the reality. To make calculations possible, but sometimes also because of lack of knowledge, we neglect certain aspects of the real world. This happens in the first step, when constructing the **physical model**, and can also happen in the second step, when we translate this model into a **mathematical model**. The latter is mainly composed of algebraic and differential equations.

The third step is the translation of the mathematical model into a computer program. Bad programming can introduce errors. Some of them may be caused by writing computer functions and commands incorrectly; we call them **syntax errors**. Fortunately, the computer objects to most of them by returning *error messages*. Other errors are due to wrongly organized calculations; they are **logic errors**, but the popular term is **bugs**.

Errors can also occur in the numerical data supplied to the computer. Such data can be obtained from measurements, which, in most cases, are error affected. More often we supply *standard data*, as required by the various codes of practice. Thus, in thermodynamical calculations we may use standard temperatures and pressures, in naval architectural calculations standard seawater densities and standard wind velocities, and in electrical calculations standard voltages, for example 110 V in the USA, 220 V in most other parts of the world. Actual values fluctuate around the standard values, but practical calculations cannot take such fluctuations into account. It is important to design algorithms and programs that are not too sensitive to input errors. Whenever necessary, we discuss this in our book, but for a more thorough treatment we refer the reader to textbooks on numerical analysis.

There is a further source of errors and it is this that we want to discuss in the following sections: the way in which numbers are represented in the computer. A simple example will help us to introduce the subject. Try for yourself the following

```
≫ A = 10^20;
≫ B = A + 1
B =
  1.0000e+020
≫ B - A
ans =
  0
```

The difference should have been 1, not 0.

Usually, books on numerical methods begin with a description of the computer representation of numbers and of computer errors. See, for example,

Wood (1999), Hultquist (1988), or Gerald and Wheatley (1994). We prefer to treat the subject at the point where the reader has mastered enough MAT-LAB tools to experiment with interesting examples and the ways of avoiding, or at least minimizing, the errors.

In the following subsections we give more examples of computer-generated errors. We encourage the reader to look also on the web where many sites treat this subject, for example Robert Piché,
http:/virtual.cvut.cz/odl/partners/tut/unit/mklunit1.html

5.2.1 A loop that does not stop

Write the following short program and run it.

```
%INFINITELOOP   Example of loop that does not terminate

x = 1; while x~= 0
    x = x - 0.2
end
```

Running this loop we obtain a non-terminating sequence of numbers that begins with:

```
x =
   0.8000
x =
   0.6000
x =
   0.4000
x =
   0.2000
x =
   5.5511e-017
x =
  -0.2000
x =
  -0.4000
```

We aborted the run by pressing simultaneously the `Ctrl` and C keys, otherwise the program would have continued to run. We expected the program to have stopped after the fifth step because

$$1 - 5 \times 0.2 = 0$$

The screen display shows that the *halting condition* was missed by a very small error.

5.2.2   Errors in trigonometric functions

We know that

$$\sin 2k\pi = 0, \ cos2k\pi = 1, \ tan2k\pi = 0$$

for any integer k. As it appears below, the computer does not know this.

```
≫ sin(10^10*pi)
ans =
 -2.23936276195592e-006
≫ cos(10^10*pi)
ans =
 0.999999999997493
≫ tan(10^10*pi)
ans =
 -2.23936276196154e-006
```

The errors are very small, but we may have expected the computer to be exact. Interestingly, the MATLAB function sin yields a result different from zero even for k as small as 1. Try for yourself sin(2*pi). The MATLAB function sind, however, yields the correct results. Try for yourself sind(1000*360). Experiment also with the functions tan and tand.

5.2.3   An unexpected root

This is a classical problem treated, for example, by Anonymous (1992), Hultquist (1988), Schneider and Eberly (2003), or Piché (no year given). We know that the second-degree equation

$$Ax^2 + Bx + C = 0 \tag{5.1}$$

has two roots given by

$$x_1 = \frac{-B + \sqrt{B^2 - 4AC}}{2A} \tag{5.2}$$

and

$$x_2 = \frac{-B - \sqrt{B^2 - 4AC}}{2A} \tag{5.3}$$

If $A = 1$, $C = 1$, and $B^2 \gg 4AC$, one approximate solution is

$$x_2 \approx -B \tag{5.4}$$

As $C = x_1 x_2$, the other approximate solution is

$$x_1 \approx -1/B \tag{5.5}$$

Trying to calculate the roots on a computer, using Equations 5.2 and 5.3, can yield an unexpected result, instead of the root approximated by Equation 5.5. To experiment with this example, write the following program in a file called Roots2Degree.m and run it.

```
%ROOTS2DEGREE Solutions of the second-degree equation
%              'Ax^2 + Bx + C = 0
%      Written by Adrian Biran, April 2006

  A  = 1; C = 1;              % coefficients of 2nd-degree equation
  fid = fopen('Roots2Deg.out', 'w'); % open file for results
  fprintf(fid, '         B          x1           x2\n');

  for k = 1: 10                          % begin loop over B
     B = 10^k;
     D  = sqrt(B^2 - 4*A*C);             % discriminant
     x1 = (-B + D)/(2*A);
     x2 = (-B - D)/(2*A);
     fprintf(fid, '%+13.5e %+13.5e %+13.5e\n', B, x1, x2)
  end                                    % end loop over B
  fclose(fid)                            % close file of results
```

indexfclose

The results, printed in the file Roots2Deg.out, are

```
      B               x1                x2
+1.00000e+001 -1.01021e-001 -9.89898e+000
+1.00000e+002 -1.00010e-002 -9.99900e+001
+1.00000e+003 -1.00000e-003 -9.99999e+002
+1.00000e+004 -1.00000e-004 -1.00000e+004
+1.00000e+005 -1.00000e-005 -1.00000e+005
+1.00000e+006 -1.00001e-006 -1.00000e+006
+1.00000e+007 -9.96515e-008 -1.00000e+007
+1.00000e+008 -7.45058e-009 -1.00000e+008
+1.00000e+009 +0.00000e+000 -1.00000e+009
+1.00000e+010 +0.00000e+000 -1.00000e+010
```

We see that for $B = 10 \ldots 10^6$ the results are as expected, for $B = 10^7 \ldots 10^8$ they are approximately as expected, but for $B > 10^8$ the root x_1 has no significant digits. The reader may modify the file Roots2Degree.m to catch the exact value of B where this unexpected result appears first.

Above we use the command fopen to open a file for printing the results. We name the file fid. The values of the variables B, x_1 and x_2 are printed with the command fprintf; see the MATLAB help for details. We use the command fclose to close the file fid.

5.2.4   Other unexpected roots

The equation

$$(x - 1)^n = 0$$

has exactly n solutions

$$x_1 = x_2 = \ldots = x_n = 1$$

If we expand the polynomial and calculate its roots by means of the MATLAB function roots, from a certain n up we are finding unexpected complex roots. To experiment with this problem write the following program in a file called polyroots.m. The polynomial coefficients of the expansion are obtained with the function conv. For each n we can visualize the polynomial function, around the solutions 1. To advance the programme, after each display press the Enter key.

```
%POLYROOTS   Shows strange roots of the polynomial equation
%                    (x - 1)^n
%     Written by Adrian Biran, April 2006

P0 = [ 1 -1 ], P  = P0;
for k = 1: 10;                  % begin loop over equation degree
    disp([ 'For the degree ' num2str(k) ' the roots are ' ])
    roots(P)
    x = 0.99: 0.0001: 1.01; % interval around x = 1
    y = polyval(P, x);
    plot(x, y), grid
    title([ 'Plot of polynomial of degree ' num2str(k) ])
    xlabel('x')
    ylabel('Polynomial values')
    pause
    P = conv(P, P0);
end                             % end loop over equation degree
```

Running the program we see that for $n = 1$, 2, 3 we obtain the correct results. The calculated roots for $n = 4$ are

```
1.0002
1.0000 + 0.0002i
1.0000 - 0.0002i
0.9998
```

Obviously, these are not roots of the given equation, even if they are very close to 1. As the degree n increases, the calculated roots deviate more and more from the correct value. Note, in the plots, the oscillations of the higher-degree polynomials.

5.2.5 Accumulating errors

The following example is inspired from Colonna (2006). Let us consider the following iterative function:

$$x_{n+1} = Ax_n - B \qquad (5.6)$$

Colonna proposes the values $A = 64.1$, $B = 63.1$, $x_1 = 1$. To calculate several iterations write the following program in a file called `IterrError`:

```
%ITERERROR  Shows how errors accumulate in the
%                iterative scheme
%                    Xn+1 = AXn - B
%    Example proposed by Colonna (2006). Script file
%    Written by Adrian Biran, April 2006

A = 64.1; B = 63.1;
X = 1;
for k = 1:10
   X = A*X - B
end
```

It is easy to see that, for the given values, all results should be equal to 1. Running the program we obtain the following unexpected numbers:

```
 0.99999999999999
 0.99999999999954
 0.99999999997039
 0.99999999810210
 0.99999987834492
 0.99999220190909
 0.99950014237250
 0.96795912607752
-1.05382001843072
-1.306498631814092e+002
```

We shall understand later what happens. In the meantime try also the following:

```
≫ A - B
ans =
  0.99999999999999
```

At this point we may ask ourselves what happens here. We have learned arithmetic and practiced it. It appears that computer arithmetic is different. Why? We explain this in the next section.

5.3   Computer representation of numbers

In everyday life we work with numbers represented in the *decimal positional system*. In this system the value of each digit depends upon its position relative to the decimal point; for example, the meaning of 124.32 is

$$1 \times 10^2 + 2 \times 10^1 + 4 \times 10^0 + 3 \times 10^{-1} + 2 \times 10^{-2}$$

We say that the above number is represented in *fixed-point format*. In MATLAB the default display is in fixed-point format with four decimal digits. In the *long format* double-precision numbers are represented with 15 decimal digits. We are also familiar with the *scientific notation* in which the number 124.32 is represented as 1.2432×10^2. Today computers work with an adaptation of the scientific format called **floating-point representation**; in MATLAB we obtain it with

```
≫ format short e, x = 124.32
x =
     1.2432e+002
```

We can return to fixed-point notation by

```
≫ format short, x
x =
    124.3200
```

To obtain more digits use

```
≫ format long, x
x =
      1.243200000000000e+002
```

or

```
≫ format long e, x
x =
      1.243200000000000e+002
```

Computers use the floating-point representation, but not the decimal system. The favored representation is that in the *binary system* in which there are only two digits, 0 and 1. As in the decimal system, the value of each digit depends upon its position relative to the decimal point — perhaps it would be better to call it *fractional point* — but it multiplies powers of 2. For example, the meaning of the binary number 101.101 is

$$1 \times 2^2 + 0 \times 2^1 + 1 \times 2^0 + 1 \times 2^{-1} + 0 \times 2^{-2} + 1 \times 2^{-3}$$

The main reason for using binary numbers is that they can be put into correspondence to the states *on* and *off* of physical components; for example 1 for a closed circuit and 0 for an open circuit. Another example: 1 for 'magnetized', 0 for 'demagnetized'.

Putting things together, numbers are represented in the computer under the form $\sigma m b^e$ where σ is the sign, that is + or -, m, the fractional part, called *significand* or *mantissa*, b the *base* of the number system, and e the *exponent*.

In 1985 the Institute of Electrical and Electronics Engineers — American institute mostly known as IEEE — issued an *IEEE Standard for Binary Floating-Point Arithmetic* (ANSI/IEEE Std 754-1985). The standard is fully implemented on many computers, partially on a few others. In the *single-precision* standard the information is stored in 32 bits. In the IEEE *double-precision* standard a number is defined by information stored in 64 bits as follows:

- 1 bit for the sign (marked above as σ);

- 11 bits for the exponent data (marked above as e);

- 52 bits for the fractional part of the mantissa (marked above as m).

To understand the anatomy of a number stored in the *IEEE 754* format let us write the number as

$$N = \sigma n \times 2^x$$

where σ, n and x are binary numbers. By convention, the number σ equals 0 for $N > 0$, and 1 for $N < 0$. The exponent x is chosen so that n has only one digit before the fractional point. That digit can only be 1. The other possible digit would 0, and this is impossible as leading zeros have no meaning. By choosing x as above the fractional point is shifted to the left. Therefore, a number represented in this way is called *floating-point number.* As the leading digit is always 1, it conveys no information and, therefore, it is neglected. The resulting number, m, represents only the fractionary part of the mantissa. The exponent is *biased* by adding 1023 to it. Defining $e = x + 1023$, we obtain the above representation $N = \sigma m \times 2^e$. Thus, the number N is stored in 64 bits in the format

$$\sigma e_1 e_2 \ldots e_{11} m_1 m_2 \ldots m_{52}$$

where $e_1 \ldots e_{11}$ are the digits of the biased exponent, and $m_1 \ldots m_{52}$ those of the fractional part of the mantissa.

The above description shows that the lowest biased exponent can be 0; however, this value is reserved for special cases, so that the lowest biased exponent is, in practice, 1. To this corresponds the unbiased exponent $1 - 1023 = -1022$. The smallest possible mantissa is obviously 1. We conclude that the smallest number that can be stored in the IEEE double-precision format is $1 \times 2^{-1022} = 2.2251 \times 10^{-308}$.

Let us find now the largest number that can be stored in the IEEE double-precision standard. The largest possible exponent, in binary representation, has 11 digits all equal to 1. Adding 1 would result in a binary number that is represented by 1 followed by 11 zeros, that is $2^{12} = 2048$. Then, the biased exponent equals $2048 - 1 = 2047$. This number is reserved for special cases. Thus, the largest meaningful biased exponent is 2046, and the largest actual exponent is $2046 - 1023 = 1023$. The maximum possible fractionary part of the mantissa, in binary form, is represented by 52 digits all equal to 1. Adding 1 to this number, we obtain a binary number that consists of 1 followed by 52 zeros. Considering the unwritten leading 1, we obtained the number 2. We conclude that the largest number that can be stored in the IEEE double-precision format is $(2 - 2^{-52}) \times 2^{1023} = 1.7977 \times 10^{+308}$.

In earlier versions of MATLAB it was possible to check if the computer worked in IEEE arithmetic:

```
≫ isieee
ans =
    1
```

If 1 was the answer, the computer used, indeed, IEEE arithmetic. Today MATLAB implements IEEE arithmetic on all computers.

Two MATLAB functions, `realmax` and `realmin`, allow us to find the largest and the smallest number that can be stored in the computer. For a computer using the IEEE standard the result of invoking these functions is

```
≫ realmax
ans =
    1.7977e+308
≫ realmin
ans =
    2.2251e-308
```

We recovered the maximum and the minimum number found above.

These are the default answers in MATLAB and they correspond to *double-precision* numbers. MATLAB allows also other representations, namely single-precision floating point, signed integers and unsigned integers. Each of these classes has its own range bounded by the respective **realmin** and **realmax** values. Any number larger than that yielded by the **realmax** command will cause an *overflow* error, while any number smaller than that yielded by the **realmin** command will cause an *underflow* error. A famous example of overflow error is the one that caused the failure of the *Ariane 5* Flight 501 on 4 June 1996 (see Example 5.3). Using different classes of numbers may result in better use of computer resources. However, as the example of *Ariane 5* shows, any lack of attention can end by catastrophic results. Therefore, we strongly recommend beginners to refrain from changing the default MATLAB representation.

An indication of the floating-point accuracy of the standard is given by the MATLAB variable eps; as set initially it corresponds to the distance from 1.0 to the next largest floating point number. The following command retrieves eps

```
≫ eps
    eps =
    2.2204e-016
```

This means that the computer cannot represent a number between 1 and 1.00000000000000022204. If the user defines a variable called eps the above meaning is changed.

We can now explain why the computer did not reach zero when extracting five times 0.2 from 1. The reason is that the binary number corresponding to the decimal fraction 0.2 is

$$0.00110011001100\ldots \tag{5.7}$$

This representation requires an infinite number of digits. The consequence is that the computer works with an approximate value of 0.2 that is closest to the true number. Subtracting five times the approximate value of 0.2 from 1 does not yield exactly 0.

Converting back from rounded binary representation, to decimal representation, can introduce other errors.

5.4   The set of computer numbers

From the preceding section we understand that the computer can deal only with a finite set of numbers; we call it the *set of computer numbers* or *machine numbers*. In single precision this set is limited between 1.1755^{-38} and 3.4028×10^{38}, and in double precision between 2.2251×10^{-308} and 1.7977×10^{308}.

The set of computer numbers is not only bounded by a minimum and a maximum number, but it is also *discrete*. To understand this fact let us consider the set of *real numbers*, $\mathbb{R}$, as defined in modern mathematics. Let $a, b \in \mathbb{R}$ be two real numbers chosen as close one to another as we want. We can always define another number, c, that lies between the chosen numbers, that is $a < c < b$. Take, for instance

$$c = \frac{a + b}{2}$$

We cannot do this with computer numbers as there are gaps between them. Because of the way in which computer numbers are defined, the gaps between them are not equal. The larger a number is, the larger also the gaps between it and the preceding or following computer number.

The MATLAB function `eps(X)` yields the positive distance from `abs(X)` to the next larger in magnitude floating point number of the same precision as X. To experiment systematically with this function and draw a plot that shows how the gaps increase with computer magnitude, write the following script file, `epsilonx.m`.

```
%EPSILONX Plots distances from abs(X) to the next larger
% in magnitude floating point number of the same precision
% as X.
% Written by Adrian Biran, April 2006
%%%%%%%%%%%%%%%%%%%%%%%%%%%% SINGLE PRECISION %%%%%%%%%%%%%%%%
n = 0:20;
X = single(10.^n);           % array of plot points
S = eps(X);                  % array of distances
loglog(X, S, 'k-'), grid
title('Plot of distances to next larger floating point number')
xlabel('Number'), ylabel('Distance')
hold on
%%%%%%%%%%%%%%%%%%%%%%%%%%%% DOUBLE PRECISION %%%%%%%%%%%%%%%%%%%%%%%%%
```

```
X = double(10.^n);          % array of plot points
S = eps(X);                 % array of distances
loglog(X, S, 'r--')
legend('Single precision', 'Double precision')
hold off
```

Above, the function `single` converts the value of its argument to its single-precision representation. The function `loglog` plots the decimal logarithm of S against the decimal logarithm of X. The resulting plot is shown in Figure 5.1.

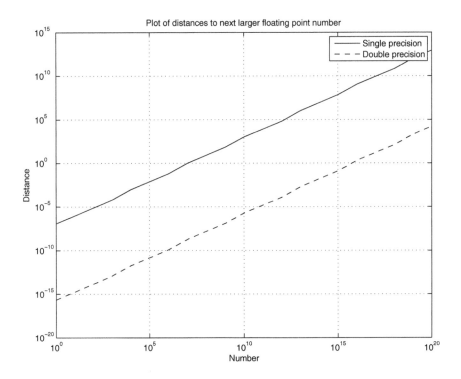

FIGURE 5.1: Distance between computer numbers

To understand the origin of the gaps we must look again at the IEEE representation. Given a machine number, M_i, the next largest machine number, M_{i+1}, is obtained by adding 1 to the mantissa. In the 64-bit format the fractionary part of the mantissa has 52 digits. Therefore, adding one to the mantissa means adding to M the number 2^{-52} times the exponent of M. This is the distance between M_i and M_{i+1}. For example

```
≫ 2^(-52)*(10^10)
g =
  2.2204e-006
≫ 2^(-52)*(10^(-100))
g =
  2.2204e-116
```

See for yourself that the above values are close to those yielded by the commands `eps(10^10)` and `eps(10^(-100))`. These simple considerations explain why the larger the number M_i, also the larger the gap between it and M_{i+1}.

To conclude this section we remark that being bounded and discrete, the set of machine numbers is *finite*.

5.5   Roundoff

Arithmetic operations can yield results with many decimal digits. For example, knowing the diameter of a circle, $d = 1.2866$, and using the approximation $\pi = 3.1416$, the circumference is

$$\pi d = 3.1416 \times 1.2866 = 4.04198256 \qquad (5.8)$$

that is a number with eight decimal digits. We get still more digits if we want to find the area of the circle by

$$\frac{\pi d^2}{4} = \frac{3.1416 \times 1.2866^2}{4} = 1.30010369042400$$

Actually, π has an infinite number of digits, and so can have rational or irrational numbers resulting from such simple operations as $1/3$, $1/9$, or $\sqrt{2}$. When performing numerical calculations we can deal only with finite numbers of decimal digits. For obvious reasons of convenience, the number of digits considered in manual calculations is small, for example $\pi \approx 3.1416$, more often $\pi \approx 3.14$. Computers can deal with more digital digits, but still with finite numbers of them. Take, for instance, the number π; the approximation displayed by MATLAB, in the `long` format, is

$$\pi \approx 3.14159265358979$$

To use an approximation with four decimal digits one could have dropped all digits to the right of the fourth and retain 3.1415. This way of dealing with numbers is called *chopping*. It can be shown that chopping leads to gross errors which, when accumulated in a long sequence of calculations, can yield catastrophically erroneous results. To minimize errors, most computers use *roundoff* instead of chopping, for example $\pi = 3.1416$ instead of 3.1415.

Rounding rules employed in computers are the same as in manual calculations. The simplest statement of the rounding rule is:

> For rounding a number to n digits, add half the base, that is $b/2$, to the digit in position $(n + 1)$ and discard all digits to the right of position n.

In the decimal system the rule for rounding to n digits becomes: if the $(n + 1)$ digit is less than 5, the $n - th$ digit does not change, and if it is 5 or more, the n-th digit is increased by one. Rounding-off as shown above yields approximate values that are sometimes smaller, sometimes larger than the true values. We can thus expect that errors cancel partially each other. Chopping yields always approximate values that are smaller than the true ones; therefore, no error cancellation is possible.

Examples of rounding off in MATLAB are:

```
>> 1/9, 2.2222222222222222/4, 2/3, -2/3
ans =
    0.1111
ans =
    0.5556
ans =
    0.6667
ans =
    -0.6667
```

The reader is invited to repeat the same examples in the long format.

5.6   Roundoff errors

Rounding off a number to n decimal places, following the rule described in the preceding section, produces an error whose absolute value is not larger than $b/2 \times b^{-n}$. Thus, the absolute value of the error in the approximation $\pi = 3.1416$ is not larger than $\rho = 0.5 \times 10^{-4}$. Assuming that the MATLAB value of π is correct, the *error* is

```
≫ E = pi - 3.1416
E =
      -7.3464e-006
```

Indeed, the error is smaller than the error bound shown above. We used here the definition

$$error = true\ value - calculated\ value$$

Some authors call this *absolute error*; others use the qualifier *absolute* for the absolute value of the error defined above. It makes sense to compare the error, or the absolute error, to the true value of the affected number. To do this we calculate the *relative error*, that is

$$relative\ error = error/(true\ value)$$

or

$$absolute\ relative\ error = (absolute\ error)/(true\ value)$$

In our example we calculate the relative error with

```
≫ E/pi
ans =
      -2.3384e-006
```

Frequently the relative error is expressed in percent, in our case:

```
≫ 100*E/pi
ans =
      -2.3384e-004
```

It can be easily shown that the absolute error of a sum or difference of two numbers is at most equal to the sum of the error bounds of the two numbers. Consider two numbers rounded to n decimal places, the absolute error in their sum or difference is not larger than

$$0.5 * 10^{-n} + 0.5 * 10^{-n} = 10^{-n}$$

When analyzing products or quotients it is the relative error that counts. It can be shown that the absolute relative error of the product or quotient of two numbers is not larger than the sum of their relative error bounds. Consider again the example in Equation 5.8

$$\pi d = 3.1416 \times 1.2866 = 4.04198256$$

or, in MATLAB

```
≫ circum = 3.1416*1.2866, format long
circum =
    4.04198256
```

The relative error bounds of the two factors are

```
≫ E1 = 0.5*10^(-4)/3.1416
E1 =
    1.5915e-005
≫ E2 = 0.5*10^(-4)/1.2866
E2 =
    3.8862e-005
```

The relative error bound of the product is given by `E1 + E2` and the error bound by

```
≫ (E1 + E2)*circum
ans =
    2.2141e-004
```

It follows that only the first three digits of the product are reliable; the others must be discarded.

A particularly bad error is produced by *cancellation*; it occurs in the addition of two numbers that are nearly equal, but have different signs. Enter, for example, the following numbers

```
≫ b = 0.543210987654321*10^2
b =
     5.432109876543210e+001
≫ c = -0.543210987650001*10^2
c =
    -5.432109876500011e+001
```

Try now the addition

```
≫ d = b + c
d =
     4.319957724874257e-010
```

The true result is 4.3199×10^{-10}. The digits common to b and c cancelled themselves. The two numbers b, c, have 16 significant digits, the result only 5. The digits beginning in the fifth decimal place of d were 'fabricated' by the computer. The following simple analysis shows that they must be discarded. The maximum error expected in the numbers b, c is 0.5×10^{-15}, that of the result is 10^{-15}. The conclusion is that the digit 5, in the fifth place of the floating-point representation, and the following are unreliable. The *relative-error bounds* are

```
≫ 0.5*10^(-15)/b
ans =
     9.2045e-018
≫ 0.5*10^(-15)/c
ans =
    -9.2045e-018
≫ 10^(-15)/d
ans =
     2.3148e-006
```

The relative-error bound of the result is by 12 orders of magnitude larger than those of the numbers b and c.

5.7 Computer arithmetic

The preceding sections show that the set of numbers that can be represented on the computer is finite. Let us note this set by M. Machine numbers are not evenly distributed. Figure 5.1 shows that the distance between two consecutive machine numbers is smaller in the vicinity of zero.

The computer can accept only data that belong to the set M. Any other number must be approximated by the nearest number in M. Less evident is the fact that operations on numbers belonging to M may yield a result that is not in M. In such cases the computer approximates the result by the nearest machine number.

Depending on the arithmetic of the computer, certain properties of real numbers do not hold within the set M. We are going to show this on a few examples run in MATLAB. Let us enter the following numbers:

```
≫ format long e, a = 0.123456789012345*10^(-4)
a =
      1.234567890123450e-005
≫ b = 0.543210987654321*10^2
b =
     5.432109876543210e+001
≫ c = -0.543210987650001*10^2
c =
    -5.432109876500011e+001
```

To show that commutativity is not fulfilled try

```
≫ d = a + b + c
d =
     1.234611089984128e-005
≫ e = c + b + a
e =
     1.234611089700699e-005
≫ d - e
ans =
     2.834290826125505e-015
```

The following calculation shows that the law of associativity is not fulfilled:

```
≫ d = (a + b) + c
d =
      1.234611089984128e-005
≫ e = a + (b + c)
e =
      1.234611089700699e-005
≫ d- e
ans =
      2.834290826125505e-015
```

Finally, distributivity is checked with

```
≫ d = a*(b + c)
d =
      5.333281093820511e-015
≫ e = a*b + a*c
e =
      5.333298906673445e-015
≫ d - e
ans =
     -1.781285293355644e-020
```

If $a + \epsilon = a$, we usually conclude that $\epsilon = 0$. This may be not true within the set of machine numbers, for example:

```
≫ format long,eps
eps =
      2.220446049250313e-016
≫ 1 + eps
ans =
      1.00000000000000
```

Throughout the book, whenever necessary, we point to potential sources of errors. The reader interested in this subject should refer to books on *numerical analysis*, for example Hartley and Wynn-Evans (1979), Hultquist (1988), or Rice (1993).

In the preceding sections we wanted to warn the reader that the computer is not omnipotent and that there are pitfalls that should be avoided. We want now to conclude this discussion by reassuring the user that careful programming may lead to good results. The following section contains some hints on how to minimize computer errors. It is also worth mentioning that MATLAB — combined with IEEE arithmetic — is much more efficient in minimizing numerical errors than some older software. We calculated in MATLAB a few problems quoted in books of numerical analysis as examples in which the computer gives wrong results; MATLAB yielded the correct ones!

Exercise 5.1 Some simple matrix operations
Enter in MATLAB the matrices

$$A = \begin{bmatrix} 67000.3 & -42010.2 & 10030.1 \\ 20075.3 & 0.001 & -6071.4 \end{bmatrix} \quad B = \begin{bmatrix} -220002 & 10001.31 & 0.2 \\ 40272.11 & -60711.11 & 1401 \end{bmatrix}$$

For $k = 1.36$ calculate

$$C = k * (A + B), \quad D = k * A + k * B, \quad C - D$$

Where are the results different from what is predicted by matrix theory?

5.8 Why the examples in Section 5.2 failed

We are now able to explain what happens in the examples shown in Section 5.2 and to indicate how to avoid some of the errors encountered there.

5.8.1 Absorbtion

Let us consider an example similar to the first one shown in Section 5.2. We show the results in the decimal system as they are displayed in MATLAB. We define a very large number, N, and we add 1 to obtain another number, M.

```
≫ N = 1.23*10^20
N =
   1.230000000000000e+020
M = N + 1
M =
   1.230000000000000e+020
```

As defined, the number N should be 123000000000000000000. Adding 1 should result in 123000000000000000001. As we see, we need 20 digits after the digit 1. The computer, however, displays only 15 digits after the decimal point. The last digit, 1, is *absorbed*. This is not just a matter of display, as we may be inclined to think at a first glance. With N and M defined as above, we find in MATLAB that

```
M - N
ans =
   0
≫ eps(N)
ans =
   16384
```

Thus, MATLAB, indeed, does not 'feel' that we added 1 to N because the result, $M = N + 1$ falls in the gap between N and the next larger machine number and the software approximates it by the closest machine number, N. We leave to an exercise the task of finding when MATLAB begins to 'feel' the addition of 1 to N.

Exercise 5.2 A case of absorbtion

Given a number N defined as in the preceding subsection, what is the smallest number, A, that added to N will produce a machine number $M \neq N$? Try, first, to answer intuitively. Next, experiment in MATLAB and confirm your intuition. Repeat with several different numbers.

5.8.2   Correcting a non-terminating loop

The *while* loop shown in Subsection 5.2.1 can be corrected as follows:

```
%TERMINATINGLOOP    while loop that terminates

x = 1; while (abs(x) - eps) >= 0
    x = x - 0.2
end
```

Write the above lines to a file **TerminatingLoop.m** and run it. This time the loop stops after the expected number of repetitions. This is one way of ensuring termination; however, it may not be universally valid. The programmer must check the terminating condition for each case in part.

5.8.3   Second-degree equation

Let us return to Subsection 5.2.3. The first root is yielded by

$$x_1 = \frac{-B + \sqrt{B^2 - 4AC}}{2A} \tag{5.9}$$

With the given values of the coefficients, $\sqrt{B^2 - 4AC} \approx B$, and an error of cancellation occurs in the numerator of Equation 5.9. It is catastrophic because we lose all significant digits. In this case a simple algebraic trick solves the problem. If we multiply and divide the fraction by $(\sqrt{B^2 - 4AC} + B)$ we obtain

$$x_1 = \frac{(\sqrt{B^2 - 4AC} - B)(\sqrt{B^2 - 4AC} + B)}{2A(\sqrt{B^2 - 4AC} + B)} = \frac{-2C}{(\sqrt{B^2 - 4AC} + B)} \quad (5.10)$$

To show that we solved the problem, write the following script to a file named `CorrectedRoots2Deg.m`.

```
%CORRECTEDROOTS2DEG  Solution of the 2nd degree equation
%             Ax^2 + Bx + C = 0
%     that avoids the error of cancellation
%     Written by Adrian Biran, April 2006

A = 1; C = 1;    % coefficients of 2nd-degree equation
fid = fopen('CorrectedRoots2Deg.out', 'w');
fprintf(fid, '       B               x1               x2\n');

for k = 1: 10
    B = 10^k;
    D  = sqrt(B^2 - 4*A*C);    % discriminant
    x1 = -2*C/(D + B)
    x2 = (-B - D)/(2*A)
    fprintf(fid, '%+13.5e %+13.5e %+13.5e\n', B, x1, x2)
end
fclose(fid)
```

The results printed on the file `CorrectedRoots2Deg.out` are

```
      B             x1               x2
+1.00000e+001 -1.01021e-001 -9.89898e+000
+1.00000e+002 -1.00010e-002 -9.99900e+001
+1.00000e+003 -1.00000e-003 -9.99999e+002
+1.00000e+004 -1.00000e-004 -1.00000e+004
+1.00000e+005 -1.00000e-005 -1.00000e+005
+1.00000e+006 -1.00000e-006 -1.00000e+006
+1.00000e+007 -1.00000e-007 -1.00000e+007
+1.00000e+008 -1.00000e-008 -1.00000e+008
+1.00000e+009 -1.00000e-009 -1.00000e+009
+1.00000e+010 -1.00000e-010 -1.00000e+010
```

Now all roots have significant digits and are close to the expected values.

5.8.4   Unexpected polynomial roots

Let us return now to Subsection 5.2.4. Running the script `polyroots` for higher-degree polynomials shows wild oscillations close to the true, multiple root. To understand what happens, let us consider, for example, the 7th-degree polynomial equation

$$x^7 - 7x^6 + 21x^5 - 35x^4 + 35x^3 - 21x^2 + 7x - 1 = 0$$

with seven roots $x_1 = x_2 = \ldots x_7 = 1$. We rearrange this equation as

$$(x^7 - 1) + (7x - 7x^6) + (21x^5 - 21x^2) + (35x^3 - 35x^4) = 0$$

When $x \to 1$ the two terms within each parentheses are close to one another. As these terms are approximated by machine numbers, cancellation errors and spurious results can occur. Let us try to solve the equation using the function `roots` recommended for polynomial equations:

```
≫ c = [ 1 -7 21 -35 35 -21 7 -1 ];
≫ r = roots(c)
r =
  1.0101
  1.0063 + 0.0079i
  1.0063 - 0.0079i
  0.9977 + 0.0098i
  0.9977 - 0.0098i
  0.9909 + 0.0044i
  0.9909 - 0.0044i
```

We call now the function `poly`, with the array of roots `r` as argument, to recover the polynomial coefficients. In the default `short` format things look well:

```
≫ poly(r)
ans =
  1.0000 -7.0000 ...   35.0000 -21.0000 7.0000 -1.0000
```

If we switch to the `long` format, we discover that the calculated coefficients are not exactly those defined in the input:

```
 ≫ format long
 ≫ poly(r)
ans =
  Columns 1 through 6
  1.00000000000000 ...   -34.99999999999998 34.99999999999996
 ...
```

To see that the elements of **r** are not the exact roots we can also use the function `polyval`

```
 ≫ polyval(c, r(1))
ans =
  1.199040866595169e-014
 ≫ polyval(c, r(2))
ans =
  1.332267629550188e-014 +1.457167719820518e-016i
```

and so on.

For the x–values obtained with the `roots` function, the value of the polynomial differs slightly from the true value 0. This may seem acceptable to some users; however, if we have to solve an engineering problem the calculated roots are unacceptable. For example, if the roots should be used to dimension a mechanical part, or to obtain the main characteristics of some engineering system, complex values are of little use. What should we do? Heuristics can help. Looking again at the results we can find some hints:

- the real root, 1.010, is close to 1;

- the imaginary parts of the complex roots are very small.

Such hints can encourage us to try the root 1. As this value proves correct, we can use deconvolution to divide the given polynomial say P_7, by $x - 1$, and obtain a 6th-degree polynomial P_6. We can apply now the function `roots` to find the zeros of P_6 and so on. We recommend to proceed in this way and we invite the reader to try this procedure on the given equation. Users who look over the list of MATLAB functions will discover the function `fzero`. Let us check how useful this function is in our case. We first build an *anonymous function* using the operator @ and specifying that the independent variable is x :

```
 ≫ f = @(x)x^7-7*x^6+21*x^5-35*x^4+35*x^3-21*x^2+7*x-1;
```

Next, we invoke the function `fzero` with two arguments, the name of the anonymous function, and an initial guess:

```
≫ fzero(f, 1.2)
ans =
    1.00800000000728
```

The result is not exact and we do not even learn that there are seven roots.
Let us try another guess, on the other side of the expected zero:

```
≫ fzero(f, 0.9)
ans =
    1.00782047101674
```

This is not much better. There is, however, a more serious disadvantage.
The function `fzero` does not find roots, but *zero crossings*. The function
worked with a 7th-degree polynomial because the corresponding curve crosses
the $y = 0$ axis. When the curve is tangent to that axis, `fzero` will fail totally.
We can experiment with the simplest possible example

$$x^2 - 2x + 1 = 0$$

Two MATLAB calculations and their results are

```
≫ f2 = @(x)x^2 - 2*x + 1;
≫ fzero(f, 0.9)
Exiting fzero:  aborting search
for an interval containing a sign change
because NaN or Inf function value encountered during search.
(Function value at -1.54458e+154 is Inf.)  Check function or
 try again with a different starting value.
ans =
NaN
fzero(f2, 1.1)
...
(Function value at -1.88782e+154 is Inf.)
Check function or try again...
ans =
NaN
```

The Optimization Toolbox® provides a more powerful function, `fsolve`.
Returning to the example of the 7th-degree polynomial we invoke the function with the initial guess 0.9 and two output arguments. The first output
argument is x, the zero we are looking for, and the second argument is `fval`,
the value of the function f at the point x.

```
≫ [ x, fval ] = fsolve(f, 0.9)
Optimization terminated:  first-order optimality is
less than options.TolFun.
x =
   0.9000
fval =
  -1.0000e-007
```

The calculated x−value is evidently not a root, and MATLAB shows, indeed, that at the above point the value of the polynomial is not zero.

We may play with other starting values, but the results are worse than those obtained with the **roots** function. Again we do not even learn that there are seven roots.

The *Symbolic Math Toolbox*® provides the true solution. Readers who have access to this toolbox can try the simple command

```
≫ solve('x^7-7*x^6+21*x^5-35*x^4+35*x^3-21*x^2+7*x-1')
ans =
 1
 1
 1
 1
 1
 1
 1
```

Above, the argument of the function **solve** is the equation supplied as a string.

5.9   Truncation error

In this section we treat an error that is not due to the computer, but to the methods we **must** use. Certain mathematical processes imply an infinite number of operations, or steps. Obviously, neither we nor the computer can perform an infinite number of operations. The error caused by cutting an infinite process after a finite number of steps is called **truncation error**. This error can be best exemplified on series. Consider, for example, the following series

$$1 = \frac{1}{2} + \frac{1}{4} + \dots \frac{1}{2^n} + \dots \tag{5.11}$$

This is, in fact, a *geometrical progression* and mathematical analysis teaches us that it *converges* to the value one. We invite the user to find this result by the well-known formula of geometric progressions, while we prefer to give here a geometric interpretation. Let the area of the square in Figure 5.2 be 1. We divide the square into two. Each half has the area 1/2. We divide one half into two halves. Each one of the resulting squares has the area 1/4. We divide one of these squares into two. Each one of the resulting rectangles has the area 1/8. Continuing in this way, and summing up the areas of the resulting figures, we reproduce the series defined by Equation 5.11. In the figure we stopped after the sixth term. The error is represented by the square left without notation.

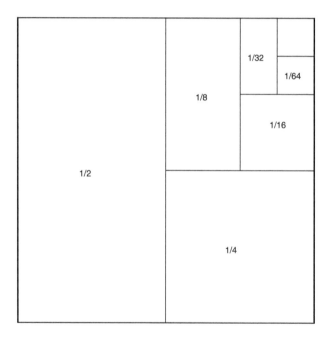

FIGURE 5.2: A series that sums up to 1

As mentioned above, we cannot sum an infinite number of terms, and we must satisfy ourselves to sum a finite number of terms, consistent with a reasonable computational time and the admissible error. To show numerically the effect of neglecting terms write the following programme to a file SeriesOne.m.

```
%SERIESONE Demonstrates truncation error on a series that
%      sums up to one.
%      Written by Adrian Biran, April 2006.
```

```
S = 0;
for k = 1:20                % begin with one term
    S = S + 1/2^k
end                         % end after adding k terms
```

The results, in format long, are

```
S =
  0.50000000000000
S =
  0.75000000000000
...
S =
  0.99996948242188
```

In the exercises we propose more series and ask the reader to write programmes that show the effect of their truncation and how we approximate
better the true values as more terms are summed up.

Exercise 5.3 Truncation error

Convert the script file `SeriesOne.m` into a function `TruncOne(n)`, that receives n as an input argument and yields the summation of n terms of the
series defined by Equation 5.11. Try to catch up the smallest number n for
which

```
≫ 1 - TruncOne(n)
```

yields 0.

EXAMPLE 5.1 A Patriot missile that did not intercept a Scud

This is a famous example of catastrophic accumulation of numerical errors
and more documentation can be found on the Internet, for example GAO
(1992), Skeel (1992), or Toich (1998). During the Gulf War, on 25 February
1991, the Iraqis fired a Scud missile against a US base in Dhahran, Saudi
Arabia. An American Patriot missile launched against the Scud failed to
intercept it. The Scud struck US Army barracks killing 28 servicemen and
wounding about 100.

The analysis of this bloody episode showed that the failure was due to the
software that controlled the Patriot system. The internal clock of the battery
measured time in tenths of seconds expressed as integers. To convert the
numbers to actual values, the software multiplied the integers by 0.1. We
already know that in binary representation the number 0.1 has an unending
number of digits. As the precision of the software was 24 bits, the system
worked with the truncated representation

$$0.00011001100110011001100$$

After 100 hours of operation the accumulated error was 0.3433 s. For a Scud speed of 1676 ms^{-1} this means an error that exceeded 570 m. Thus, the Scud was outside the tracking range of the Patriot battery. Following recommendations of the Israeli Army, the software was soon corrected to work with a precision of 48 bits.

To get a better quantitative appreciation of what happened, we can make a quick calculation as shown below. Here T is a decimal approximation of the truncated binary representation of the number 0.1, E is the error with respect to the MATLAB approximation of 0.1, and we calculate the accumulated error by taking into account that there are 10 tenths of a second in a second, 3600 seconds in an hour, and the system operated 100 hours.

```
>> format long
>> T = 1/(2^4) + 1/(2^5) + 1/(2^8) + 1/(2^9)
+ 1/(2^12) + 1/(2^13) + 1/(2^16) + 1/(2^17)
+ 1/2^(20) + 1/(2^21)
T =
   0.09999990463257
>> E = 0.1 - T
E =
   9.536743164617612e-008
>> Accumulated = 10*100*3600*E
Accumulated =
   0.34332275392623
>> RangeError = 1676*Accumulated
RangeError =
   5.754089355803682e+002
```

5.10   Complexity

5.10.1   Definition, examples

We have learned how to write programs and functions. A program or function can consist of one or several algorithms. MATLAB functions appear to us as simple expressions; however, calling one of them causes the execution of a particular algorithm. Often a computational problem can be solved by different algorithms. A useful way of choosing the 'best' algorithm is to compare the alternatives in terms of required resources, such as execution time, number of elementary operations, or computer memory. The **complexity** of an algorithm is the difficulty of executing it, as measured in terms of the above resources. Execution time depends on hardware and it can also change with input data. The number of elementary operations appears to be a more convenient measure. In fact, frequently complexity is defined as the relationship between the number of input data and the number of elementary operations required.

To illustrate the idea of counting the number of operations let us consider the two vectors

$$\mathbf{A} = [a_1, a_2, \ldots, a_n]$$
$$\mathbf{B} = [b_1, b_2, \ldots, b_n]$$

The calculation of the sum

$$\mathbf{A} + \mathbf{B} = [a_1 + b_1, a_2 + b_2, \ldots, a_n + b_n]$$

requires n additions. In other words, the number of operations grows linearly when the number of input data, n, increases.

The form of the relationship can change in other cases. In general it is less important to know what happens when the number of input data is small. Then, even if the number of operations grows quickly, it still may be an acceptable cost. What is important is to appreciate how quickly the number of required operations grows when the set of input data is large. One way of expressing the growth rate is to say that it is like that of a known function, say $g(n)$. Usual examples of $g(n)$ are $\ln n$, n, n^2, or 2^n. A formal definition of this concept of complexity follows below.

Let A be an algorithm, and $f(n)$ the number of elementary operations required to execute A when the number of input data is n. If we can find a function $g(n)$, a constant $C > 0$ and a number $m > 0$ such that

$$\frac{f(n)}{g(n)} \leq C$$

for all $n \geq m$, we say that $f(n)$ is **of order** $g(n)$, or, in other words, *of the order large O with respect to g* and we write

$$f(n) = O(g(n))$$

Returning to the sum of two vectors of length n, we can write

$$\frac{f(n)}{n} = 1$$

for all $n \geq 1$, and say that the addition of the vectors is of order n.

The scalar product of the same vectors, $\mathbf{A}$, $\mathbf{B}$, requires n multiplications plus the addition of the resulting n products $a_i b_i$, in total $2n$ operations. As

$2n/n = 2$ for all $n \geq 1$, we can write that the scalar product of two vectors of length n is $O(n)$.

Let us consider now two n-by-n matrices, $\mathbf{A}$, $\mathbf{B}$; each has n^2 elements. Therefore, n^2 operations are required to calculate the sum $\mathbf{S} = \mathbf{A} + \mathbf{B}$ (remember, the general element of the sum is $s_{ij} = a_{ij} + b_{ij}$). We conclude that the sum of two n-by-n matrices is $O(n^2)$.

If we multiply the same matrices $\mathbf{A}$ and $\mathbf{B}$ we obtain a third n-by-n matrix, $\mathbf{P}$, such that each element p_{ij} is the scalar product of the i-th row of $\mathbf{A}$ by the j-th column of $\mathbf{B}$. We already know that each scalar product requires $2n$ operations. There are n^2 elements in $\mathbf{P}$; therefore the product of the two matrices implies $2n^3$ operations and we conclude that this product is of order n^3.

So far we have only considered real numbers. When applied to complex numbers, the same algorithms require more floating-point operations. More specifically, the sum or the difference of two complex numbers requires two floating-point operations, and the product or quotient, six.

Complexity analysis allows the development and selection of algorithms that work faster and require less computer memory. Algorithms that are $O(\ln n)$ work faster than algorithms that are $O(n)$. Also, algorithms that are $O(n)$ are faster than those that are $O(n^2)$, and so on. Algorithms of exponential order, like $O(2^n)$, require amounts of time that can become prohibitive for large sets of input data. Complexity analysis can indeed identify classes of problems that cannot be solved by even the fastest computers available today.

The following examples are commonly cited:

- the summation of n numbers has complexity $O(n)$;

- optimal sorting algorithms have complexity $O(n \log(n))$;

- the multiplication of two *n-by-n* matrices has complexity $O(n^3)$;

- the algorithm for solving the problem of *Hanoi towers* has the complexity $O(2^n)$. For the definition and solution of this problem see, for example, Corge (1975), or Standish (1994).

The subject of complexity belongs to computer science and is treated in specialized books such as Baase (1983), Horowitz and Sahni (1983), Kronsjö (1987), or Standish (1994). An instance of application is shown in Section 5.11 which compares the complexity of two algorithms for the evaluation of polynomials.

5.11 Horner's scheme

Let us suppose that we want to evaluate the polynomial

$$p(x) = c_1 x^n + c_2 x^{n-1} + \cdots + c_m x^{n-m+1} + \cdots + c_{n+1}$$

at a real point x_0. In this example we shall program two algorithms that perform this evaluation and count the number of multiplications and additions required by each algorithm. In both we shall use MATLAB's form of describing the polynomials by storing their coefficients, c_i, in an array, say c, in order of descending powers of x, that is

$$c = [c_1 \; c_2 \; \ldots \; c_m \; \ldots \; c_{n+1}]$$

MATLAB provides an efficient function for the evaluation of polynomials; it is called polyval and it is described in this book in Subsections 3.8.1 and 4.4.1. This function is used in this example to check the exactitude of the two algorithms. The first algorithm exemplified here evaluates the polynomial term by term; its listing is

```
function   p = evalpol1(c, x)
%EVALPOL1 polynomial evaluation, term by term scheme.
%          EVALPOL1(C, X) evaluates at X the polynomial whose
%          coefficients are contained in the array C.
%          The elements of C are ordered according to
%          descending powers of X.
%          Written for Biran & Breiner (1995).
%          Reproduced by courtesy of Pearson Education.

l = length(c); p = c(1); pow = 1;
for k = l-1: -1: 1
        pow = pow*x;
        p   = p + c(k)*pow;
end
```

Write this function to a file evalpol1.m. It is easy to see that each loop execution requires two multiplications and one addition, that is, three elementary operations. The loop is repeated $l - 1$ times, that is, a total of $3(l - 1)$ elementary operations. Add to this one evaluation of the number $l - 1$ and we get $3 \times (l - 1) + 1$ as the total number of floating-point operations. As $l = n + 1$, we conclude that we need $3n + 1$ floating-point operations.

The second algorithm proposed here is based on a special factorization of the polynomial known as **Horner's scheme**:

$$p(x) = (\ldots ((c_1 x + c_2)x + c_3)x \cdots + \cdots)x + c_{n+1}$$

A simple example of such a factorization is

$$3x^2 + 5x + 3 = (3x + 5)x + 3$$

This time only n multiplications and n additions are needed, in total $2n$ floating-point operations. To implement this scheme write the following function to a file evalpol2.m:

```
function  p = evalpol2(c, x)
%EVALPOL2 Polynomial evaluation by Horner's scheme.
%        EVALPOL1(C, X) evaluates at X the polynomial whose
%        coefficients are contained in the array C.
%        The elements of C are ordered according to
%        descending powers of X.
%        Written for Biran & Breiner (1995).
%        Reproduced by courtesy of Pearson Education.

l = length(c); p = c(1);
for k = 2:l
        p = p*x + c(k);
end
```

Older MATLAB versions provided a function, flops, that counted floating-point operations. With the help of this function we obtained Figure 5.3. The function flops is no longer available. The interpretation of the plot in Figure 5.3 is simple; for example, for a polynomial of degree $n = 7$, the term-by-term algorithm requires 22 elementary operations, while Horner's scheme requires only 14. The results confirm the analysis carried out above.

Exercise 5.4 Polynomial evaluation

To check the functions evalpol1 and evalpol2 described in Section 5.11, write a function, evalpol(n), that produces the array of random coefficients of an n-th degree polynomial and evaluates this polynomial by means of

1. the MATLAB function polyval;

2. the function evalpol1;

3. the function evalpol2.

Compare the results for several values of n.

5.12   Problems that cannot be solved

Complexity analysis reveals that the general solution of certain problems may require a number of operations that is so large that, in practice, we cannot

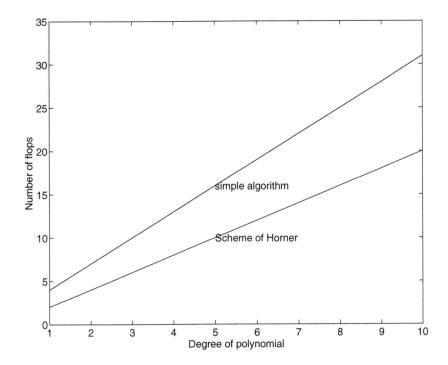

FIGURE 5.3: Number of floating-point operations required for polynomial evaluation

solve those problems. For some of these problems it is possible to devise intelligent ways of choosing a few directions among all possible ones and solve the problems without exhaustive searches. We say then that we use *heuristics*. Heuristic methods do not succeed always. Sometimes they yield approximate solutions, other times they may fail completely.

There are also problems for which no solution can be found, or, in other terms, no algorithm can be designed. Such problems were identified before the advent of the computers. The revolutionary breakthrough in this direction was that of the Austrian mathematician Kurt Gödel (1906-78). Decisive contributions were made by the British mathematician Alan Turing (1912-54) who is also known for the *Enigma* affair. Turing devised in 1936 a hypothetical computing device known since then as the *Turing machine*. As *Encyclopedia Britannica* (2008, Ultimate Reference Suite DVD) puts it, 'The Turing machine is not a machine in the ordinary sense but an idealized mathematical model that reduces the logical structure of any computing device to its essentials'. Although the Turing machine is a very simple model and can assume only a finite number of states, mathematicians and computer specialists agree that it can solve any problem that any computer can solve. This assertion is known as the *Church-Turing thesis* (Alonzo Church, American logician, 1903-1995). Here, under the term computer we include even the most powerful machines known today and machines that will be built in the future. A computational problem that cannot be solved by any Turing machine is called an *unsolvable problem*. Mathematicians and computer scientists have established a hierarchy of unsolvable problems and their study led to the development of important theories in mathematical logic. A few relatively accessible books that treat this subject are those of Douglas Hofstadter (1979), Martin Davis (2000), David Harel (2003), or Harel with Feldman (2004).

5.13   Summary

Numbers are represented in the computer as binary, floating-point numbers. The set of computer, or machine, numbers is finite. In the default double-precision format this set is bounded at the smallest end by the number 2.2251×10^{-308}, and at the largest end by the number 1.7977×10^{308}. There are gaps between computer numbers and these gaps are smaller between small numbers and larger between large numbers.

The computer can work only with numbers belonging to the set of machine numbers. Any other number, defined in the input or resulting from a computer operation, is approximated by the closest machine number. This process and the rounding off of numbers can lead to errors that can have serious effects in particular cases or when accumulated after many repeated calculations. Another source of errors is the truncation of infinite processes that are replaced by finite processes.

In general, a computational problem can be solved by different algorithms; these can be compared in terms of required resources such as execution time, number of elementary operations, or computer memory. In a general sense, the **complexity** of an algorithm is the difficulty of executing it as measured by the amount of necessary resources. In a restricted sense, the complexity of an algorithm is the relationship, $f(n)$, between the number of input data, n, and the number of elementary operations required. One way of appreciating the behaviour of this relationship for large sets of data is to compare its evolution with that of some known function. Formally, we say that $f(n)$ is **of order** $g(n)$ if we can find a function $g(n)$, a constant $C > 0$ and a number $m > 0$ such that

$$\frac{f(n)}{g(n)} \leq C$$

for $n > m$.

Complexity analysis helps in the development of more efficient algorithms and in the identification of problems that require prohibitive execution times.

5.14   More examples

EXAMPLE 5.2 Truncation of the series of the exponential function

In Section 5.9 we exemplify the effect of truncation of a series that yields one value. It is interesting to see the effect of truncation of a series that yields a function. As as an example we choose the Taylor series of the exponential function

$$e^x = 1 + x + \frac{x^2}{2!} + \frac{x^3}{3!} + \frac{x^4}{4!} + \cdots$$

Write the following M-file:

```
%EXPONENTIAL Shows the effect of truncation on the shape
%            of the exponential function
%            Written by Adrian Biran, May 2006

x = 0: 0.01: 5;
y = exp(x);
plot(x, y, 'k-', 'LineWidth', 1.5)
xlabel('x')
ylabel('Approximation of e^x')
hold on
plot([ 0 5 ], [ 1 1 ], 'r--')
S = 1;
for k = 1:7
    S = S + x.^k/factorial(k);
    plot(x, S, 'b-.')
    text(5, S(end), [ ' ' num2str(k + 1) ' terms' ])
end
legend('True function', 'One term')
```

The resulting plot is shown in Figure 5.4.

EXAMPLE 5.3 The explosion of Ariane 5

On 4 June 1996 the European Space Agency launched an *Ariane 5* rocket that was to put satellites in orbit. About forty seconds after lift-off the rocket left the designed path and destroyed itself. The analysis of the event showed that the horizontal velocity of the rocket, measured as a 64-bit floating-point

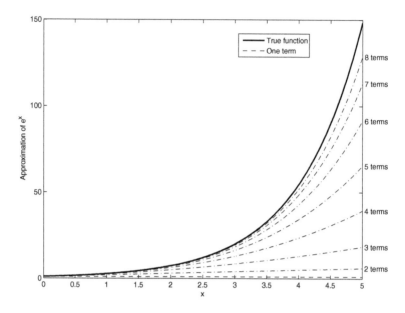

FIGURE 5.4: The effect of truncation on the series that yields the exponential function

number, was converted into a 16-bit signed integer resulting in overflow. Losing information, the computer could not correctly control the vehicle. The software was based on that used for the previous *Ariane 4* model that reached in the corresponding phase a much lower horizontal velocity.

Let us see how MATLAB would respond in such a case. The largest number that can be stored as a 16-bit signed integer is $2^{15} - 1 = 32767$. We define the next larger integer and convert it into a 16-bit signed integer with the function int16:

```
≫ N = 32768
N =
   32768
int16(N)
ans =
   32767
```

In conclusion, any number larger than 32767 will be converted into this number, the limit of the range of 16-bit signed integers. We invite the reader to experiment with a few more numbers.

5.15 More exercises

Exercise 5.5 Inexact sums
We know that the following expressions should yield the number 1

$$\sum_{i=1}^{100} 0.01, \ \sum_{i=1}^{50} 0.02, \ \sum_{i=1}^{32} 0.03125$$

Write a function

```
S = InexactSum(x, n)
```

that sums n times the number x. Calculate the three sums shown above and check, in the long format, how much is $S - 1$. You will find out that two of the sums are not exact, while the third is. Explain why.

Exercise 5.6 Strange polynomial roots
Modify the programme in Subsection 5.2.4 to catch the exact value of B for which one of the solutions of the second-degree equations has no significant digits.

Exercise 5.7 Again strange polynomial roots
The equation

$$(x - 2)^7 = 0$$

has the root 2 with multiplicity 7. Try to solve it with the function `roots`. Do you know the explanation of the discrepancies? Try also the function `fzero` and see again that it does not yield the correct answer and that it does not even mention that there are seven roots.

Exercise 5.8 Binary representation of 0.1
The binary representation of the decimal number 0.1 is 0.00011001100110... Mind the group 0011 that repeats itself indefinitely. The conversion to decimal representation is

$$\frac{1}{2^4} + \frac{1}{2^5} + \frac{1}{2^8} + \frac{1}{2^9} + \frac{1}{2^{12}} + \frac{1}{2^{13}} + \dots$$

Write a loop that adds at each iteration two of the above terms and see how rapidly the result approaches the value 0.1

Exercise 5.9 Truncation of the series that yields 1
Modify the script file in Section 5.9 so that it runs long enough to show the number 1 on the screen.

Exercise 5.10 Truncation of a series that yields $\pi/4$.

Let us consider the series

$$1 - \frac{1}{3} + \frac{1}{5} - \frac{1}{7} + \frac{1}{9} - \ldots = \frac{\pi}{4}$$

To show how this series tends to the limit value, and to see the effect of truncation, write a script file, `SeriesPi4`, based on a `For` loop that

1. sums the first, the first two, and so on up to the first 20 terms of the series;

2. plots the results against the number of summed terms;

3. shows the limit line $\pi/4$.

Exercise 5.11 Truncation of a series that yields $1/2$.

Let us consider the series

$$1 + \frac{1}{1 \times 3} + \frac{1}{3 \times 5} + \frac{1}{5 \times 7} + \frac{1}{7 \times 9} + \ldots = \frac{1}{2}$$

To show how this series tends to the limit value, and to visualize the effect of truncation, write a script file, `Series1_2`, based on a `For` loop that

1. sums the first, the first two, and so on up to the first 20 terms of the series;

2. plots the results against the number of summed terms;

3. shows the limit line $1/2$.

Exercise 5.12 Truncation of a series that yields $3/4$.

Let us consider the series

$$1 + \frac{1}{1 \times 3} + \frac{1}{2 \times 4} + \frac{1}{3 \times 5} + \frac{1}{5 \times 7} + \ldots = \frac{3}{4}$$

To show how this series tends to the limit value, and to visualize the effect of truncation, write a script file, `Series3_4`, based on a `For` loop that

1. sums the first, the first two, and so on up to the first 20 terms of the series;

2. plots the results against the number of summed terms;

3. shows the limit line $3/4$.

Exercise 5.13 Truncating the series that yields e

One of the most famous series in mathematics is that defining the number e, the base of natural logarithms. The series that yields e^x, that is the *exponential function*, is

$$e^x = 1 + x + \frac{x^2}{2!} + \frac{x^3}{3!} + \frac{x^4}{4!} + \cdots$$

Obviously, to obtain the number e one should substitute 1 for x. In MATLAB the number e is retrieved with

```
≫ format long
exp(1)
ans =
   2.71828182845905
```

To visualize the effect of truncation in the above series, write a function that begins with the first term of the series and at each successive iteration adds one more term to the sum and calculates the difference between the result and the e−value produced by MATLAB.

Exercise 5.14 Approximation of the sine function

Following Example 5.2, write an M-file that shows the effect of truncation of the series that yields the sine function. Plot the curves against arc values in radians. In this example the effect of truncation appears also in the approximations of π.

The series to be used in this and several of the following exercises can be found in books of mathematical tables and formulas such as Spiegel and Liu (2001).

Exercise 5.15 Approximation of the cosine function

Following Example 5.2, write an M-file that shows the effect of truncation of the series that yields the cosine function. Plot the curves against arc values in radians. In this example the effect of truncation appears also in the approximations of π.

Exercise 5.16 Approximation of the tangent function

Following Example 5.2, write an M-file that shows the effect of truncation of the series that yields the tangent function. Plot the curves against arc values in radians. In this example the effect of truncation appears also in the approximations of π.

Exercise 5.17 Approximation of the arcsine function

Following Example 5.2, write an M-file that shows the effect of truncation of the series that yields the $\sin^{-1}$ function.

Exercise 5.18 Approximation of the number e

If we define truncation as the substitution of an infinite process by a finite one, we can include under this term also the following example. The number e, the base of natural logarithms, is defined by

$$e = \lim_{n \to \infty} \left(1 + \frac{1}{n}\right)^n$$

We cannot use this formula to calculate exactly the number e because the computer does not know what infinity is. However, we may try to approximate the number e by choosing a very large value of n. Write a function that uses this formula and see how for increasing values of n the calculated value approaches the value of e yielded by the MATLAB command `exp(1)`.

Exercise 5.19 A truncated Fourier series

Consider the function

$$f(x) = x(\pi - x)(\pi + x), \quad -\pi \leq x \leq pi$$

The extension of this function for the whole real axis has the Fourier expansion

$$F(x) = 12\left(\frac{\sin x}{1^3} - \frac{\sin 2x}{2^3} + \frac{\sin 3x}{3^3} - \frac{\sin 4x}{4^3} + \cdots\right)$$

Write a function according to the following specification

1. plot in black the function f;

2. open a `For` loop that adds at each iteration one term of the Fourier expansion of the function F;

3. at each iteration plot in red the approximation of F. If there was a previous red plot, delete it before plotting the new graph. Two successive iterations should be separated by the command `pause`.

How many terms are necessary to produce an acceptable approximation?

6

Data types and object-oriented programming

In this chapter we discuss some data types, classes and object-oriented programming. We can introduce the **data type** as a *set of values defined by the operations that apply to them*. In the most recent documentation provided by The MathWorks we read,

> 'There are many different data types, or classes, that you can work with in the MATLAB software. You can build matrices and arrays of floating-point and integer data, characters and strings, and logical true and false states. Function handles connect your code with any MATLAB function ... Structures and cell arrays provide a way to store dissimilar types of data in the same array.
>
> There are 15 fundamental classes in MATLAB.'

By adding the term **classes** the above description includes **object-oriented** terminology. To read more in the MATLAB documentation click on `Help` in the toolbar and in the pull-down menu that opens click on `Product Help`. In the left-hand window navigate through

`MATLAB` → `User Guide` → `Programming Fundamentals` → `Classes`.

We begin the chapter by introducing the notions of **structure** and **cell** array, and showing a few ways of working with them. We continue with a short introduction to *object-oriented programming* using structures in Subsections 6.3.3 and 6.3.5, and cell arrays in Subsection 6.3.4. As examples we show how to program functions that work with dimensioned quantities. In this way, together with the value of physical quantities we can assign them units and obtain after calculation the units of the result. The examples are simple and do not constitute a full system. Such an enterprise goes beyond the limits of the book.

6.1   Structures

6.1.1   Where structures can help

Let us assume that a mass-spring-damper system is described in *state space* form by the equations

$$\acute{x} = \mathbf{A}x + \mathbf{B}u$$
$$y = \mathbf{C}x + \mathbf{D}u$$

where

$$\mathbf{A} = \begin{bmatrix} -0.7 & -0.05 \\ 1.0 & 0.00 \end{bmatrix}, \quad \mathbf{B} = \begin{bmatrix} 1 \\ 0 \end{bmatrix}, \quad \mathbf{C} = \begin{bmatrix} 0 & 0.5 \end{bmatrix}, \quad \mathbf{D} = [0]$$

We can store all the above data in one variable of the type **structure**. We name this variable **spring** and define four **fields**. Using the syntax

structure_name.field

we write:

```
≫ spring.A = [ -0.7 -0.05; 1 0 ];
≫ spring.B = [ 1; 0 ];
≫ spring.C = [ 0 0.5 ];
≫ spring.D = 0;
```

We can operate on the individual fields of a structure exactly as we operate on the individual elements of an array. For example, to change one value:

```
≫ spring.A(1, 1) = -0.65;
```

To see the various fields of the structure let us write a function **showsys** to a file **showsys.m**

```
function showsys(sys)

%Show the matrices that represent a system
% in state space form.
sys.A, sys.B, sys.C, sys.D
```

and we call the function with the name of the structure as input argument:

```
≫ showsys2(spring)
ans =
  -0.6500  -0.0500
   1.0000  0
ans =
   1
   0
ans =
   0  0.5000
ans =
   0
```

Using structures reduces the amount of housekeeping. If we use four matrices to describe a system, it is the responsibility of the user to keep them together and not mix the **A**-matrix of one system with the **B**-matrix of another. This danger is avoided by storing the four matrices under one name that is different from the names we assign other systems. The individual fields of a structure can contain different data types. In other words, there are no restrictions against collecting under the same name numerical arrays and strings. As shown in the next subsection, it is possible to add an explanation to the structure `spring`.

6.1.2 Working with structures

Structures can be defined either by direct assignment, as we have seen in the preceding subsection, or using the command `struct`:

```
≫ spring = struct('A', [ -0.7, -0.05; 1, 0 ],...
'B', [ 1; 0 ],....
'C', [ 0, 0.5 ],...
'D', 0)
spring =
A: [2x2 double]
B: [2x1 double]
C: [0 0.5000]
D: 0
```

The function `struct` takes as input a list of parameters consisting of strings defining field names and corresponding values. In object-oriented terminology `struct` is a *constructor*. More complex examples of constructors are given in Subsections 6.3.3, 6.3.4 and 6.3.5.

Some assignment and retrieval operations of the field values, which can be performed by means of the equal sign, can also be carried on by built-in MATLAB functions. For example, use `getfield` to retrieve one filed of the structure spring

```
>> blockA = getfield(spring, 'A')
blockA =
  -0.7000  -0.0500
   1.0000   0
```

which is equivalent to

```
>> blockA = spring.A
blockA =
  -0.7000  -0.0500
   1.0000   0
```

The function `rmfield` does not actually remove a field from a structure, as the name may suggest, but makes a copy of the variable, without a specific field, for example

```
>> noD = rmfield(spring, 'D')
noD =
A: [2x2 double]
B: [2x1 double]
C: [0 0.5000]
```

Above we left the structure **spring** unchanged, and built a new variable, noD, identical to **spring**, but without the field D. Analogously, the function `setfield` does not add a filed to a structure, but produces a copy of the given structure with an additional field. The syntax is

```
newstruct = setfield(givenstruct, 'newfield', value);
```

For example,

```
>> newspring = setfield(spring,...
'info', 'Spring in Chapter 11')
newspring =
  A: [2x2 double]
  B: [2x1 double]
  C: [0 0.5000]
  D: 0
  info:  'Spring in Chapter 11'
```

6.2 Cell arrays

The MATLAB documentation defines *cell arrays* as 'a collection of containers called *cells* in which you can store different types of data'.

Like structures, cell arrays can be defined by *direct assignment*:

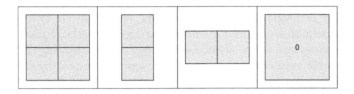

FIGURE 6.1: Graphic description of the cell array `cspring`

```
≫ cspring1 = [ -0.7, -0.05; 1, 0 ];
≫ cspring{2} = [ 1; 0 ];
≫ cspring{3} = [ 0 0.5 ];
≫ cspring{4} = 0;
≫ cspring
≫ cspring =
  [2x2 double]   [2x1 double]   [1x2 double]   [0]
```

In this case curly brackets, { }, are placed around the index. As mentioned in the MATLAB User's Guide, 'curly brackets are cell array constructors, just as square brackets are numeric array constructors'. An alternative way of defining the same cell array is by enclosing the indexes between regular parentheses and placing the curly brackets in the right-hand side of the equality sign:

```
≫ cspring(1) = {[ -0.7, -0.05; 1, 0 ]};
≫ cspring(2) = {[ 1; 0 ]};
≫ cspring(3)= {[ 0 0 .5 ]};
≫ cspring(4)= { 0 };
≫ cspring
≫ cspring =
  [2x2 double]   [2x1 double]   [1x3 double]   [0]
```

Let us examine how MATLAB interprets our sequence of commands. When we write

```
≫ cspring(1) = {[ -0.7, -0.05; 1, 0 ]};
```

MATLAB requests from the operating system the amount of memory necessary to store the array $[-0.7, -0.05; 1, 0]$ and the space to store a cell array of
length 1 that points to it. We continue with

```
≫ cspring(2) = {[ 1; 0 ]};
```

MATLAB requests now the system to store $[1; 0]$ and finds out that `cpsring`
is a cell array of length 1, which is not sufficient to contain two references.
Therefore, it requests memory for a new `cspring`, of length 2, makes a copy of
the existing array into a new array, assigns the value of `cspring(2)`, calls the
new cell variable `cspring`, and releases the memory occupied by the previous
cell array. These operations are repeated each time we add a new cell to an
existing variable, a time-consuming procedure. For efficient code we should
pre-allocate the space for a cell array of four rows and one column using the
command

```
≫ cspring = cell(1, 4);
```

This creates a cell array with empty contents that we can fill later without
having to interact so much with the operating system. After having defined
`cspring` we can display it with the command: `celldisp`

```
≫ celldisp(cspring)
cspring1 =
  -0.7000   -0.0500
   1.0000  0
cspring2 =
   1
   0
cspring3 =
   0  0  0.5000
cspring4 =
   0
```

The function `cellplot` produces the graphic description of the cell shown
in: Figure 6.1

```
≫ cellplot(cspring)
```

The contents of individual cells can be extracted by calling them with the
cell index enclosed between curly brackets. Thus we can also operate on the

contents of individual cells

```
≫ AA = 2*cspring{1}
AA =
   -1.4000   -0.1000
    2.0000   0
```

6.3 Classes and object-oriented programming

6.3.1 What is object-oriented programming?

An **object** is an instance of a class. Conversely, as Weisfeld (2000) puts it, 'Classes can be thought of as templates ... for objects ... A class is used to create an object'. To exemplify these concepts in plain MATLAB try the following.

```
≫ a = pi;
≫ b = [ 2.5 4 9.1 ];
≫ c = single(pi);
≫ t = 'MATLAB classes';
≫ whos
  Name  Size  Bytes  Class   Attributes

  a     1x1     8    double
  b     1x3    24    double
  c     1x1     4    single
  t     1x14   28    char
```

We see that *a* and *b* are objects of the class *double* precision, *c*, of the class *single* precision, and t, of the class *character* string. While MATLAB has used for a long time classes, objects and other features of object-oriented programming (further called *OPP*), it did it to a lesser extent than languages dedicated to OOP, and in a way that conserved the original orientation of MATLAB that is built around the basic concept of array. With the appearance of the 2008a version, MATLAB has come closer to other object-oriented languages and new powerful OO features have been introduced. The following examples are built in the 2008b version.

6.3.2   Calculations with units

In the next subsections we introduce the main notions of OO programming by means of an example. This is done in several textbooks and other documentation on the subject. Most examples, however, have no connection to engineering, or no serious importance in engineering. In this book, written for engineering students and practising engineers, we are picking up a basic and ubiquitous problem of engineering and science, that of calculating with units. Engineers and scientists work with **physical quantities** that have *dimensions* such as length, mass, time or electric current. Another term for such quantities is **dimensioned quantities**. A dimensioned quantity is defined by one number, that we further call **value**, and by **units**. As formulated by Maxwell, a *unit* is a 'quantity of the same kind as the quantity to be expressed' and 'is to be taken as a standard of reference. The other component is the number of times the standard is to be taken in order to make up the required quantity' (quoted in Emerson, 2008). Thus, we can talk about a length of 1 meter, a mass of 2 kilograms, or a time interval of 6 seconds. A quantity that can be completely described without reference to units is called **non-dimensional** or **dimensionless**.

Several systems of units have been used during the centuries. In this section we refer only to the most recent one, the **SI** system. The letters *SI* are an acronym of *Système International*. The SI system is widely accepted today and used in most textbooks. A quick, but rigorous and official introduction can be found in Anonymous (2000). There are seven **SI basic units**, as detailed in Table 6.1. Further units, called **SI derived units**, are calculated from the basic units by means of simple equations. A few examples are shown in Table 6.2. These tables are based on more detailed and extended tables presented in the document cited above. We have included many units that are not used in our examples, but may allow the reader to design more examples.

Table 6.1: SI basic units

Quantity	Unit	Symbol
length	meter	m
mass	kilogram	kg
time	second	s
electric current	ampere	A
thermodynamic temperature	kelvin	K
amount of substance	mole	mol
luminous intensity	candela	cd

For the continuation of this discussion it is important to note that a derived unit is the product of integer powers of basic units.

Table 6.2: A few SI derived units

Quantity	Unit	SI basic units	Symbol
area	square meter	m^2	m^2
volume	cubic meter	m^3	m^3
speed, velocity	meter per second	m/s	m/s
acceleration	meter per second squared	m/s^2	m/s^2
mass density	kilogram per cubic meter	kg/m^3	kg/m^3
current density	ampere per square meter	A/m^2	A/m^2
plane angle	radian	$m \cdot m^{-1} = 1$	rad
frequency	hertz	s^{-1}	Hz
force	newton	$m \cdot kg \cdot s^{-2}$	N
pressure, stress	pascal	$m^{-1} \cdot kg \cdot s^{-2}$	Pa
energy, work	joule	$m^2 \cdot kg \cdot s^{-2}$	J
power	watt	$m^2 \cdot kg \cdot s^{-3}$	W
electric charge	coulomb	$s \cdot A$	C
electric potential difference	volt	$m^2 \cdot kg \cdot s^{-3} \cdot A^{-1}$	V
capacitance	farad	$m^{-2} \cdot kg^{-1} \cdot s^4 \cdot A^2$	F
electric resistance	ohm	$m^2 \cdot kg \cdot s^{-3} \cdot A^{-2}$	Ω
electric conductance	siemens	$m^2 kg^{-1} \cdot s^3 \cdot A^2$	S
inductance	henry	$m^2 \cdot kg \cdot s^{-2} \cdot A^{-2}$	H
Celsius temperature	degree Celsius	K	°C

Calculations with physical quantities are subject to certain restrictions and follow a few simple rules. First, the operations of addition, subtraction and comparison are defined only between quantities of the same kind. For example, the operations

$$c = a + b$$
$$c = a - b$$

can be carried on only if a and b have the same dimensions and the result, c, inherits the dimensions of a and b. Thus, we can add two lengths and obtain a length, but we cannot add a length to an area.

Second, the operations of division and multiplication are not subject to constraints like those mentioned above. The value of the result of a multiplication is the product of the values of the multiplicands, and the units of the result are the product of the units of the multiplicands. The multiplication of units is carried on by adding the powers of the basic units. For example, multiplying an area of 5 m^2 by a height of 2 m results in a volume of 10 m^3. It is easy to deduce the rule for division. Elevating to a given power is an extension of multiplication. Certain expressions used in engineering or science can contain non-integer powers of units. These powers are rational numbers and their set is rather limited.

A third rule can be formulated for trigonometric and exponential functions. The argument of such functions must be **non-dimensional**. To explain this restriction let us consider, for example, the Taylor-series expansion of the sine function

$$\sin x = x - \frac{x^3}{3!} + \frac{x^5}{5!} + \dots$$

Let us assume that we measure x in meters. Then, to obtain the value of the sine we would subtract cubic meters from meters, add to the result meters at the fifth power a.s.o. This contradicts the first rule. In Table 6.2 the unit of angle, *radian*, is defined as $m \cdot m^{-1} = 1$. Measuring the arguments of trigonometric function in radians does not lead to a contradiction of the first rule. We let the reader check that the *degree* is also an acceptable angle unit.

There is an additional, more general rule: when formulating an expression containing dimensioned quantities, the units of the left-hand side must be the same as those of the right-hand side. The rule is even more stringent: on the two sides we should have quantities of the same kind. Two quantities can be of different kinds even if measured in the same units. For example, moments of area are not of the same kind as volumes, although both are measured in m^3. Another example is that of torque and work that are of different kinds even if both are measured in $m^2 \cdot kg \cdot s^{-2}$.

Our explanation is short and far from complete. We still hope that it is sufficient to show that there may be interest in software that carries on various computations only if they are legal and that yields the units of the results. Several such systems have been implemented in a few computer languages, and not a few papers have been written on the subject. In this section we are going to exemplify a simple implementation in MATLAB. It is just an example of OOP, not a complete system of calculating with units.

6.3.3   Defining a class

In this subsection we are going to create a class called PhQuant, an acronym for *Physical quantity*. That class will consists of objects measured in units derived from the first three basic units: meter, kilogram second. Thus we limit the scope of the class to mechanical quantities, but keep the example as simple as possible. We design four properties: the value, and the exponents of the meter, kilogram and second units. We also design the functions that can be applied to the objects of the class; in OOP parlance we call them **methods**. One of these methods modifies the MATLAB `display` function to show the values of the objects together with their units. We say that our function **overloads** the built-in MATLAB function. Further methods should overload MATLAB arithmetic operations. As the operations of addition, subtraction and comparison can be performed only between quantities of the same kind, we design a function that compares the units of the objets involved.

To begin, in the directory you want to work in, open a subdirectory called `@PhQuant`. Please note the character `@`. In this subdirectory open the editor to write a file `PhQuant.m`. The listing of the file we wrote appears on the next page. The first line contains the word `classdef` followed by the name of the class `PhQuant`. Note that the last line of the file should contain the command `end`.

Each class may have its own *attributes*; we call them **properties**. The properties are declared in a block contained between the words `properties` and `end`. As said, there are four properties, all initialized to zero.

The block of properties is followed by the block of methods. This block is contained between the words `methods` and `end`. The first method is the **constructor** of the class. Called with four arguments, the constructor creates an object of the `PhQuant` class and sets its properties to the values defined by the arguments.

The next method is `display`; it overloads the MATLAB built-in display function and is activated when the user calls the constructor without ending the call with a semicolon. The `if` constructs build a string composed by the names of the basic units, each one followed by the power symbol, '`^`', and the corresponding exponent. If the exponent of a basic unit is zero, its name is not included in the string. Thus, an area of 8 m^2 would be displayed as `8 m^2`, and a force of 12 N would be displayed as `12 m kg s^-2`. To keep the example simple and readable we did not add the statements that would display N instead of `m kg s^-2`. Anonymous (2000) allows two style conventions for separating the basic units that must be multiplied: either a space or a half-high dot. As the latter cannot be displayed in the Command Window, we choose the former. The last conditional construct of the function causes the display to be in the format declared for the Command Window, that is either `loose` or `compact`.

The third method is `SameUnit`; it checks if the two arguments have the same units and yields as output a boolean variable.

The fourth method overloads the function `plus`, usually entered as the operator '`+`'. While the MATLAB built-in function carries on the addition of any numerical summands, the overloaded function is restricted to summands having the same units. The restriction is implemented after calling the method `SameUnit`.

The fifth method overloads the function `times`, usually called by the array operator '.*'. While the original function yields only a numerical value, the overloaded function adds also the exponents of the corresponding units and yields the units of the result.

Including more overloaded functions, for division, power, trigonometric functions, etc, would lengthen too much the example. We invite the reader to experiment with more overloaded functions.

```
classdef PhQuant
% Creates the class PhQuant of physical quantities.
% Call: PhQuant(value, power of m, power of kg, power of s)

    properties
        value    = 0;
        meter    = 0;
        kilogram = 0;
        second   = 0;
    end                 %  properties

    methods

        function obj = PhQuant(value, meter, kilogram, second)
        % class constructor
            if (nargin > 0)
                obj.value    = value;
                obj.meter    = meter;
                obj.kilogram = kilogram;
                obj.second   = second;
            end
        end                 % constructor

        function display(obj)
            t = [ ' ' num2str(obj.value) ' ' ]; % print value
            if obj.meter ~= 0
                if obj.meter == 1
                    t = [ t ' m' ];
                else
                    t = [ t ' m^' num2str(obj.meter) ];
                end
            end
            if obj.kilogram ~= 0
                if obj.kilogram == 1
                    t = [ t ' kg' ];
                else
                    t = [ t ' kg^' num2str(obj.kilogram) ];
                end
```

```
        end
        if obj.second ~= 0
            if obj.second == 1
                t = [ t ' s' ];
            else
                t = [ t ' s^' num2str(obj.second) ];
            end
        end    % display
            if isa(obj, 'PhArea') % to use also specific display
                disp(t)
            elseif isequal(get(0, 'FormatSpacing'), 'compact')
                disp([ inputname(1) ' =']);
                disp(t)
            else        % that is format loose
                disp(' ' )
                disp([ inputname(1) ' =']);
                disp(' ' );
                disp(t)
            end
end                % class display function

function s = SameUnit(p, q)
% checks if two objects have the same units
    if  (p.meter ==  q.meter) &...
        (p.kilogram == q.kilogram) &...
        (p.second == q.second)
            s = true;
    else
            s = false;
    end
end                         % SameUnit

function s = plus(p, q)
% addition function for PhQuant class
    if SameUnit(p, q) == true % summands have the same units
        s          = PhQuant;
        s.value    = p.value + q.value;
        s.meter    = p.meter;
        s.kilogram = p.kilogram;
        s.second   = p.second;
    else
        error('Summands have different units')
    end
end  % function plus
```

```
function s = times(p, q)
% multiplication function for PhQuant class
    if (isa(p, 'PhQuant') & isa(q, 'PhQuant'))
        s            = PhQuant;
        s.value    = p.value.*q.value;
        s.meter    = p.meter + q.meter;
        s.kilogram = p.kilogram + q.kilogram;
        s.second   = p.second + q.second;
    elseif  isnumeric(p) & isa(q, 'PhQuant')
        s            = PhQuant;
        s.value    = p.*q.value;
        s.meter    = q.meter;
        s.kilogram = q.kilogram;
        s.second   = q.second;
    else
        if isnumeric(q) & isa(p, 'PhQuant')
            s            = PhQuant;
            s.value    = q.*p.value;
            s.meter    = p.meter;
            s.kilogram = p.kilogram;
            s.second   = p.second;
        else
            error('Incompatible multiplicands')
        end
    end
end     % multiplication function

end         % methods
end
```

To create objects of the class **PhQuant** and calculate with them, return to the parent directory of **@PhQuant**. Below are a few examples.

```
≫ a = PhQuant(2, 1, 0, 0)
a =
  2 m
≫ b = PhQuant(3, 1, 1, 0);
c = a + b
???  Error using ==> <a href="matlab:
...
Summands have different units
```

To create the object *a* we called the constructor without ending with a semicolon. The display shows the value of *a* and the unit m. To create the object *b* we ended the call with a semicolon. There is no display. The attempt to add a volume to a length failed. Otherwise happens below where we first added two lengths and afterwards multiplied them.

```
≫ b = PhQuant(3, 1, 0, 0)
b =
   3 m
≫ c = a + b
c =
   5 m
≫ a.*b
ans =
   6 m^2
```

Finally, let us try two operations not included in the methods of the class. MATLAB does not allow us to carry them on. The multiplication function, however, allows us to multiply a dimensioned quantity by a scalar.

```
≫ d = a - b
???  Undefined function or method 'minus' for input...
≫ a/b
???  Undefined function or method 'mrdivide' for input...
≫ 25.*a
ans =
   50 m
≫ 5.*a.*b
ans =
   30 m^2
```

6.3.4 Defining a subclass

We have mentioned above pairs of quantities that bear the same units, but are of different kinds. Certain operations allowed between two numbers, have no meaning between two dimensioned quantities of different kind. Thus, it makes no sense to add a torque to a mechanical work, and neither can we decide which one is greater. It may be interesting to find a way to distinguish between different kinds of dimensioned quantities and define the operations allowed between them. OOP programming offers an elegant and efficient solution: **subclasses** and **inheritance**. Let us remark, for example, that *area* is a physical quantity, but of a particular kind; it has the units m^2. Similar statements can be made for *length, volume, work, torque* and so on. Thus, after having built a class for physical quantities, it may make sense to build **subclasses** for *length, volume, work, torque* and so on. For example, let PhArea be a subclass of PhQuant. In OO parlor we may also say that PhQuant is a

parent of `PhArea`, or, alternatively, `PhArea` is a **child** of `PhQuant`.

Let us assume now that we have created more subclasses of physical quantities. Each of those classes **inherits** the attributes and methods of the parent class, while having something special. Thus it is not necessary to repeat for each subclass the attributes and methods that are common with the **superclass**. It is necessary to define only what is particular. As an example, let us define a subclass `PhArea`. As the particular method of the subclass we design a function that calculates the areas of certain plain figures. The choice of figures is presented in a **predefined dialog box**, that is a simple graphical user interface, shortly **GUI**, of a type built-in in MATLAB.

To show the relationship to the parent class, we have kept in the name the first two letters, 'Ph'. We have to open a subdirectory of the same directory that includes `@PhQuant`. We name the new subdirectory `@PhArea`. In this subdirectory we write the following file called `PhArea.m`.

```
classdef PhArea < PhQuant
% Creates PhArea as a child of class PhQuant, prompts the user
% to input values in a dialog box and calculates the 'area' value
% that is passed to the parent class together with the unit m^2.
% Call: PhArea('Type') where possible types are:
%         Triangle, Rectangle,  Trapezoid, Circle, Ellipse.

   properties
        type  = ' ';
        Area  = 0;
   end    %  properties

   methods

     function obj = PhArea(type)
     % class constructor
        Area      =   PhArea.CalcValue(type);
        obj       = obj@PhQuant(Area, 2, 0, 0); % calls parent class
        obj.type = type;
        obj.Area = Area;
     end % constructor


     function display(obj)
     % display function for PhArea subclass
        t = [ 'Area of ' obj.type ];
        disp(t)
        disp([ inputname(1)  ' = '])
        display@PhQuant(obj)   % display of  parent class
      end % class display function
```

```
end     % methods

methods(Static = true)

    function Area = CalcValue(type)
    % dialog box for input and calculates 'area' value
      switch type

          case 'Triangle'
              prompt = {'Base in meters',...
                          'Altitude in meters'};
              name  = 'Area of triangle';
              rdata = inputdlg(prompt, name);
              b     = str2num(rdata{1});
              a     = str2num(rdata{2});
              Area  = b*a/2;

          case 'Rectangle'
              prompt = {'Length in meters',...
                          'Width in meters'};
              name  = 'Area of rectangle';
              rdata = inputdlg(prompt, name);
              l     = str2num(rdata{1});
              w     = str2num(rdata{2});
              Area  = l.*w;

          case 'Trapezoid'
              prompt = {'Side a in meters',...
                          'Side b in meters',...
                          'Altitude in meters'};
              name  = 'Area of trapezoid';
              rdata = inputdlg(prompt, name);
              a     = str2num(rdata{1});
              b     = str2num(rdata{2});
              h     = str2num(rdata{3});
              Area  = h*(a + b)/2;;
          case 'Circle'
              prompt = {'Radius in meters'};
              name  = 'Area of circle';
              rdata = inputdlg(prompt, name);
              r     = str2num(rdata{1});
              Area  = pi*r^2;
          case 'Ellipse'
              prompt = {' Semi-axis major in meters',...
```

```
                        'Semi-axis minor in meters'};
            name    = 'Area of ellipse';
            rdata   = inputdlg(prompt, name);
            a       = str2num(rdata{1});
            b       = str2num(rdata{2});
            Area    = pi*a*b;
        otherwise
            error('Undefined type for PhArea quantities')

      end       % switchyard
    end       % CalcValue
  end       % static methods

end                       % subclass definition
```

The opening line of the new file is composed of the key word `classdef`, the name of the subclass, and the string `< PhQuant`, which states that *PhArea* is a subclass of *PhQuant*. To the opening line corresponds a closing line with the key word `end`.

Next we define two properties that belong only to the subclass: *type* and *Area*, and initialize them.

The first method is again the constructor. It calls the constructor of the superclass using the syntax `obj@Phquant`. The first argument is the value to be calculated by the method `Area` described a little later. The other three arguments set the units to m². The following two lines set the properties particular to the subclass.

The display method begins with statements that characterize the result as the area of a certain figure. Next, the method calls the display of the superclass using the syntax `display@PhQuant(obj)`.

The function `Area` is called without reference to an object. Therefore, it is defined as a **Static** method. It has only one argument that can be anyone of the strings `'Triangle'`, `'Rectangle'`, `'Trapezoid'`, `'Circle'` or `'Ellipse'`. The method calls `inputdlg`, a *predefined dialog box* built-in in MATLAB. The argument `prompt` defines the questions to be answered by the user, and the argument `name`, the title of the dialog box. If there is more than one question, the user has to use the `Tab` key to advance from one line to the next. The answer of the user is stored in the cell array `rdata`. The function converts the strings to numerical values that are used in the calculation of the desired area. To try the function, enter, for example, the following command and answer the prompt.

```
>> a =PhArea('Rectangle')
```

For the length 3 and the width 2 the result will be that shown below. Create another object, *b*, which will calculate the area of an ellipse.

```
Area of Rectangle
a =
   6 m^2
≫ b = PhArea('Ellipse')
```

As shown in continuation, we can add the two areas. Asking for the area of a parabola we get an error message.

```
Area of Ellipse
b =
   779.115 m^2
≫ a + b
ans =
   785.115 m^2
a =PhArea('Parabola')
...
Undefined type for PhArea quantities
```

6.3.5 Calculating with electrical units

In this subsection we give an example of the application of OOP to calculations with electrical quantities. To keep the example simple we consider only those quantities that play a role in dc circuits including only resistors. The units involved are the ampere, A, the volt, V, the ohm, Ω, and the siemens, S. We include the unit of conductance to allow calculations with resistors connected in parallel. As MATLAB cannot display the Greek letter Ω in the Command Window, we chose to display the unit of electric resistance as ohm. Table 6.2 defines the derived units as products of various powers of the basic units introduced in Table 6.1. If we would want to calculate not only with electrical quantities, but also with mechanical quantities, we should use those definitions. For example, this would be the case when calculating the resistance of a conductor starting from its diameter, length and resistivity. The length and complexity of the resulting OO functions would be beyond the aims of this book. Therefore, in the following example, instead of the expressions based on basic units we use the symbols A, V, ohm, and S. In this subsection we also illustrate the use of a *set function* and of a second predefined dialog box.

We start the design by defining a multiplication table; it is shown in Table 6.3. We note the first multiplicand by p and the second by q. We place a dash, '-', where the multiplication is not defined. The table shows, for example, that multiplying a current, measured in A, by a voltage, measured in V, results in a power, measured in w.

Table 6.3: A multiplication table for electrical quantities

q	p			
	A	V	ohm	S
A	-	w	V	-
V	w	-	-	A
ohm	V	-	-	-
S	-	A	-	-

We invite the reader to create a subdirectory, @ElQuant, of the working directory. The following file, ElQuant.m, should be written in the new directory. Explanations follow the listing of the file.

```
classdef ElQuant
% Creates the class of electrical quantities ElQuant
% Use: ElQuant(value, unit)
   properties
      value = 0;
      unit  = ' ';
   end     % of properties

   methods
    function obj = ElQuant(value, unit)
    % class constructor
       % define set of acceptable units
       S = {'A', 'V', 'ohm', 'S', 'w'};
       if ismember(unit, S)
         if (nargin > 0)
           obj.value = value;
           obj.unit  = unit;
         end
       else
         errordlg('Invalid unit', 'Input error')
       end
    end     % of constructor

    function  display(obj)
    % displays value and electrical unit
       t = [ ' ' num2str(obj.value) ' ' obj.unit ];
       if isequal (get(0, 'FormatSpacing'), 'compact')
         disp([ inputname(1) ' =' ]);
         disp(t)
       else % format loose
         disp(' ')
```

```
            disp([ inputname(1) ' =' ]);
            disp(' ' )
            disp(t)
      end % of conditional construct
   end    % of quantity display

   function s = SameUnit(p, q)
   % verifies if two objects have the same unit
      if (p.unit == q.unit)
         s = true;
      else
         s = false;
      end      % of unit check
   end         % of function SameUnit

   function r = plus(p, q)
   % addition function for class ElQuant
      if SameUnit(p, q) == true % summands have same unit
         Value = p.value + q.value;
         Unit  = p.unit;
         r     = ElQuant(Value, Unit);
      else
         errordlg('Summands have different units',...
                  'Input error')
      end     % of conditional construct
   end        % of function plus

   function r = times(p, q)
   % multiplication function for class ElQuant
      if  isnumeric(p) & isa(q, 'ElQuant')
         Value = p.*q.value;
         Unit  = q.unit;
         r     = ElQuant(Value, Unit);
      elseif  isnumeric(q) & isa(p, 'ElQuant')
         Value = q.*p.value;
         Unit  = p.unit;
         r     = ElQuant(Value, Unit);
      else       % both  multiplicands electrical quantities
         switch p.unit
            case 'A'     % current
               if (q.unit == 'V')
                  Value = p.value.*q.value;
                  Unit  = 'w';
                  r     = ElQuant(Value, Unit);
               elseif (q.unit == 'ohm');
```

```
                    Value = p.value.*q.value;
                    Unit  = 'V';
                    r     = ElQuant(Value, Unit);
                else
                    errordlg('Incompatible units of multiplicands',...
                            'Input error')
                end
            case 'V'      % voltage
                if(q.unit == 'A')
                    Value = p.value.*q.value;
                    Unit  = 'w';
                    r     = ElQuant(Value, Unit);
                elseif (q.unit == 'S');
                    Value = p.value.*q.value;
                    Unit  = 'A';
                    r     = ElQuant(Value, Unit);
                else
                    errordlg('Multiplicands have incompatible units', ..
                            'Input error')
                end
            case 'ohm'    % electrical resistance
                if (q.unit == 'A')
                    Value = p.value.*q.value;
                    Unit  = 'V';
                    r     = ElQuant(Value, Unit);
                else
                    errordlg('Multiplicands have incompatible units', ..
                            'Input error')
                end
            case 'S'         % electrical conductance
                if (q.unit == 'V')
                    Value = p.value.*q.value;
                    Unit  = 'A';
                    r     = ElQuant(Value, Unit);
                else
                    errordlg('Multiplicands have incompatible units', ..
                            'Input error')
                end
        end           % of switchyard
      end             % of conditional construct
    end               % of multiplication function

  end           % of methods
end
```

The class constructor has two input arguments: the value and the unit. The set of acceptable units is S. It is defined as a cell array because its members are character strings of different lengths. The *set function* `ismember` checks if the second input argument, the unit, is a member of the set S. In the affirmative the constructor creates the desired object. Otherwise, the constructor displays a predefined dialog box called with the command `errordlg`. This command has two arguments, the text to be displayed and the name of the box.

As in Subsection 6.3.3, the display function distinguishes between *format compact* and *format loose*. The displayed string consists of a number, which is the value, a blank, and the characters of the unit.

The function `SameUnit` plays the same role as the function `SameUnit` of class *PhQuant*; in this example it checks if addition is legal. The reader can extend the use to subtraction and comparison.

The function `plus` overloads the usual addition function and adapts it to the electrical quantities involved in our example. If the summands, p and q, have different units, the function `plus` calls the error dialog box with the title 'Input error' and the text 'Summands have different units'.

The function `times` overloads the array operation '.*' and adapts it to the electrical quantities involved in our example. This function is structured as a conditional construct of the form *if, elseif, else*. The statements following *if* and *elseif* allow the multiplication of an electrical quantity by a pure number. Thus, for example, it is possible to calculate, by a simple multiplication, the equivalent resistance of two resistors of 2 Ω connected in series. The statements following *else* constitute a switch construct. The expression that classifies the *cases* to be considered is the unit of the first multiplicand, p. To each unit defined in the set S, in the class constructor, corresponds a conditional structure that implements Table 6.3. If the unit of the second multiplicand, q, does not allow a multiplication defined in the above mentioned table, the function `times` calls the error dialog box.

To experiment with the new file the reader has to exit the directory `@ElQuant` and work in the parent directory of `@ElQuant`. Below are a few examples.

```
≫ a = ElQuant(2, 'A')
 a =
  2 A
≫ b = ElQuant(3, 'V')
 b =
  3 V
≫ c = ElQuant(5, 'ohm')
 c =
  5 ohm
≫ d = ElQuant(10, 'S')
 d =
  10 S
≫e = ElQuant(4, 'ohm')
 e =
  4 ohm
≫ a.*b
ans =
  6 w
≫ a.*c
ans =
  10 V
≫ c + e
ans =
  9 ohm
≫ b.*d
ans =
  30 A
```

We invite the reader to try the addition $a + b$, the multiplication $c. * e$, and see the reaction of the computer. Try also $2. * a$, $b. * 3$.

6.4   Summary

Structures and *cell arrays* are two *data types* that allow the user to store under one name data of different natures. Using these data types simplifies housekeeping and eliminates potential sources of error. On the other hand, structures and cell arrays require more computer resources than simple arrays. MATLAB provides functions for changing, querying, or displaying structures and cell arrays. In this chapter we used these data types to design objected-oriented functions that allow calculations with *dimensioned quantities*. These functions allow the user to assign units to physical quantities, to perform a few calculations, and to retrieve the units of the result. The software allows only operations defined for the variables involved. In other words, it prevents the user to add, subtract or compare quantities that have different units, for example meters and square units. The scope of the functions illustrated in this chapter is limited; a fully-fledged system would be beyond the aims of the book.

In our examples we have shown three features of OOP that are advantageous in certain situations, such as the ones illustrated above:

Encapsulation: the data of physical quantities have been packed together with the methods that are allowed to operate on them. No user can apply operations that make no sense for the given objects.

Overloading: operations that perform in a certain way on pure numbers, have been modified to work also on the units.

Inheritance: By defining subclasses that inherit some of the attributes and methods of the superclass, the designer spares work.

We have also shown how to use *predefined dialog boxes* that are simple, MATLAB built-in GUIs. To learn more about predefined dialog boxes type `doc dialog`, or look for `Predefined dialog boxes` in the MATLAB Help.

The commands introduced in this chapter include

celldisp - `celldisp(C)` displays the contents of the cell array `C`;

cellplot - `cellplot(C)` yields a graphic description of the structure of the cell array `C`;

errordlg - `errordlg(error,name)` displays a dialog box with the title `name` and the message contained in the string `error`.

getfield - `getfield(s,'field')` returns the contents of the specified field of the structure `C`;

ismember - given two arrays, A and S, `ismember(A, S)` returns an array containing logical 1 (true) where the elements of A are members of the set S, and logical 0 (false) where not;

rmfield - `S1 = rmfield(S, 'f'` copies the structure S into the structure S1, but without the field `f`;

setfield - `S1 = setfield(S, 'newfield', x` produces a new structure, S1, with the additional field `newfield` assigning it the value x

struct - `S = struct('f1', x, 'f2', y, ...)` produces a structure, S, with the fields `f1`, `f2`, ..., and the values x, y, ...

6.5   Exercises

Exercise 6.1 The mass of a steel plate
Consider a steel plate of length L = 20 m, height H = 6 m, thickness t = 6 mm, and density $7.85 \text{ kg} \cdot \text{m}^{-3}$. Declare the above properties as objects of the class *PhQuant* and calculate the area, the volume, the mass and the weight of the plate. Check that the units of the weight are, indeed, those corresponding to the newton in Table 6.2.

Exercise 6.2 Defining subtraction for the class PhQuant
You are required to define the method of subtraction for the class `PhQuant`. The MATLAB function to be overloaded is `minus`. The file `PhQuant.m` is already rather long. To avoid lengthening it too much it is possible to write the subtraction method on a separate file, `minus.m`, belonging to the directory `@PhQuant`. Then, in the block of methods of the file `PhQuant.m` insert

```
s = minus(p, q)
% subtraction function for PhQuant class
```

Try the modified class in a few examples.

Exercise 6.3 Defining division for the class PhQuant
You are required to define division for the class `PhQuant`. To solve this exercise read the text of Exercise 6.2 and look in the MATLAB help for 'Implementing Operators for Your Class'.
Try the modified class in a few examples.

Exercise 6.4 Simple calculation with electrical objects
Consider three resistors connected in series; $R_1 = 3 \ \Omega$, $R_2 = 5 \ \Omega$, $R_3 = 10 \ \Omega$. Assuming that the current passing through those resistors is 10 A, calculate: a) the equivalent resistance of the connection; b) the voltage across the connection; c) the power dissipated through that connection.

Exercise 6.5 Parallel connection of resistors
To make possible the calculation of parallel connections of resistors add two methods to *ElQuant*. The first method, *ohm2S*, should be called with an argument having the unit *ohm* and should:

1. check if the units of the argument are, indeed, *ohm*;

2. calculate the corresponding conductance and output the result with the unit *S*.

The second method, *S2ohm*, should be called with an argument having the unit *S* and should:

1. check if the units of the argument are, indeed, S;

2. calculate the corresponding resistance and output the result with the unit *ohm*.

Using the above functions calculate the equivalent resistance of the two resistances, R1 $= 2\ \Omega$, R2 $= 3\ \Omega$, connected in parallel.

Part III

Progressing in MATLAB®

7

Complex numbers

7.1 The introduction of complex numbers

In this book we first met complex numbers in Subsection 3.8.2 where they appeared in the solutions of certain algebraic equations. This is also the way in which complex numbers appeared in mathematics. Initially considered as *impossible*, with time the complex numbers have become *indispensable* in many fields of engineering and science. Using complex numbers it is easier to understand and solve problems that involve vibrations, alternating current, or control. And, as Penrose (2004) writes, understanding complex numbers is essential for understanding modern mathematics.

In MATLAB, working with complex numbers is almost as simple as working with real numbers. We prove this first by showing how to perform elementary operations over complex numbers. In continuation, we introduce a few elementary functions of complex numbers and exemplify their proprieties or ask the reader to check them in several exercises. Two further applications are conformal mapping and phasors. Phasors simplify to a large extent the study of oscillatory systems; therefore, they are extensively used in the analysis of alternating-current circuits. In this chapter we use phasors to analyze two classic systems, one in mechanical, the other in electrical engineering.

Rotating vectors represented by complex numbers can also be used in the study of certain mechanisms. This chapter includes an extended example of such an application.

7.2 Complex numbers in MATLAB

The imaginary unit, $i = \sqrt{-1}$, is entered in MATLAB as i, or j. To see this, try

```
≫ i^2
ans =
  -1
≫ j^2
ans =
  -1
```

The character 'j' is often preferred in electrical engineering where 'i' is reserved for the intensity of the current. To define complex numbers in MAT-LAB, for example $z_1 = 3 + 4i$, $z_2 = -11 + 2i$, it is sufficient to type:

```
≫ z1 = 3 + 4i
z1 =
≫ 3.0000 + 4.0000i
≫ z2 = -11 + 2i
z2 =
 -11.0000 + 2.0000i
```

Above we just concatenated the character 'i', or 'j', to a number. If instead of numbers we use letters, it is necessary to insert the multiplication symbol, '*':

```
≫ a = 3; b = 4;
≫z3 = a + b*i
z3 =
   3.0000 + 4.0000i
≫ c = -11; d = 2;
≫ z4 = c + 2*j
z4 =
   -11.0000 + 2.0000i
```

Once complex numbers are defined, we can work with them as simply as we work with real numbers. Thus, for addition and subtraction we type:

```
≫ z1 + z2
ans =
 -8.0000 + 6.0000i
≫ z1 - z2
ans =
 14.0000 + 2.0000i
```

The multiplication

$$z_1 \cdot z_2 = (3 + 4i)(-11 + 2i) = -33 + 6i - 44i - 8 = -41 - 38i$$

is carried on with one command:

```
≫ z1*z2
ans =
  -41.0000 -38.0000i
```

Similarly, the division

$$\frac{z_1}{z_1} = \frac{-11 + 2i}{3 + 4i} = \frac{(-11 + 2i)(3 - 4i)}{(3 + 4i)(3 - 4i)} = \frac{-25 + 50i}{9 + 16} = \frac{-25 + 50i}{25}$$

is performed with one command:

```
≫ z2/z1
ans =
  -1.0000 + 2.0000i
```

Given a complex number, $z = a + bi$, we can retrieve the *real part*, a, with the command `real(z)`, and the *imaginary part*, b, with the command `imag(z)`

```
≫ real(z1)
ans =
   3
≫ imag(z1)
ans =
   4
```

In textbooks it is usual to note the real part of a complex number, z, by $\Re(z)$, and the imaginary part by $\Im(z)$. We are going to use these notations when plotting complex numbers and functions of complex numbers.

There is an alternative way of defining complex numbers. It uses the command `complex` and it is helpful when dealing with arrays of complex numbers:

```
≫ a = [ 1; 2; 3 ];
≫ b = [ 10; 20; 30 ];
≫ c = complex(a, b)
c =
   1.0000 +10.0000i
   2.0000 +20.0000i
   3.0000 +30.0000i
```

Given a complex number, $z = a + ib$, its **conjugate** is $\bar{z} = a - ib$. In MAT-LAB, the conjugate of a complex number is retrieved with the command `conj`. For example, with z_1 defined as above:

```
≫ conj(z1)
ans =
   3.0000 - 4.0000i
```

At the end of this subsection it is worth pointing out that the use of i or j to define the imaginary unit is built-in in MATLAB, but can be overruled if the user assigns another value to one of those letters.

```
≫ i = 2, j = 3
...
≫ i^2
ans =
   4
≫ j^2
ans =
   9
```

Many programmers use the letter i, or j, to index array elements, or in *For loops*. As complex numbers are ubiquitous in modern engineering, in this book we avoid the overruling of the default meaning of i and j, and use instead the letter k for indexing iterations in *For loops*.

7.3   Geometric representation

To give complex numbers a geometric interpretation, we consider a plane in which the cartesian coordinates of points are their real and imaginary parts. For example, in Figure 7.1 the complex number, $z1 = 3 + 4i$, is represented by the point z_1. Alternatively, a complex number, $z = a + bi$, can be inter-

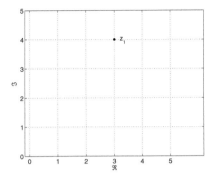

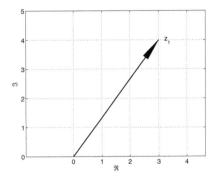

FIGURE 7.1: Complex number represented as a point

FIGURE 7.2: Complex number represented as a vector

preted as a vector whose components parallel to the cartesian axes of a plane are the real and the imaginary parts, a, b. This interpretation of the above number, z_1, is shown in Figure 7.2. We say that the plane introduced here is the **complex plane**, also called the *Argand plane* after the mathematician Jean-Robert Argand (1768-1822) who described it. Actually the Norwegian-Danish Caspar Wessel (1745-1818) had the idea nine years earlier. In German literature the complex plane is known as the *Gauss plane*. The axis along which we measure the real parts, a, is the **real axis**, and the axis along which we measure the imaginary parts, b, is the **imaginary axis**. In the complex plane, the distance of a point from the origin, or the magnitude of a vector, can be obtained with the command **abs**, for example

```
≫ abs(z1)
ans =
   5
```

This value is known as the **absolute value**, or **modulus** of z and is noted in textbooks as *mod z*.

The angle between a complex vector, Z, and the real axis is called the **argument**, or **phase** of z, and is noted in textbooks as *arg z*. This value can be retrieved in MATLAB with the command **angle(Z)**, for example

```
≫ angle(z1)
ans =
   0.9273
```

The result is measured in radians and is the same that could have been obtained as below:

```
≫ atan2(imag(z1),real(z1))
ans =
   0.9273
```

7.4   Trigonometric representation

Given the *absolute value*, ρ, and the *argument*, θ, of a complex number, z, the **trigonometric**, or **polar form**, of the number is

$$z = \rho\cos\theta + i\rho\sin\theta \qquad (7.1)$$

For example, in MATLAB we can write

```
≫ rho = 5;
≫ theta = pi/12;
≫ z = rho*cos(theta) + rho*sin(theta)*i
z =
   4.8296 + 1.2941i
```

We invite the experiment to check the commands `abs(z)` and `angle(z)` on this representation and interpret the results.

7.5   Exponential representation

Using the relation

$$e^{i\theta} = \cos\theta + i\sin\theta \qquad (7.2)$$

we can convert from the trigonometric representation (Equation 7.1) to the **exponential representation**

$$z = \rho e^{i\theta} \qquad (7.3)$$

Thus, continuing the example in the preceding subsection we write in MAT-LAB:

```
>> z = rho*exp(i*theta)
z =
   4.8296 + 1.2941i
```

Using the exponential representation of complex numbers, we can give a geometric interpretation of multiplication. To show this, let us consider the complex numbers and their product

$$Z_1 = \rho_1 e^{i\theta_1}$$
$$Z_2 = \rho_2 e^{i\theta_2}$$
$$Z_1 Z_2 = \rho_1 e^{i\theta_1} \cdot \rho_2 e^{i\theta_2} = \rho_1 \rho_2 e^{\theta_1 + \theta_2} \tag{7.4}$$

Thus, the product of two complex numbers is a complex number whose module is the product of the modules of the multiplicands, and is rotated from the first multiplicand by the angle of the second multiplicand. To give an example in MATLAB, let us define two complex numbers, multiply one by the other, and find the module and the argument of the result:

```
>> z1 = 3 + 4i;
>> z2 = 1 + 2i;
>> z = z1*z2
Z =
  -5.0000 +10.0000i
>> abs(z)
ans =
   11.1803
>> angle(z)
ans =
    2.0344
```

The absolute values of the two numbers, and the product of the absolute values are, indeed,

```
>> rho1 = abs(z1)
rho1 =
    5
>> rho2 = abs(z2)
rho2 =
    2.2361
>> rho1*rho2
ans =
   11.1803
```

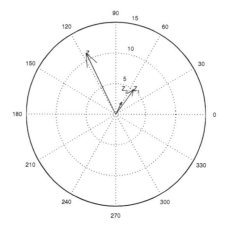

FIGURE 7.3: The multiplication of two complex numbers

The arguments of the two multiplicands and their sum are

```
≫ theta1 = angle(Z1)
≫ theta1 =
   0.9273
≫ theta2 = angle(Z2)
theta2 =
   1.1071
≫ theta1 + theta2
ans =
   2.0344
```

Thus, the multiplication of a complex number, Z_1, by another complex number, Z_2, causes the counterclockwise rotation of Z_1 by the argument of Z_2.

To plot the above multiplicands and their product we collect the definitions of the three vectors in one array and use the function compass as follows:

```
≫ V = [ Z1, Z2, Z ];
≫ compass(V)
≫ text(real(Z1), imag(Z1), 'Z_1')
≫ text(real(Z2), imag(Z1), 'Z_2')
≫ text(real(Z), imag(Z), 'Z')
```

The result is shown in Figure 7.3 where the angles are measured in degrees. We invite the reader to check that the various values are those calculated above.

As a particular, important case, let us consider the multiplication of a vector by i. The argument of i is $\pi/2$. Therefore, multiplication by i amounts to counterclockwise rotation by $\pi/2$. We invite the reader to check this in Exercise 7.9. We also invite the reader to interpret the division of a complex number by another complex number and to try this interpretation on a few concrete examples.

7.6 Functions of complex variables

Given a set of complex numbers, z, let us associate each member of it with one or more complex numbers, w. We say that the set z is a **complex variable**, and the set $w = f(z)$ a **function of** z. An important example is that of the **exponential function**, $w = e^z$. Similarly to the real-valued exponential function of a real variable, there are several possibilities of defining the complex exponential function. For example, the exponential function can be defined by the power series

$$e^z = \sum_{n=0}^{\infty} \frac{z^n}{n!}$$

With $z = x + iy$ we can write

$$w = e^z = e^{x+iy} = e^x e^{iy} = e^x(\cos y + i\sin(y))$$

To give an example in MATLAB, let us consider the complex number $z_1 = 3 + 4i$ defined in the preceding subsection and calculate in two ways:

```
≫ f1 = exp(z1)
f1 =
   -13.1288 -15.2008i
≫ f2 = exp(3)*(cos(4) + i*sin(4))
f2 =
   -13.1288 -15.2008i
≫ f1 - f2
ans =
   0
```

The complex exponential function has the same properties as the real-valued exponential function. For example, considering also the number $z_2 = 1 + 2i$ defined in the previous section, we can check the multiplication:

```
>> exp(z1)*exp(z2) - exp(z1 + z2)
ans =
   0 -3.5527e-015i
```

The imaginary part of the result should have been 0. As the computer performs the calculations over the set of computer, and not real, numbers (see Chapter 5), we obtain a very small number and not zero. Now, let us check the division:

```
>> exp(z1)/exp(z2) - exp(z1 - z2)
ans =
   4.4409e-016 -8.8818e-016i
```

Again, and for the same reasons, we obtain a very small number instead of the expected zeroes.

If $w = e^z$, then $z = \ln w$, that is the **natural logarithm** of z. An example in MATLAB is

```
>> w1 = exp(z1)
w1 =
   -13.1288 -15.2008i
>> z = log(w1)
z =
   3.0000 - 2.2832i
```

At the first glance it seems that we have not recovered, as expected, the value of z_1. The result, however, is correct. The complex natural logarithm is a **multiple-valued function**. In fact, if we use the exponential notation for z_1, then

$$w = \ln z_1 = \ln \rho + i(\theta + 2k\pi), ...k = 0, \pm 1, \pm 2, ...$$

Checking in MATLAB, we find that, in this case, $k = 2$:

```
>> (imag(w1) - imag(z))/pi
ans =
   2
```

The complex natural logarithm has the same properties as the natural logarithm of real numbers. As an example in MATLAB let us check:

```
≫ w2 = exp(z2);
≫ log(w1*w2) - log(w1) - log(w2)
ans =
    0
```

The trigonometric functions of complex variables can be defined in terms of the exponential function, for example,

$$\sin z = \frac{e^{iz} - e^{-iz}}{2i}, \quad \cos z = \frac{e^{iz} + e^{-iz}}{2} \tag{7.5}$$

The trigonometric functions of complex variables have the same properties as the trigonometric functions of real variables, for example,

```
≫ cos(z1)^2 + sin(z1)^2
ans =
    1
```

In the section of exercises the reader may find more examples of the properties of functions of complex variables.

7.7 Conformal mapping

Consider the complex variable $z = x + iy$ and a function of this variable, $w(z) = u(x, y) + iv(x, y)$. The equations

$$u = u(x, y)$$
$$v = v(x, y)$$

define a **transformation**, or **mapping**, between the z−plane $(x, y$-plane) and the w−plane $(u, v$-plane).

Let us assume that two curves, $C_1(z)$, $C_2(z)$ of the z−plane, are transformed by $w(z)$ into two curves, K_1, K_2 of the w−plane. If the angle between C_1 and C_2, at their point of intersection, equals the angle between the curves K_1, K_2, at their intersection, we say that the mapping $w(z)$ is **conformal**. If the derivative

$$w'(z) = \lim_{\Delta \to 0} \frac{f(z + \Delta z) - f(z)}{\Delta z}$$

exists in all points of a region, D, of the complex plane, the function $w(z)$ is said to be *analytic* in the region D. An important theorem states that if $w(z)$

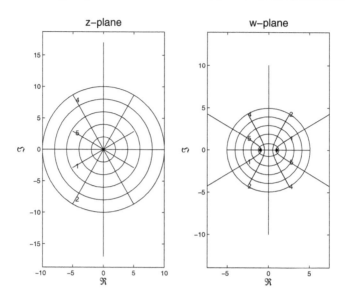

FIGURE 7.4: The conformal mapping $w = (z + 1/z)/2$

is analytic and $w'(z) \neq 0$ in a region D, then the mapping $w(z)$ is conformal at all points of D.

To give an interesting example of the use of MATLAB in performing a conformal mapping, we consider in Figure 7.4 a set of straight lines passing through the origin of the z−plane, and a set of circles with center in the origin of the same plane. The two sets of curves are, obviously, *orthogonal*, that is they intersect at right angles. The transformation

$$w = \frac{1}{2}\left(z + \frac{1}{z}\right)$$

maps the circles onto ellipses, and the straight lines onto hyperbolas. The two sets are orthogonal. Moreover, the ellipses and the hyperbolas have the same *foci*, which are shown in the right-hand figure as the points

$$F_1 = 1 + 0i, \ F_2 = -1 + 0i$$

We say that the two families of curves are *confocal*. To carry on this mapping, we have written a script file, ConfMap.m, whose listing appears below. Some explanations follow the listing.

```
%CONFMAP   First example of conformal mapping
%          Circles are mapped onto ellipses, and straight
%          lines onto hyperbolas
```

```
%%%%%%%%%%%%%%%% z-plane %%%%%%%%%%%%%%%%%%%%%%%%%
subplot(1, 2, 1)
% plot coordinate axes
plot([ -10 10 ], [ 0 0 ], 'k-', [ 0 0 ], [ -17 17 ], 'k-')
axis equal
ht = title('z-plane');
set(ht, 'FontSize', 16)
hl = xlabel('\Re');
set(hl, 'FontSize', 14)
hl = ylabel('\Im');
set(hl, 'FontSize', 14)
% define curves in the z-plane
x1 = linspace(-5, 0);
x2 = linspace(0, 5);
% straight lines
a   = 0;          % initialize slope of straight lines
y1 = zeros(6, length(x1)); % allocate space
y2 = zeros(6, length(x2));
% plot curves in the z-plane
% circles
t   = 0: 2.5: 360;         % degrees
r   = 0;                   % initialize circle radii
xc = zeros(6, length(t)); % allocate space
yc = zeros(6, length(t));
hold on
for k = 1:5
    % straight lines
    a = a + 30;                % increase slope by 15 degrees
    y1(k, :) =  tand(a)*x1; % left-hand part of lines
    text(x1(10), y1(k, 10), num2str(k))
    plot(x1, y1(k, :), 'm-')
    y2(k, :) = tand(a)*x2;   % right-hand part of lines
    plot(x2, y2(k, :), 'm')
    % circles
    r  = r + 2;                % increase circle radius by 1
    xc(k, :) = r*cosd(t);
    yc(k, :) = r*sind(t);
    plot(xc(k, :), yc(k, :), 'c-')
    % text(xc(10), yc(k, 10),  num2str(k) )
end
x1 = repmat(x1, 6, 1);
x2 = repmat(x2, 6, 1);
z1 = x1 + y1*i;
z2 = x2 + y2*i;
z3 = xc + yc*i;
```

```
hold off
%%%%%%%%%%%% CONFORMAL MAPPING %%%%%%%%%%%%%%
w1 = (z1 + 1./z1)/2;
w2 = (z2 + 1./z2)/2;
w3 = (z3 + 1./z3)/2;
%%%%%%%%%%%%%% z-plane %%%%%%%%%%%%%%%%%%%%%%
subplot(1, 2, 2)
% plot coordinate axes
plot([ -5 5 ], [ 0 0 ], 'k-', [ 0 0 ], [ -10 10 ], 'k-')
axis equal
ht = title('w-plane');
set(ht, 'FontSize', 16)
hl = xlabel('\Re');
set(hl, 'FontSize', 14)
hl = ylabel('\Im');
set(hl, 'FontSize', 14)
hold on
point([ -1; 0 ], 0.15)
point([  1; 0 ], 0.15)
for k= 1:5
    plot(real(w1(k, :)),    imag(w1(k, :)), 'm-')
    text(real(w1(k, 2)),    imag(w1(k, 2)), num2str(k))
    plot(real(w2(k, :)),    imag(w2(k, :)), 'm-')
    text(real(w2(k, end)), imag(w2(k, end)), num2str(k))
    plot(real(w3(k, :)),    imag(w3(k, :)), 'c-')
  end
hold off
```

In the above listing we use the command `subplot` to generate two plots, one of the $z-$plane, the other of the $w-$plane. The first use is `subplot(1, 2, 1)`, which means one row, two columns, and 'this is the first plot'. The second use is `subplot(1,2 2)`, which means the same arrangement of two plots in one row, but also 'this is the second plot'. We begin by plotting a bundle of straight lines passing through the origin of the $z-$ plane. The slope of the lines is a, and their equation ax. We start the plotting with $a = 30°$. We avoid thus the line $\Im(z) = 0$ whose mapping is not interesting. For the $x-$coordinate we define two semi-axes, `x1` for negative values, and `x2` for positive values. We do not use one axis from negative to positive values because it would cause a problem in the $w-$plane. As initially defined, `x1` and `x2` are row vectors of length 100. The arrays of $y-$coordinates, `y1` and `y2`, are defined, however, as *6-by-100* arrays. To define the corresponding $z-$functions we must first convert the `x1` and `x2` arrays into arrays of the same size as the $y-$arrays. More specifically, `x1` and `x2` must be converted into arrays with six identical rows. To do this, we use the MATLAB command `repmat`. Calling it as `repmat(x1, 6, 1)` means that we want to have the array `x1` repeated in six rows, one column.

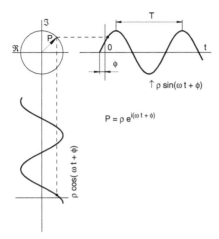

FIGURE 7.5: To the definition of *phasor*

The circles in the z-plane are plotted with the help of their parametric equations. The real and imaginary parts of the lines and circles are generated in a For loop that is iterated five times. These parts are combined, however, in complex functions only outside the loop. We first define three functions, z1, z2, z3, that are mapped by the transformation $w = (z + 1/z)/2$ into the functions, w1, w2, w3. The latter functions are plotted in the w-plane after the command subplot(1, 2, 2). We use the TEX commands \Re and \Im to label the axes.

Conformal mappings have important applications in science and engineering, for example in fluid flow, heat transfer and electrostatics. One special case of fluid mechanics is in the calculation of ship motions. The idea is to solve a problem by representing the involved figure as a conformal mapping of a simpler figure for which the solution is known. The method is useful in problems that can be solved by potential theory.

7.8 Phasors

7.8.1 Phasors

The response of simple oscillating or vibrating systems can often be represented by a sine or cosine function. A general form of such a response is

$$V = \rho \cos(\omega t + \phi) \tag{7.6}$$

where ρ is the *amplitude*, ω, the *angular frequency*, t, the time independent variable, and ϕ, the *phase angle*. The expression between parentheses, $\omega t + \phi$, is the *argument* of the response. The angular frequency is measured in rad s^{-1}. Quantities derived from those defined above are the *frequency*, $f = \omega/2\pi$, measured in *hertz* (Hz), and the `period`, $T = 1/f$, measured in seconds, s.

From Subsection 7.5 we know that Equation 7.6 is equivalent to

$$V = \Re\left(\rho e^{i(\omega t + \phi)}\right) \tag{7.7}$$

Based on this, we show in the upper, left side of Figure 7.5 a part of the complex plane and in it a vector, P, of magnitude ρ, and making with the real axis an angle equal to $(\omega t + \phi)$. If this vector is rotating counterclockwise with the angular velocity ω, starting from the angle ϕ at $t = 0$, then its projection on the real axis describes the function $V = \rho \cos(\omega t + \phi)$, while the projection on the imaginary axis describes the function $\rho \sin(\omega t + \phi)$.

The equation that defines the vector P in exponential form is

$$P = \rho e^{i\phi} e^{i\omega t} \tag{7.8}$$

The quantity

$$\rho e^{i\phi}$$

is constant and contains all the information about the amplitude and phase of P; we say that it is the **phasor** representation of P. The quantity

$$e^{i\omega t}$$

is variable. As we are going to see soon, this separation is useful because it allows us to factor out the variable part of the rotating vector and deal only with its constant part. In this way we can solve sone linear differential equations by simple algebra.

Differentiating Equation 7.8 with respect to time yields

$$\frac{dP}{dt} = i\omega\rho e^{i(\omega t + \phi)} = i\omega e^{i\omega t}\rho e^{i\phi} \tag{7.9}$$

that is the vector P multiplied by ω and rotated counterclockwise by the angle $\pi/2$.

Differentiating a second time we obtain

$$\frac{d^2 P}{dt^2} = -\omega^2 \rho e^{i(\omega t + \phi)} = -\omega^2 e^{i\omega t}\rho e^{i\phi} \tag{7.10}$$

which means a second multiplication by ω and a second counterclockwise rotation by $\pi/2$.

In electricity we are also interested in integration with respect to time. Thus

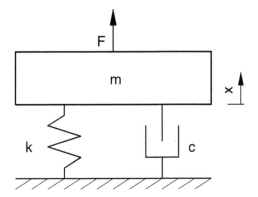

FIGURE 7.6: A mechanical system

$$\int P dt = \frac{\rho}{i\omega} e^{i(\omega t + \phi)} = \frac{e^{i\omega t}}{i\omega} \rho e^{i\phi} \qquad (7.11)$$

that is clockwise rotation by $\pi/2$ and division by ω.

7.8.2 Phasors in mechanics

In this section we describe a simple mechanical system that is modeled by a second-order, linear differential equation. Figure 7.6 shows the physical model. A mass, m, is supported by a spring characterized by the *spring constant* k. The motion is damped by a *dashpot* that produces *viscous* friction. We measure the displacement of the mass m in the direction of the x-axis shown at the right of the mass. The forces acting on the mass are:

1. inertia;

2. the force of the extended or compressed spring;

3. the viscous friction generated by the dashpot;

4. an external, periodic force of amplitude F_0 and angular frequency ω.

The mathematical model is obtained by writing that the sum of the first three forces mentioned above is equal to the external force

$$inertia\ force\ +\ spring\ force\ +\ friction\ force\ =\ external\ force$$

The external force is also known as the *driving force*. We assume the following simplifications:

Linear spring - the amplitude of the oscillations is limited to the domain within which the force developed by the spring is proportional to its extension or compression. We write this force as kx;

Linear damping - the damping force is proportional to the velocity of the mass, $\dot{x}$. We write this force as $c\dot{x}$.

The resulting equation is

$$m\frac{d^2x}{dt^2} + c\frac{dx}{dt} + kx = F_0\cos\omega t \tag{7.12}$$

The theory shows that the *steady-state* response of the system is a periodic function with the same angular frequency as the driving force, but with another amplitude and phase. Without loss of generality, we have assumed that the phase of the driving force is zero. Let the response be

$$X = X_0\cos(\omega t + \phi) = \Re(X_0 e^{i\omega t}e^{i\phi}) \tag{7.13}$$

Using phasors similar to those developed in the preceding subsection, we can rewrite Equation 7.12 as

$$-m\omega^2 X_0 e^{i\phi}e^{i\omega t} + icX_0 e^{i\phi}\omega e^{i\omega t} + kX_0 e^{i\phi}e^{i\omega t} = F_0 e^{i\omega t} \tag{7.14}$$

The theory shows that if the system is undamped ($c = 0$), and there is no driving force ($F_0 = 0$), the system will oscillate freely with the frequency $\omega_0 = \sqrt{k/m}$. Using this notation, and simplifying Equation 7.14 we obtain

$$X_0 e^{i\phi} = \frac{F_0}{m(\omega_0^2 - \omega^2) + ic\omega} \tag{7.15}$$

The expression $X_0 e^{i\phi}$ in the left-hand side of the equation is the *complex amplitude* of the response.

For a concrete example let us assume the following values: mass equal to $m = 2$ Kg, damping equal to $c = 0.4$ N·m^{-1}·s, spring constant equal to $k = 0.1$ N·m^{-1}, amplitude of the driving force equal to $F_0 = 1$ N, and the driving frequency $\omega = 0.3$ rads^{-1}. To analyze this system let us write the script file, MechSys.m, whose code is listed below.

```
%MECHSYS Example of mechanical system solved with phasors

% define parameters
m       = 2;
c       = 0.4;
k       = 0.1;
F0      = 1;
omega   = 0.3;  % natural frequency
omega0  = sqrt(k/m)
```

```
% calculate transfer function
T       = F0/(m*(omega0^2 - omega^2) + i*c*omega);
X0      = abs(T); % response amplitude
r       =omega/omega0
phi     = omega/omega0;  % phase
phideg = 180*phi/pi
% calculate phasors
spring  = k*T
damping =  i*omega*c*T
inertia = -omega^2*m*T
sum     = spring + damping + inertia - F0
% define frame
plot([ -0.5 1.5 ], [ 0 0 ], 'k-', [ 0 0 ], [ -0.5 1.5 ], 'k-')
axis equal, axis off
ht = text( 1.4, 0.08, '\Re');
set(ht, 'FontSize', 16)
ht = text( 0.05, 1.4, '\Im');
set(ht, 'FontSize', 16)
hold on
% calculate points and plot arrows
O  = [ 0; 0 ];
P1 = [ F0; 0 ];
arrow(O, P1, 0.05)
ht = text(P1(1)/2, 0.1, 'F_0');
set(ht, 'FontSize', 16)
P2 = [ real(inertia); imag(inertia) ];
arrow(O, P2, 0.05)
ht = text(P2(1)/2, P2(2)/2, 'inertia');
set(ht, 'FontSize', 16)
P3 = P2 + [ real(damping); imag(damping) ];
arrow(P2, P3, 0.05)
ht = text((P2(1) + P3(1))/1.8, P2(2)/1.2, 'damping');
set(ht, 'FontSize', 16)
P4 = P3 + [ real(spring); imag(spring) ];
arrow(P3, P4, 0.05)
ht = text(P4(1)/1.1, P3(2)/1.5, 'spring');
set(ht, 'FontSize', 16)
hold off
```

The resulting phasor diagram is shown in Figure 7.7.

The polygon of phasors is closed, which means that the solution satisfies Equation 7.12. The commands not ended with a semicolon, ';', printed the values

```
omega0  =  0.2236
```

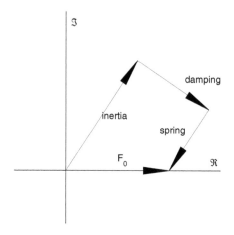

FIGURE 7.7: Phasor diagram of the mechanical system

```
r        =   1.3416
phideg   = 76.8704
spring   = -0.3846 - 0.5769i
damping  =  0.6923 - 0.4615i
inertia  =  0.6923 + 1.0385i
sum      =  0 -2.2204e-016i
```

Due to numerical errors (see Chapter 5), the printed sum is a very small number, while we expected a zero vector. The factor $e^{i\omega t}$ disappeared during our analysis; however, its interpretation is simple. As this factor multiplies each component of the polygon and multiplication by a complex number means rotation of the first multiplicand by the angle of the second, we can say that the polygon shown in Figure 7.7 is rotating with the angular velocity ω around the origin of the complex plane.

We analyzed the system for one value of the driving frequency, $0.3 \text{ rad} \cdot \text{s}^{-1}$. It is interesting to study the behavior of the system over a range of driving-force frequencies that encompasses the natural frequency, ω_0. We obtain thus the *frequency response*; it is usually represented by two graphs, one of the ratio of the amplitude of the response to that of the driving force, the other of the phase. Both graphs are usually plotted against the ω/ω_0 ratio. To give an example, let us write the following script file that we call MechExample.m. The main part of the program is a **For** loop that calculates the amplification and the phase for three values of the damping c. We use the TEX command \omega to print the Greek letter ω.

```
%MECHEXAMPLE1  Frequency response of mechanical system comprising
%             a mass, a dashpot, a spring  and a driving force
```

```
m   = 2;                  % mass, kg
c   = 0.4;                % damping, Ns/m
k   = 0.1;                % spring, N/m
F0 = 1;                   % driving force amplitude
% derived data
omega0 = sqrt(k/m)
% write transfer function
omega = linspace(0.01, 0.5);
c = 0.1;
for k = 1:3
    subplot(2, 1, 1)
        T = 1./(m*(omega0.^2 - omega.^2) + i*c*omega);
        plot(omega/omega0, abs(T), 'k-')
        grid
        ylabel('Amplitude')
        text(1, abs(T(40)), [ '   c = ' num2str(c) ])
        hold on
    subplot(2, 1, 2)
        plot(omega/omega0, 180*angle(T)/pi, 'k-')
        grid
        ylabel('Phase, deg')
        text(2.25, 180*angle(T(end))/pi, [ 'c = ' num2str(c) ])
        hold on
        c = c + 0.3;
end
xlabel('\omega/\omega_0')
hold off
```

Running the file `MechExample.m` we obtain Figure 7.8. The graph reveals the phenomenon of *resonance*, an increase of the amplification for $\omega \simeq \omega_0$. We also see that the effect of resonance increases while c decreases. For $c = 0$ the response would be infinite.

7.8.3 Phasors in electricity

Phasors are helpful in the understanding of electrical circuits. To show this on an example, let us consider in Figure 7.9 a series connection of an ac (alternating-current) voltage source, U, a resistor, R, an inductor, L, and a capacitor, C. We note the current intensity in the circuit by $i(t)$ to show that it is a function of time. As governing equation of the circuit we write that the sum of voltage drops across the three elements, R, L, and C, equals the voltage U. Let the voltage of the source be

$$U(t) = u_0 \sin \omega t = \Im(u_0 e^{j\omega t}) = u_0 \Im(e^{j\omega t}) \tag{7.16}$$

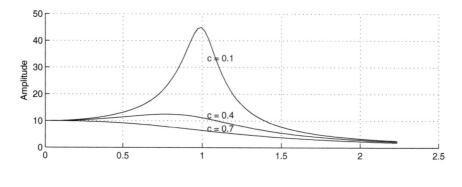

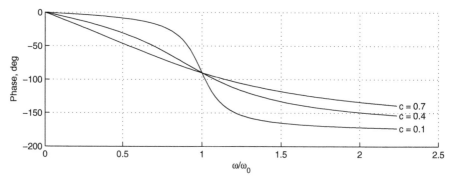

FIGURE 7.8: The frequency response of the mechanical system

As explained in Subsection 7.2, for the imaginary unit we use now 'j' instead of 'i'. The theory shows that the current through the circuit is another sine function, but its amplitude and phase differ from those of the voltage. We can describe this current by

$$i = i_0 \sin(\omega t + \phi) = i_0 \Im(e^{j\phi} e^{j\omega t} \tag{7.17}$$

Let us write the phasors that represent the voltage drops across the passive elements of the circuit. Thus, for the resistor the equation is

$$v_R(t) = R i(t) = R i_0 e^{j\phi} e^{j\omega t} \tag{7.18}$$

We obtain a phasor 'in phase' with the voltage phasor. For the inductor we write

$$v_L(t) = L\frac{di(t)}{t} = j\omega L i_0 e^{j\phi} e^{j\omega t} \tag{7.19}$$

a phasor rotated counterclockwise by $\pi/2$, that is leading the voltage. Finally, for the capacitor we write

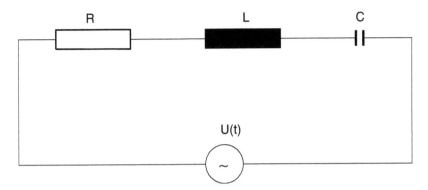

FIGURE 7.9: A series RLC circuit

$$v_C(t) = \frac{1}{C}\int i(t)dt = \frac{1}{j\omega C}i_0e^{j\omega\phi}e^{j\omega t} = -\frac{j}{\omega C}i_0e^{j\omega\phi}e^{j\omega t} \qquad (7.20)$$

a phasor rotated clockwise by $\pi/2$, that is lagging behind the voltage.

The governing equation is

$$v_R + v_L + v_C = U$$

Substituting Equations 7.16 and 7.18 to 7.20, collecting the terms that multiply $i_0e^{j\phi}$, and dividing both sides by $e^{j\omega t}$ we obtain

$$i_0e^{j\phi}\left(R + j\omega L - \frac{j}{\omega C}\right) = u_0 \qquad (7.21)$$

which can be rewritten as

$$i_0e^{j\phi} = \frac{1}{R + j\left(\omega L - \frac{1}{\omega C}\right)} \qquad (7.22)$$

The denominator

$$Z = R + j\left(\omega L - \frac{1}{\omega C}\right) \qquad (7.23)$$

is the *series impedance* of the circuit.

To give an example in MATLAB let us assume that the effective value of the voltage u is 110 V and the frequency of the source is 60 Hz. For the other values we choose the resistance $R = 50$ Ω, inductance $I = 10$ mH (milli henrys), and the capacitance $L = 400$ μF (micro farad). We begin by calculating at the command line, but intend later to put the commands on a script file. Therefore, we recommend the reader to start with the command `format compact`. We have the effective, or rms value, of the voltage $U(t)$, but we need its maximum value to write the corresponding sinusoidal function.

Therefore, we calculate this maximum value and continue by defining the other parameters of the circuit and calculating the angular frequency of the voltage:

```
≫ u0 = sqrt(2)*(110)
u0 =
   155.5635
≫ R = 50;
≫ L = 100/10^3;
≫ C = 400/10^6;
≫ f = 60;
≫ om = 2*pi*f
om =
   376.9911
```

We calculate now the series impedance of the circuit, Z, and the transfer function, TF:

```
≫ imp = om*L - 1/(om*C);
≫ Z = R + i*imp
Z =
   50.0000 +31.0677i
≫ TF = u0/Z
TF =
   2.2447 - 1.3947i
```

The amplitude of the current and its phase are obtained with

```
≫ i0 = abs(TF)
i0 =
   2.6427
≫phi = angle(TF)
phi =
   -0.5560
```

The complex amplitude of the current is given by

```
≫ I = i0*exp(j*phi)
I =
   2.2447 - 1.3947i
```

We calculate now the voltage drop, vR, across the resistor, the voltage drop, vL, across the inductor, and vC, across the capacitor.

```
≫ vR = R*I
vR =
   1.1223e+002 -6.9736e+001i
≫ vL = j*om*L*I
vL =
   52.5798 +84.6215i
≫ vC = -j*I/(om*C)
vC =
   -9.2490 -14.8853i
```

We easily check that the sum of the voltage drops equals the impressed voltage:

```
≫ vR + vL + vC - u0
ans =
   2.8422e-014 -1.7764e-014i
```

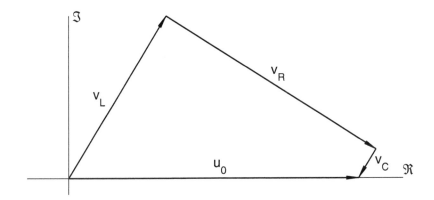

FIGURE 7.10: The phasor diagram of the RLC circuit

To get more insight into this analysis we may want to plot a phasor diagram and plot the voltage and the current in the time domain. It is convenient to do this in a script file that we can correct when necessary, and possibly reuse it for other values. We must begin the script file with the commands listed above, but how can we avoid entering them a second time? Simply, copy the screen and paste it to a new file called, for example, RLCseries.m. Delete the unnecessary, like answers, and leave only the commands. Add a semicolon after each command. Add to the file the lines shown below.

```
b =  max([ real(vR),  real(vL), real(vC) ]);
h = max([ imag(vR), imag(vL), imag(vC) ]);
plot([ -0.2*b 1.6*b ], [ 0 0 ], 'k-', [ 0 0 ], [ -0.1*h  h ], 'k-')
axis equal, axis off
ht = text(1.6*b, 0.05*h, '\Re');
set(ht, 'FontSize', 16);
ht = text(0.02*b, h, '\Im')
set(ht, 'FontSize', 16);
hold on
O  = [ 0; 0 ]
PL = [ real(vL); imag(vL) ]
arrow(O, PL, 1.3)
ht = text(0.5*PL(1)/2, PL(2)/2,  'v_L');
set(ht, 'FontSize', 16);
PR       = PL + [ real(vR); imag(vR) ]
arrow(PL, PR, 1.3)
ht = text((PL(1) + PR(1))/2, (PL(2) + PR(2))/1.8, 'v_R');
set(ht, 'FontSize', 16);
PC = PR + [ real(vC); imag(vC) ]
arrow(PR, PC, 1.3)
ht = text((PR(1) + PC(1))/1.95, (PR(2) + PC(2))/1.8, 'v_C');
set(ht, 'FontSize', 16);
arrow(O, [ u0; 0 ], 1.3)
ht = text(u0/2, 0.08*h, 'u_0');
set(ht, 'FontSize', 16);
```

The resulting diagram is shown in Figure 7.10. Continue the file with the following lines.

```
pause
clf
% plot voltage and current
T   = 2*pi/om;        % period, s
t   = 0: T/50: 3.5*T; % time scale, s
U   = u0*sin(om*t);
arg = om*t + ones(size(t))*phi;
I   = i0*sin(arg);
subplot(2, 1, 1)
    plot(t, U, 'k-'), grid
    ylabel('Voltage, V')
subplot(2, 1, 2)
    plot(t, I, 'k-'), grid
    xlabel('t, s')
    ylabel('Current, A')
```

The plot is shown in Figure 7.11.

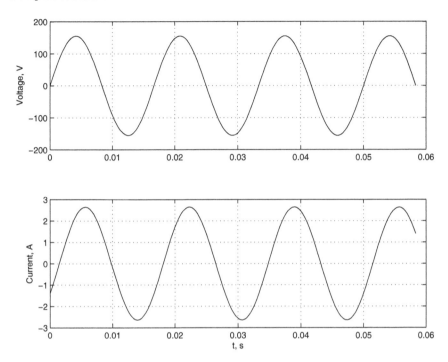

FIGURE 7.11: The time response of the RLC circuit

7.9 An application in mechanical engineering — a mechanism

MATLAB can be easily used in the analysis and simulation of mechanisms. One can even think about developing the whole theory of mechanisms using only MATLAB. This section contains only one example of application to mechanisms, but, by developing several variants of it we also show how to improve programmes and how to produce animation.

7.9.1 A four-link mechanism

Our example refers to the **four-link mechanism** shown in Figure 7.12. The links P_0 and P_3 are fixed to the ground. The mechanism is put into motion by the rotation of the bar $\overline{P_0P_1}$ around the link P_0. Let the length of the bar $\overline{P_0P_1}$ be ℓ_1, that of $\overline{P_1P_2}$ be ℓ_2, that of $\overline{P_2P_3}$, ℓ_3, and that of $\overline{P_0P_3}$, ℓ_4.

As explained above, the bar P_0P_3 is fixed. The positions of the other bars are determined by the angles they make with the horizontal. Thus, let the angle between $\overline{P_0P_1}$ and the horizontal be θ_1, the angle between $\overline{P_1P_2}$ and the

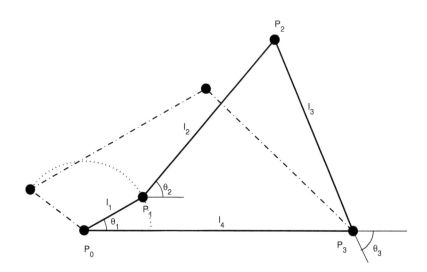

FIGURE 7.12: The four-link mechanism

horizontal be θ_2, and that between $\overline{P_2 P_3}$ and the horizontal be θ_3.

We are interested in the position, velocity, and acceleration of the bar $\overline{P_2 P_3}$ when $\overline{P_0 P_1}$ is rotating at constant angular velocity ω_1.

7.9.2   Displacement analysis of the four-link mechanism

In this subsection we are going to find the position of the bar $\overline{P_2 P_3}$ for a given position of the bar $\overline{P_0 P_1}$. In other words, the input of our problem is the angle θ_1, and the output, the angle θ_3. To complete the solution and draw the instant configuration we are also going to find the angle θ_2.

The graphical solution of this problem is simple and immediate. As shown in Figure 7.13, for a given position of the bar $\overline{P_0 P_1}$, that is of the point P_1, it is sufficient to draw a circle with the centre in P_1 and the radius ℓ_2, then a circle with the centre in P_3 and the radius ℓ_3, and find the point P_2 as the intersection of the two circles. There are two intersections, Figure 7.13 shows only one.

As in many engineering problems, the graphical solution is extremely simple; however, it cannot be continued to obtain more information, such as a continuous representation of velocities and accelerations. Therefore, we must revert to analytical solutions that require not-always simple calculations and some mathematical 'tricks'. A fruitful approach is to consider the bars as vectors represented by complex numbers (see, for example, Soni, 1974). In our case the vectors are

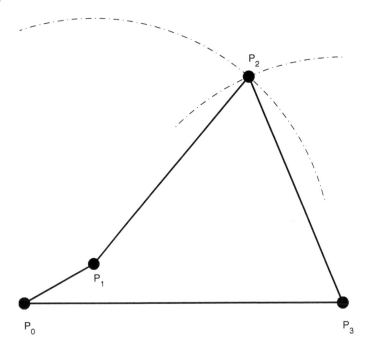

FIGURE 7.13: The four-link mechanism — two solutions

$$\overline{P_0P_1} = \mathbf{V_1}, \ \overline{P_1P_2} = \mathbf{V_2}, \ \overline{P_2P_3} = \mathbf{V_3}, \ \overline{P_0P_3} = \mathbf{V_4}$$

From Figure 7.12 we obtain

$$\mathbf{V_1} + \mathbf{V_2} + \mathbf{V_3} = \mathbf{V_4} \tag{7.24}$$

We write the above vectors as complex numbers

$$\ell_1 e^{i\theta_1} + \ell_2 e^{i\theta_2} + \ell_3 e^{i\theta_3} = \ell_4 e^{i\theta_4} \tag{7.25}$$

Using Euler's relationship, $e^{i\theta} = \cos\theta + i\sin\theta$, and knowing that $\theta_4 = 0$, we separate the real and the imaginary parts of the vectors

$$\ell_1 \cos\theta_1 + \ell_2 \cos\theta_2 + \ell_3 \cos\theta_3 = \ell_4$$
$$\ell_1 \sin\theta_1 + \ell_2 \sin\theta_2 + \ell_3 \sin\theta_3 = 0 \tag{7.26}$$

To establish a relationship between the input angle θ_1 and the output angle θ_3 we eliminate θ_2. To do so, we rearrange Equations 7.26 as

$$\ell_2 \cos\theta_2 = \ell_4 - (\ell_1 \cos\theta_1 + \ell_3 \cos\theta_3)$$
$$\ell_2 \sin\theta_2 = -(\ell_1 \sin\theta_1 + \ell_3 \sin\theta_3) \tag{7.27}$$

Squaring both sides of the two equations and adding them side by side we obtain

$$
\begin{aligned}
\ell_2^2 = \ell_4^2 &- 2\ell_4(\ell_1 \cos\theta_1 + \ell_3 \cos\theta_3) + \ell_1^2 \\
&+ 2\ell_1\ell_3(\cos\theta_1 \cos\theta_3 + \sin\theta_1 \sin\theta_3) + \ell_3^2
\end{aligned}
\tag{7.28}
$$

To obtain an equation in trigonometric functions of θ_3 we let

$$
\begin{aligned}
A &= 2(\ell_1\ell_3 \cos\theta_1 - \ell_3\ell_4) \\
B &= 2\ell_1\ell_3 \sin\theta_1 \\
C &= \ell_1^2 - \ell_2^2 + \ell_3^2 + \ell_4^2 - 2\ell_1\ell_4 \cos\theta_1
\end{aligned}
$$

With these notations we rewrite Equation 7.28 as

$$
A\cos\theta_3 + B\sin\theta_3 + C = 0
$$

or

$$
A\cos\theta_3 = -(B\sin\theta_3 + C)
$$

Squaring both sides we finally obtain

$$
(A^2 + B^2)\sin^2\theta_3 + 2BC\sin\theta_3 + (C^2 - A^2) = 0
\tag{7.29}
$$

The solution of this equation is

$$
\sin\theta_3 = \frac{-BC \pm A\sqrt{A^2 + B^2 - C^2}}{A^2 + B^2}
\tag{7.30}
$$

As explained above, there are, indeed, two solutions. The MATLAB implementation of the above analysis is shown in the next subsection.

7.9.3   A MATLAB function that simulates the motion of the four-link mechanism

In this subsection we develop a function that implements the calculations explained in the previous subsection. We follow a *top-down approach* in which we start from a first specification and refine it in steps. Let the first-level specification be

Input: the lengths of the four bars, ℓ_1, ℓ_2, ℓ_3, ℓ_4.

Function body

Output: successive plots of the mechanism configuration for θ_1 values between 0 and 2π radians.

In the second step we structure the body of the function as follows:

Define frame: plot the bar $\overline{P_0P_3}$, and the articulations P_0 and P_1, and define the size of the figure to accommodate the given bar lengths.

For loop: calculate the coordinates of the points P_1 and P_2; plot these points and the bars $\overline{P_1P_2}$ and $\overline{P_2P_3}$.

We further refine the *For loop*:

```
For 0: rotation step; one full rotation
        If step not the first
                Delete the previous image
        End
        Calculate angle of first bar
        Calculate expressions A, B, C, D
        Calculate position of point P1 and plot the point
        Calculate angle of second bar
        Calculate position of point P2 and plot the point
        Pause to show the instantaneous configuration
    End
```

The full program is shown below. Mind the letter 'l' (el); it is not the number '1'.

```
function    g = FourLink(l1, l2, l3, l4)

%FOURLINK    Simulation of four-link mechanism
%           The input consists of the lengths of the four bars,
%           l1, l2,l3, l4, where l4 connects the two fixed
%           points P0 and P3. Internal variables are the angles
%           t1, t2, t3. As an example call the function with
%           FourLink(1, 3, 3, 4).
%           Developed by Adrian Biran

P0 = [  0; 0 ]; P3 = [ 14; 0 ];
% define frame
Hp = plot([ 0 14 ], [ 0 0 ], 'k-');
set(Hp, 'LineWidth', 1.5)
axis([ -1.1*l1 1.1*(14 + 13) -1.1*l3 1.1*l3 ])
```

```
axis equal, axis off

hold on
point(P0, 0.07)
point(P3, 0.07)
t0  = 0: pi/30: 2*pi; % parameter of articulation circles
r0  = 0.07;              % radius of articulation circles
a1  = 0: pi/60: pi;    % parameter of circular trajectories
x1  = 11*cos(a1);
y1  = 11*sin(a1);
hp = plot(x1, y1, 'r--');
set(hp, 'LineWidth', 1.3)
a3  = pi/2: pi/60: pi;
x3  = P3(1) + 13*cos(a3);
y3  = P3(2) + 13*sin(a3);
hp = plot(x3, y3, 'r--');
set(hp, 'LineWidth', 1.3)

for k = 1:9
    t = (k -1)/8;
    if t > 0
        delete(Hp0)
        delete(Hp1)
        delete(Hp2)
    end
    t1 = pi*t;
    A  = 2*(11*13*cos(t1) - 13*14);
    B  = 2*11*13*sin(t1);
    L  = 11^2 - 12^2 + 13^2 + 14^2;
    C  = L - 2*11*14*cos(t1);
    D  = A^2 + B^2 - C^2 % discriminant of 2nd-degree equation
    t3 = asin((-B*C + A*sqrt(D))/(A^2 + B^2));
    P1 = 11*[ cos(t1); sin(t1) ]
    x = P1(1) + r0*cos(t0);
    y = P1(2) + r0*sin(t0);
    Hp1 = patch(x, y, [ 0 0 0 ]);
    t2 = atan((11*sin(t1) + 13*sin(t3) )/(11*cos(t1)...
        + 13*cos(t3) - 14));
    P2 = P1 + 12*[ cos(t2); sin(t2) ]
    x = P2(1) + r0*cos(t0);
    y = P2(2) + r0*sin(t0);
    Hp2 = patch(x, y, [ 0 0 0 ]);
    Hp0 = plot([ P0(1) P1(1) P2(1) P3(1) ],...
                [ P0(2) P1(2) P2(2) P3(2) ], 'k-' );
    set(Hp0, 'LineWidth', 1.5)
```

```
    pause(0.5)
end
hold off
```

To run the function call it, for example, as

```
>> FourLink(1, 3, 3, 4)
```

Exercise 7.1 Modified function `FourLink`
Modify the function `FourLink` according to the following specification:

1. plot only for the angle values $t = [0, \pi/2, \pi, 3\pi/2]$;

2. do not delete intermediate plots;

3. to distinguish between the different phases use the following sequence of line types and colors:

t	Line type	Colour
0	-	k
$\pi/2$	–	r
π	:	b
$3\pi/2$	-	g

Hint. To answer requirement 3 you must define an array of line-type values and an array of colors. Within the loop 'For k =' you call in each iteration the values identified by the index k.

7.9.4 Animation

In MATLAB there are several possibilities of playing an animation. For example, in the function `FourLink` we inserted the command **pause(0.5)** in a loop and this caused the program to stop for half a second and show a frame. Another possibility, better for display, is to use the pair of commands **getframe** and **movie**. To exemplify this facility, copy the file `FourLink.m` to another file, `FourLinkMovie.m`, and insert the command M(k) = getframe before the end of the loop, and the command movie(M, -10) after the loop. The scheme below shows how to do it.

```
function    g = FourLinkMovie(l1, l2, l3, l4)
...
    pause(0.5)
    M(k) = getframe;
end
hold off
movie(M, -10)
```

Run now `FourLinkMovie(1, 3, 3, 4)`. Change the parameter `-10` and see the effect.

7.9.5   A variant of the function `FourLink`

In the preceding subsection we did not use fully the complex-number facilities of MATLAB. We are going to do so in this subsection. From the formulation of the problem and Figure 7.14 we write

$$\mathbf{V_2} + \mathbf{V_3} = \mathbf{V_4} - \mathbf{V_1} \tag{7.31}$$

Each side of Equation 7.31 is equal to the vector that connects $\mathbf{P_1}$ to $\mathbf{P_3}$. Let this vector be $\mathbf{V_5}$; it is shown in Figure 7.14.

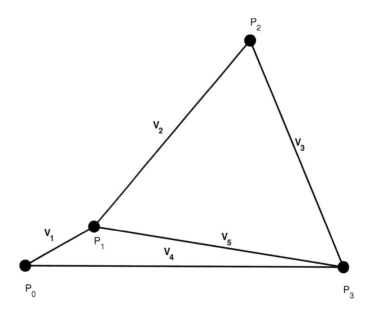

FIGURE 7.14: The four-link mechanism

Applying the cosine formula in the triangle $P_1P_2P_3$, and calling tv_2 the angle opposed to V_2, we write

$$\cos tv_2 = \frac{l_3^2 - l_2^2 - l_5^2}{2l_2l_5} \tag{7.32}$$

A revised programme listing based on the above relationships is shown below.

```
function    g = FourLinkNew(l1, l2, l3, l4)

%FOURLINKNEW  Simulation of four-link mechanism,
%            complex-number notation. The input consists
%            of the lengths of the four bars, l1, l2, l3, l4,
%            where l4 connects the two fixed points P0 and P3.
%            Internal variables are the angles t1, t2, t3,
%            and t5. As an example call the function
%            with FourLinkNew(1, 3, 3, 4).
%            Developed by Adrian Biran

P0 = [  0; 0 ]; P3 = [ l4; 0 ];
% define frame
Hp = plot([ 0 l4 ], [ 0 0 ], 'k-');
set(Hp, 'LineWidth', 1.5)
axis([ -1.3*l1 1.1*(l4 + l3/2) -1.1*l3 1.1*l3 ])
axis equal, axis  off
hold on
point(P0, 0.07)
point(P3, 0.07)
a1  = 0: pi/60: pi; % parameter of circular trajectories
x1  = l1*cos(a1);
y1  = l1*sin(a1);
hp = plot(x1, y1, 'r--');
set(hp, 'LineWidth', 1.3)
a3  = pi/2: pi/60: pi;
x3  = P3(1) + l3*cos(a3);
y3  = P3(2) + l3*sin(a3);
hp = plot(x3, y3, 'r--');
set(hp, 'LineWidth', 1.3)
t0 = 0: pi/30: 2*pi; % parameter of articulation circles
r0 = 0.07;           % radius of articulation circles
   for k = 1:9
       t = (k - 1)/8;
      if t > 0
          delete(Hp0)
          delete(Hp1)
          delete(Hp2)
      end
      t1    = pi*t;
      V1    = l1*exp(i*t1);
      P1    = [ real(V1); imag(V1) ];
      x1    = P1(1) + r0*cos(t0);
      y1    = P1(2) + r0*sin(t0);
      Hp1   = patch(x1, y1, [ 0 0 0 ]);
```

```
V4     = 14*exp(i*0);
V5     = V1 - V4;
15     = abs(V5);
tv2    = acos(13^2 +15^2 - 12^2)/(2*13*15);
t3     = pi - tv2;
V3     = V4 + 13*exp(i*t3);
P2(1)  = real(V3;
P2(2)  = imag(V3);
x2     = P2(1) + r0*cos(t0);
y2     = P2(2) + r0*sin(t0);
Hp2    = patch(x2, y2, [ 0 0 0 ]);
Hp0    = plot([ P0(1) P1(1) P2(1) P3(1) ],...
              [ P0(2) P1(2) P2(2) P3(2) ], 'k-' );
set(Hp0, 'LineWidth', 1.5)
F(k) = getframe;
pause(0.5)
end
hold off
movie(F, -5)
```

Exercise 7.2 Calculate velocities in the four-link mechanism

In this exercise we are going to complete the analysis of the four-link mechanism by adding an analysis of velocities. To do so we differentiate Equation 7.25 with respect to time and obtain

$$i\ell_1 \frac{d\theta_1}{dt} e^{i\theta_1} + i\ell_2 \frac{d\theta_2}{dt} e^{i\theta_2} + i\ell_3 \frac{d\theta_3}{dt} e^{i\theta_3} = i\ell_4 \frac{d\theta_4}{dt} e^{i\theta_4} \qquad (7.33)$$

Now, $d\theta_1/dt$ is the angular velocity of the bar $\overline{P_0 P_1}$ around the point P_0; we note it by ω_1. With similar considerations for the other derivatives we write

$$\frac{d\theta_1}{dt} = \omega_1, \ \frac{d\theta_2}{dt} = \omega_2, \ \frac{d\theta_3}{dt} = \omega_3, \ \frac{d\theta_4}{dt} = 0$$

as the bar $\overline{P_0 P_4}$ is fixed. The vector

$$i\ell_1 \frac{d\theta_1}{dt} e^{i\theta_1}$$

represents the linear velocity of the point P_1 around the point P_0. The factor i shows that this vector is perpendicular to the bar $P_0 P_1$. We can say similar things about the other velocities. Substituting the values of the angular velocities we write

$$(\ell_1 \omega_1) e^{i\theta_1} + (\ell_2 \omega_2) e^{i\theta_2} + (\ell_3 \omega_1) e^{i\theta_3} = 0 \qquad (7.34)$$

Separating the real and the imaginary parts we obtain a system of two equations linear in ω_2 and ω_3 :

$$(\ell_2 \cos \theta_2)\omega_2 + (\ell_3 \cos \theta_3)\omega_3 = -(\ell_1 \cos \theta_1)\omega_1 \qquad (7.35)$$
$$(\ell_2 \sin \theta_2)\omega_2 + (\ell_3 \sin \theta_3)\omega_3 = -(\ell_1 \sin \theta_1)\omega_1$$

The solutions of this system are

$$\omega_2 = \frac{\ell_1}{\ell_2} \cdot \frac{\sin(\theta_1 - \theta_3)}{\sin(\theta_3 - \theta_2)}\omega_1$$
$$\omega_3 = \frac{\ell_1}{\ell_3} \cdot \frac{\sin(\theta_2 - \theta_1)}{\sin(\theta_3 - \theta_2)}\omega_1 \qquad (7.36)$$

Your task is to complete the programme described in Subsection 7.9.3 by adding

1. statements that calculate the velocities using Equations 7.36 and store them in the arrays omega2 and omega3;

2. statements that plot the values ω_2/ω_1 and ω_3/ω_1 against θ_1.

The required plots show one nondimensional variable against another nondimensional variable. This is is a desirable representation any time it is possible. It is also meaningful when the input angular speed, ω_1, is constant.

Exercise 7.3 Calculate accelerations in the four-link mechanism
We assume that you have done the previous exercise. In this exercise we are going to complete the analysis of the four-link mechanism by adding an analysis of accelerations. To do so we differentiate Equation 7.34 with respect to time. Your task is to complete the programme obtained in the previous exercise by adding

1. statements that calculate the accelerations and store them in the arrays a2 and a3;

2. statements that plot the values of the accelerations against θ_1. Assume ω_1 constant.

7.10 Summary

We first met complex numbers in Chapter 3, in the solution of algebraic equations. In this chapter we show that in MATLAB one can deal with complex numbers as simply as with real numbers. In MATLAB the imaginary unit, $\sqrt{-1}$, can be represented by either the letter 'i', or the letter 'j'. The latter is preferred in electrical engineering where 'i' is usually reserved for current intensity. These built-in definitions of MATLAB can be overruled by assigning other values to the letters 'i' and 'j'. Therefore, care must be exercised when dealing with such definitions.

Complex numbers can be represented in

algebraic form as $z = a + ib$;

trigonometric form as $\rho \cos\theta + i\rho \sin\theta$;

exponential form as $\rho e^{i\theta}$.

Considering the algebraic form, a is the *real part* of the number, noted as $\Re z$, and b, the imaginary part, noted as $\Im z$. In the geometric or exponential form ρ is the *absolute value*, or *modulus*, and *theta*, the *argument*, or *phase* of the number. MATLAB provides commands for retrieving the values corresponding to theses notions; they are listed at the end of this section.

Complex numbers can also be represented as points or vectors in a plane in which the cartesian coordinates are the real and the imaginary parts of the numbers. The resulting plane is alternatively called the *complex plane*, *Argand plane*, or, in German literature, the *Gauss plane*. A plot of complex numbers represented by vectors can be obtained with the command `compass`, other plots by separating the real and the imaginary parts and using the usual `plot` command.

Examples of elementary operations include addition, subtraction, multiplication and division.

Given a set of complex numbers, Z, and another set of complex numbers, W, such that to each member of Z corresponds one or more members of W, we say that $W = f(Z)$ is a function of Z. Examples of functions of complex numbers are the exponential, the logarithm, the sine and the cosine.

A periodic function can be described by complex numbers represented as vectors. Important examples are

$$\rho \sin(\omega t + \phi) = \rho \Im(e^{i(\omega t + \phi)}) = \rho \Im(e^{i\omega t} e^{i\phi})$$
$$\rho \cos(\omega t + \phi) = \rho \Re(e^{i(\omega t + \phi)}) = \rho \Re(e^{i\omega t} e^{i\phi})$$

The quantity $\rho e^{i\phi}$ is called *phasor* and can be regarded as a vector rotating with the angular speed ω.

The commands introduced in this chapter include:

abs - given a complex number, Z, `abs(Z)` returns the modulus, or magnitude $\sqrt{(\Re z)^2 + (\Im Z)^2)}$. For a real number, x, this command returns its absolute value, $|x|$.

angle - given a complex number, Z, the command `angle(Z)` returns its phase angle in radians.

compass - plots vectors emanating from the origin of the plot.

complex - given two arrays, a and b, `complex(a, b)` generates the array of complex numbers $a + ib$.

conj - given the complex number $z = a + ib$, the command `conj(z)` yields the *conjugate of z*, $\overline{z} = a - ib$.

getframe - `M(k)= getframe` grabs screen for animation and stores it in M.

imag - given the complex number $z = a + ib$, the command `imag(z)` yields the *imaginary part b*.

\Im - command that prints $\Im$, the symbol for the imaginary part of a complex number

linspace - `x = linspace(a, b)` generates a vector of 100 numbers equally spaced between and including a and b.

movie - `movie(M, n)` plays n times the animation of frames stored in M.

\Re - command that prints $\Re$, the symbol for the real part of an imaginary number.

real - given the complex number $z = a + ib$, the command `real(z)` yields the *real part a*.

repmat - the command `B = repmat(A, m, n)` generates an array in which the array A is repeated in m rows and n columns.

7.11   Exercises

Exercise 7.4 Euler's formula
  The famous formula of Euler is

$$\cos\theta + i \cdot \sin\theta = e^{i\theta} \tag{7.37}$$

Check this formula for $\theta = 0$, $\pi/4$, $\pi/2$.

Exercise 7.5 Another Euler formula
  Check in MATLAB

$$e^{\pi i} + 1 = 0$$

Exercise 7.6 Moivre's formula
  Moivre's formula states that

$$(\cos\theta + i\cdot\sin\theta)^n = \cos n\theta + i\cdot n\theta \tag{7.38}$$

Check this formula for $\theta = 30^0$ and 60°.

Exercise 7.7 Trigonometric functions
  Let $z1 = 1 + i; z2 = 2 + 3i$. Show that the following equations hold within the numerical precision of the computer;

$$\cos(z1 + z2) = \cos(z1)\cos(z2) - \sin(z1)\sin(z2)$$
$$\sin(z1 + z2) = \sin(z1)\cos(z2) - \cos(z1)\sin(z2)$$

Exercise 7.8 Division of complex numbers
  Given the two complex numbers

$$z_1 = 1.9319 + 0.5176i,\ z_2 = 2.7716 + 1.1481$$

calculate $z_3 = z_1/z_24$ and explain why the modulus of the result equals $2/3$ and the argument is approximately $15^\circ - 22.5^\circ$.

Exercise 7.9 Multiplication by i
  Consider the complex number that has the module 2 and the argument $\pi/4$. Multiply this number by i, $i^2; i^3, i^4$. Using the command **compass** plot the vectors that represent the given number and the results of the required multiplications.

Exercise 7.10 Conjugates of complex numbers
  Define in MATLAB the two complex numbers $z_1 = 3 + 4i$, $z_2 = 6 + 8i$, calculate their conjugates, $\overline{z_1}$, $\overline{z_2}$, and verify the following proprieties:

$$z_1\overline{z_1} = 25 = |z_1|^2$$
$$\Re(z_1) = \frac{z_1 + \overline{z_1}}{2}$$
$$\Im(z_1) = \frac{z_1 - \overline{z_1}}{2i}$$
$$\overline{z_1 + z_2} = \overline{z_1} + \overline{z_2}$$
$$\overline{z_1 z_2} = \overline{z_1}\cdot\overline{z_2}$$
$$\overline{\left(\frac{z_1}{z_2}\right)} = \frac{\overline{z_1}}{\overline{z_2}}$$

Exercise 7.11 Series RLC circuit

We refer to the script file `RLCseries` developed in Subsection 7.8.3. The source voltage and frequency suit values used in the U.S. Run this file with the values common in Europe, that is effective voltage 220 V, frequency 50 Hz.

8

Numerical integration

8.1 Introduction

Many engineering problems require the calculation of a definite integral of a function bounded in the finite interval $[a, b]$

$$\int_a^b f(x)dx$$

Often it is difficult, or even downright impossible, to find the answer analytically. This can happen for several reasons:

- the indefinite integral of the given function is known, but its expression is very complex;

- the definite integral cannot be expressed in terms of elementary functions;

- it is not known if there exists an *antiderivative* of the given integrand;

- no analytical expression of the integrand is known, but only a table of $x, f(x)$ pairs.

A list of functions belonging to the second case can be found, for example, in Wikipedia, at the entry *List of integrals*; it includes some famous examples, such as the *Gaussian integral*. The fourth case is frequent in certain engineering fields, such as naval architecture or topography. In all cases we can approximate the definite integral by the weighted sum of a number of function values, $f(x_1)$, $f(x_2)$, ... $f(x_n)$, measured or evaluated at n points in the given interval, that is

$$\int_a^b f(x)dx \simeq \sum_{i=1}^n \alpha_i f(x_i)$$

In this chapter we describe two methods of integration that can be directly applied to data given in tabular form: the *trapezoidal* and *Simpson's rules*. These methods can also be applied to functions given analytically. In this case it is necessary to use the analytical expression to calculate a table of

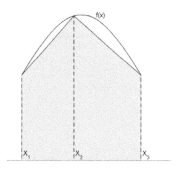

FIGURE 8.1: The trapezoidal rule

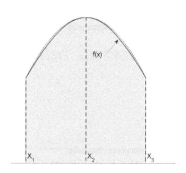

FIGURE 8.2: Simpson's Rule

function values spanning the given integral of integration, and next apply the integration rules to those values. For the trapezoidal rule we use the MATLAB built-in function **trapz**; for Simpson's rule we reuse, with minor modifications, a function written for a previous book of the authors. MATLAB provides several functions for calculating numerically the definite integrals of functions described by analytical expressions. We introduce in this book only one, **quadl**. We conclude this chapter by giving a few examples of *symbolic integration* that can be carried on with the help of the *Symbolic Math Toolbox*.

8.2 The trapezoidal rule

8.2.1 The formula

In Figure 8.1 we consider a function $f(x)$ whose values are known, or can be evaluated or measured at the points x_1, x_2, x_3. The integral of the function between the limits x_1, x_3 represents the area under the curve in this interval. We can approximate this area as the sum of two trapezes

$$\int_{x_1}^{x_2} f(x)dx \simeq (x_2 - x_1)\frac{f(x_1) + f(x_2)}{2} + (x_3 - x_2)\frac{f(x_2 + f(x_3)}{2} \qquad (8.1)$$

The area corresponding to this approximation is shown in grey in Figure 8.1 and the error of this approximation corresponds to the white surfaces between the trapezes and the given curve, $f(x)$.

Usually the available data consists in more values, $f(x_1)$, $f(x_2)$, ... $f(x_n)$, evaluated at the points x_1, x_2, ... x_n. The trapezoidal rule can be simplified if the points are equally spaced, that is $x_2 - x_1 = x_3 - x_2 = ... x_n - x_{n-1} = \delta x$

$$\int_{x_1}^{x_2} f(x)dx \simeq \delta x \left(\frac{1}{2}x_1 + x_2 + \ldots x_{n-1} + \frac{1}{2}x_n \right)$$

It can be shown that the upper boundary of the error of integration is proportional to δx^2. This means that halving the interval of integration reduces the error approximately to one quarter.

8.2.2 The MATLAB trapz function

To give an example, let us return to the kinematic viscosity of fresh water, as shown in Table 4.1, Subsection 4.4.1. In that subsection we stored the data in the file kvisc.m. Let us calculate the mean viscosity in the interval $0°$ to $28°C$. Noting temperature by t, and kinematic viscosity by ν, the exact definition of the mean value is

$$\nu_m = \frac{\int_0^{28} \nu dT}{28 - 0} \tag{8.2}$$

We do not have a function that relates ν to T, but only data in tabular form. Therefore, we cannot carry on the integration analytically; we can do this only numerically. Below are the calculations in MATLAB. We first call the file kvisc.m to import the data into our workspace. Afterward we separate the variables. For an indication we calculate the arithmetic mean using the MATLAB function **mean**. Finally, we call the MATLAB function **trapz** with two input arguments, the array of independent-variable values, T and the array of the dependent-variable values, visc.

```
≫ kvisc
≫ T = nu(:, 1);
≫ visc = nu(:, 2);
≫ mean(visc)
ans =
    1.2209
≫ nu_m = trapz(T, visc)/(T(end) - T(1))
nu_m =
    1.2176
```

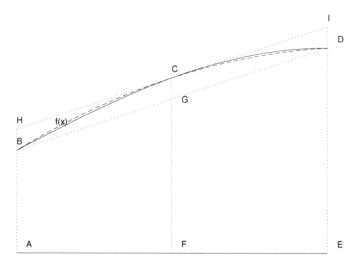

FIGURE 8.3: To Simpson's Rule

8.3 Simpson's rule

8.3.1 The formula

In Figure 8.2 we consider again a function $f(x)$ whose values are known or can be evaluated or measured at the points x_1, x_2, x_3. The integral of the function between the limits x_1, x_3 represents the area under the curve in this interval. This time we want to approximate this area as that under a parabola of the form

$$y = ax^2 + bx + c$$

This equation has three coefficients. To define them we need three data points. Moreover, the derivation of Simpson's law is based on three equally-spaced points. Therefore, while the trapezoidal rule imposed no restriction on the given data, Simpson's rule imposes one: the data should be given as pairs of values evaluated at equally-spaced points.

In Figure 8.3 the solid line passing through the points B, C and D represents the integrand $f(x)$. We want to calculate the integral of $f(x)$ between $x = A$ and $x = E$, that is, the area $ABCDEFA$. The approximating parabola is represented by a dashed line in Figure ??. We need three points to define this curve; therefore, in addition to the values of $f(x)$ calculated at the two extremities, that is, at the points B and D, we shall also evaluate $f(x)$ at the half-interval, obtaining the point C. Let

$$\overline{AB} = f(x_1), \ \overline{FC} = f(x_2), \ \overline{ED} = f(x_3)$$

and

$$h = \overline{AE}/2 = (x_3 - x_1)/2$$

We decompose the total area under $f(x)$ into two partial areas:

1. the trapezoid $ABDEA$;

2. the parabolic segment $BCDGB$.

The first area equals

$$\overline{AE} \cdot \frac{\overline{AB} + \overline{ED}}{2} = 2h \cdot \frac{f(x_1) + f(x_3)}{2}$$

For the second area we use a result from geometry which says that the area of a parabolic segment equals two thirds of the area of the circumscribed parallelogram. Correspondingly we calculate the second area as $2/3$ of the circumscribed parallelogram $BHID$, that is

$$\frac{2}{3} \cdot \overline{AE} \cdot \overline{CG} = \frac{2}{3} \cdot 2h \left(f(x_2) - \frac{f(x_1) + f(x_3)}{2} \right)$$

Adding the two partial sums yields

$$\int_{x_0}^{x_2} f(x)dx \approx \frac{h}{3} [f(x_1) + 4f(x_2) + f(x_3)] \tag{8.3}$$

which is the elementary form of **Simpson's rule.**

Usually we have to integrate the function $f(x)$ over a larger interval $[a, b]$. Then we achieve a better approximation by dividing the given interval into more subintervals. From the way we derived Equation 8.3 we see that the number of subintervals must be even, say $n = 2k$, where k is a natural number. Let

$$h = \frac{a - b}{n} = x_2 - x_1 = x_3 - x_2 = \cdots = x_{n+1} - x_n$$

Applying Equation 8.3 for each pair of subintervals, and adding all partial sums, we get

$$\int_{x_1}^{x_{n+1}} f(x)\, dx = \frac{h}{3}[f(x_1) + 4f(x_2) + 2f(x_3) +$$

$$4f(x_4) + \cdots + 4f(x_n) + f(x_{n+1})] \tag{8.4}$$

which is the extended form of Simpson's rule, for equal subintervals. This form was very helpful when calculations were carried out manually.

It can be shown that the upper boundary of an integral evaluated with Simpson's rule is proportional to the fourth power of the interval of integration, *deltax*. This means that if we halve the interval, the upper boundary is reduced in the $1/16$ ratio.

8.3.2   A function that implements Simpson's rule

Below we reproduce, with the kind permission of Pearson Education U.K., a function that we wrote for Biran and Breiner (1995). We made a few changes in the first comments and changed error messages to error dialog boxes.

```
function z = simp(x, y)

%SIMP Simpson integration of tabular data y(x).
%    Z = simp(x, y) integrates y with respect to x
%    using Simpson's first rule.
%  Input:
%        x and y, column vectors of the same size, or
%        x is a column vector of length n
%        and y a matrix of size n-by-m. In the latter
%        case Z is a row vector containing the integrals
%        of each column of y.  If x or y are row vectors
%        they are converted to column vectors.
%        The vector x should contain pairs of
%        equally spaced values.
%    Z = simp(y) integrates y assuming that all x-intervals
%    have the same value. The user must multiply
%    the result by this constant interval.
%    Written for Biran and Breiner (1995) and reproduced,
%    with a few changes, by kind permission of
%    Pearson Education U.K.

if nargin < 2, y = x; end        % only one input argument
[ n, m ] = size(y);
if n == 1, y = y(:); n = m; end % convert to column vector
x = x(:);
if length(x) ~= n
    errordlg('Input arguments are not of the same length'...
             'Input error')
end
n = length(x);
if rem(n, 2) == 0
    errordlg('Uneven number of intervals', 'Input error')
end
c        = [ 1 4 1 ];        % Simpson multipliers
dx      = diff(x) ;          % intervals of integration
z        = 0;                % initialize output value
flag =   true;
for i = 1: 2: (n - 2)   % loop over pairs of intervals
    if abs(dx(i) - dx(i + 1)) > 0.0001
            flag = false;   % intervals not equal
```

```
            break
      else
            z = z + c*[ y(i, :); y(i + 1, :); y(i + 2, :) ]*dx(i);
      end
end
if flag == false
      errordlg('Two sequential intervals not equal',...
                  'Input error')
else
      z = z/3;
end
```

To check whether the input x consists in pairs of equally- spaced points, we use the command rem. Given two numbers, a and b, rem(a, b) returns the remainder of the division of a by b. Thus, if the number, n, of pairs x, y in the input of the function simp is even, then rem(n, 2) = 0, and the number of $x_{i+1} - x_i$ intervals is odd. We invite the reader to write the function on a file called simp.m. To give an example of application we return to the kinematic viscosity data used in Subsection 8.2.2. We calculate again the mean value of the kinematic viscosity

```
≫ simp( T, visc)/(T(end) - T(1))
ans =
   1.2170
```

The result is slightly less than that obtained by the trapezoidal rule. According to theory it is also closer to the exact value.

8.4 The MATLAB quadl function

MATLAB provides several *quadrature functions* that can be used for the numerical integration of functions defined analytically. In this section we are going to illustrate one of the them, the quadl functions. The integrand will be defined as an *anonymous function*, a possibility of creating functions at the command line. It is a good solution when we have to use the function several times during a single session. On the other hand, it works only with functions that can be defined in one line. The syntax is handlename = @ (variable) expression. It is not necessary to define the expression as a string of characters. The following example illustrates the syntax to be employed for defining the function and calling it.

```
>> hp = @ (x) -3*x.^2 + 3
hp =
   @(x)-3*x.^2+3
area = quadl(hp, -1, 1)
area =
   4.0000
fplot(hp, [ -1, 1 ])
```

The command `fplot(hp, [ -1 1 ]` plots the function whose handle is `hp`, in the interval $[-1\ 1]$.

As said, it makes sense to define anonymous functions when we need them only for the current computer session. What happens, however, if the user has to close suddenly the session and wants to keep the definition to resume the work in another session? It is possible to save the function handle as a `mat` file and load it in the next session. Continuing the above example, write `save hparabola.mat` at the command line and exit MATLAB. Open again MATLAB, in the same directory you worked previously, and type `load hparabola` followed by `hp`.

EXAMPLE 8.1 Mean value of kinematic viscosity by `quadl`

In Subsection 8.2.2 we calculated the mean value of kinematic viscosity using the trapezoidal rule, and in Subsection 8.3.2 using Simpson's rule. The input of those calculations was tabular data. If we have an equation approximating the given data, we can use the MATLAB `quadl` function. In Subsection 4.4.1 we found such an equation by fitting a cubic curve. Of course, we can use that equation, but prefer to perform here a new regression and use the full precision of the resulting coefficients. In other words, we get more decimal digits than by copying the equation shown in Figure 4.4. In the calculations detailed below we

1. import the data by writing the name of the file containing them;

2. separate the variables;

3. fit a cubic polynomial;

4. write an anonymous function that uses the polynomial coefficients found in the preceding step;

5. calculate the mean by calling the `quadl` function.

```
≫ kvisc;
≫ T = nu(:, 1);
≫ visc = nu(:, 2);
≫ C = polyfit(T, visc, 3);
≫ h = @(T) C(1)*T.^3 + C(2)*T.^2 + C(3)*T + C(4);
≫ Vmean = quadl(h, 0, 28)/(T(end) - T(1))
Vmean =
   1.2175
```

8.5 Symbolic calculation of integrals

The *Symbolic Math Toolbox®* of The Mathworks provides tools for analytic integration. As not all users have this toolbox, we do not treat the subject extensively, but give in this section only a few simple examples. We begin with

$$\int_0^\pi \sin\alpha \, d\alpha = |-\cos\alpha|_0^\pi = 2$$

If the toolbox is installed, it is sufficient to declare the *symbolic* variables we are going to use. This is done by writing the term **syms** followed by the list of variables. In continuation we define the function, call the function **int** with the name of the function to obtain the indefinite integral, and call **int** with the function name and the limits of integration to obtain the definite integral.

```
≫ syms alpha x y
≫ y = sin(alpha);
≫ Ii = int(y)
Ii =
-cos(alpha)
≫ Id = int(y, 0 , pi)
Id =
2
```

As a second example we consider the integral

$$\int_0^1 \frac{dx}{1+x}$$

We have already defined the symbolic variable x; therefore, we can go directly to the definition of the integrand and call **int** to obtain the indefinite and the definite integrals.

```
>> y = 1/(1 + x)
>> Ii = int(y)
Ii =
log(x + 1)
>> Id = int(y, 0 , 1)
Id =
log(2)
>> eval(Id)
ans =
0.6931
```

Within the symbolic toolbox we can perform also analytic differentiation by means of the command `diff`. Thus, applying `diff` to the indefinite integral found above, we recover the integrand.

```
>> diff(Ii)
ans =
1/(x + 1)
```

The *Symbolic Math Toolbox* cannot solve all cases because there are functions the integral of which cannot be expressed in terms of elementary functions, or is unknown. An example is the the perimeter of the ellipse; it cannot be expressed in terms of elementary functions. We can, however, calculate it easily by numerical integration. Thus, given an ellipse with semi-axes a and b, we can write the parametric equations of the ellipse as

$$x = a \cos 2\pi t$$
$$y = b \sin 2\pi t$$

where the *parameter* t runs from 0 to 1. The perimeter of the ellipse is then given by

$$L = \int_0^1 \sqrt{\dot{x}^2 + \dot{y}^2} dt$$

Let us see what happens if we try to integrate symbolically this function. We begin by declaring the symbolic variables. Next we write the parametric equations of the ellipse and call `diff` to differentiate them. We define the integrand I0 and try to integrate it.

```
≫ syms x y t
≫ x = a*cos(2*pi*t);
≫ y = b*sin(2*pi*t);
≫ xdot = diff(x)
xdot =
(-2)*pi*a*sin(2*pi*t)
≫ ydot = diff(y)
ydot =
2*pi*b*cos(2*pi*t)
≫ I0 = sqrt(xdot^2 + ydot^2)
I0 =
2*(pi^2*a^2*sin(2*pi*t)^2 + pi^2*b^2*cos(2*pi*t)^2)^(1/2)
≫ I = int(I0)
Warning:  Explicit integral could not be found.
I =
int(2*(pi^2*a^2*sin(2*pi*t)^2 +
pi^2*b^2*cos(2*pi*t)^2)^(1/2), t)
```

As we see, the software issued a warning message and returned the integrand. This problem is connected to that of *elliptic functions* and MATLAB has a function, `ellipke`, that calculates them.

8.6   Summary

The MATLAB commands introduced in this chapter are

eval - evaluates the function given as argument.

fplot - `fplot(hp, lim)` plots the function whose handle is `hp`, between the limits specified in the array `lim`.

load - calls variables from disk and loads them as workspace variables. Examples of calling syntax: `save filename, save('filename', 'x', 'y', ...)`.

mean - `M = mean(A)` returns the mean values of the elements of `A`.

quadl - `quadl(h,a,b)` approximates the integral of the function `h` from `a` to `b`.

rem - `rem(a, b)` yields the remainder of the division of `a` by `b`. In other words, for $a = mb + r$, `r = rem(a, b)`.

save - saves space variables to disk.

syms - declares variables for constructing symbolic expressions.

trapz - integration of tabular data by the trapezoidal rule. Given an array of values y, `trapz(y)` computes an approximation of the integral of y for unit spacing. To compute the integral for another spacing, multiply the result by the interval of integration. Given a table of data $y = f(x)$, `trapz(x, y)` computes the integral of y with respect to x.

8.7   Exercises

Exercise 8.1 Integral of the cosine function
a) Calculate the integral

$$\int_0^{pi/2} \cos x dx$$

in the following ways:

1. analytically;

2. by the trapezoidal rule, with two intervals of integration, in other words, using cosine values evaluated at 0, $\pi/4$, $\pi/2$;

3. by the trapezoidal rule, with four intervals of integration;

4. by Simpson's rule, with two intervals of integration;

5. by Simpson's rule, with four intervals of integration.

b) Calculate the errors of integration of the four results obtained by numerical methods. Verify that halving the interval by two reduces the error approximately in the ratio 1/4 for the trapezoidal rule, and 1/16 for Simpson's rule.

Exercise 8.2 Ellipse perimeter
As mentioned in Section 8.5, the perimeter of the ellipse cannot be expressed in terms of elementary functions. We can, however, calculate it easily by numerical integration. Thus, given an ellipse with semi-axes a and b, we can write the parametric equations of the ellipse as

$$x = a \cos 2\pi t$$
$$y = b \sin 2\pi t$$

where the *parameter t* runs from 0 to 1. The perimeter of the ellipse is then given by

$$L = \int_0^1 \sqrt{\dot{x}^2 + \dot{y}^2}\, dt$$

Your assignment includes the following tasks:

1. write explicitly the equation that yields the perimeter of the ellipse;

2. write an anonymous function based on the above equation and assuming $a = 4$, $b = 3$. Let the function handle be **he** and use the trigonometric functions that accept arguments in degrees;

3. to check your function run it first for limit cases, such as $a = b = 3$, $a = b = 0$, or $a = 4$, $b = 0$. Assess the plausibility of the results;

4. define `t = 0: 1/90: 1`, evaluate the anonymous function at these points, and calculate the integral by means of `quadl`;

5. calculate the integral by calling `trapz(x, y)`. Why do you get the result 0?

6. calculate the integral by calling `trapz(t, y)`;

7. try to find the ellipse perimeter calling `simp(x, y)`. Why do you receive an error message?

8. calculate the integral calling `simp(t, y)`;

9. calculate the integral using the approximation $L = 2\pi\sqrt{(a^2 + b^2)/2}$.

Exercise 8.3 Gaussian integral

The *error function* is defined as

$$\frac{2}{\sqrt{\pi}} \int_0^x e^{-u^2}\, du$$

This integral has no solution that can be expressed in terms of elementary functions. In other words, it has no antiderivative. Put into another form, there is a solution for one particular case and there are elegant proofs that

$$\int_{-\infty}^{\infty} e^{-u^2}\, du = \sqrt{\pi}$$

See, for example, Wikipedia, the entry *Gaussian integral*. Calculate the integral

$$\int_{-n}^{n} e^{-u^2}\, du = \sqrt{\pi}$$

using the functions `quadl`, `trapz` and `simp`. Compare the results with $\sqrt{\pi}$ and see how your results approach the true value as you assume larger $n-$values.

Exercise 8.4 Symbolic integration of the error function

Let us return to the *error function* defined as

$$\frac{2}{\sqrt{\pi}} \int_0^x e^{-u^2} du$$

Try to integrate it symbolically. The answer contains `erf`, a MATLAB built-in function.

9

Ordinary differential equations

9.1 Introduction

Many physical models can be described by an ordinary differential equation, shortly ODE. In this chapter we show how to analyze in this way a mass-spring-damper system. The corresponding equation is

$$m\ddot{\ell} + c\dot{\ell} + h\ell = F \tag{9.1}$$

where ℓ represents the displacement of the mass, c, the damping, and h, the spring constant. By F we indicate the sum of all the external forces that act on the mass. For more explanations on such a system see Subsection 11.5. We assume the following constants:

$$m = 2 \text{ kg}, \ c = 1.4 \text{ N} \cdot \text{s} \cdot \text{m}^{-1}, \ h = 0.1 \cdot \text{m}^{-1}$$

It is shown how to integrate Equation 9.1 with solvers provided by MATLAB and several methods of passing parameters to the solvers are discussed. A short theoretical treatment of ODEs follows, for those who want to get more insight. At the end, the chapter contains a short treatment of *stiff* differential equations.

9.2 Numerical solution of ordinary differential equations

9.2.1 Cauchy form

A large number of dynamic systems can be represented in a standard mathematical form, called the **Cauchy form**, as a system of n first-order differential equations that looks like

$$\begin{cases} w'_{(1)}(t) = f_{(1)}(t, w) \\ \quad \vdots \\ w'_{(n)}(t) = f_{(n)}(t, w) \end{cases} \tag{9.2}$$

where t is the independent variable, usually the time, $w(t)$ is a vector of n components, and the derivative with respect to the independent variable is indicated by a prime sign as in $w'_{(1)}$. In vector notation, this system can be written as

$$\mathbf{w}'(t) = \mathbf{f}(t, \mathbf{w}) \tag{9.3}$$

This form of describing a system is probably the most common in the simulation community.

The function $\mathbf{f}(t, \mathbf{w})$ is usually contained in a separate MATLAB M-file. The MATLAB ode23 and ode45 functions, which we discuss in the next section, require the model to be described in this form. There are many reasons why this form is so popular. First of all it is very general, as $\mathbf{f}(t, \mathbf{w})$ can be any reasonably smooth function of t and $\mathbf{w}$. There are no requirements on linearity, time invariance, and so on. Secondly, once the model is defined, it can often be submitted in this form to different solving routines, based on very different strategies, such as Runge–Kutta, predictor-corrector, Gear, and so on, which we discuss later in this chapter.

9.3   Numerical solution of ordinary differential equations

Numerical strategies for solving differential equations are often selected according to the underlying physical properties of the problem, and while most of them *do often produce* a solution, it is not uncommon to encounter real-life problems where one routine is faster than another by several orders of magnitude.

We are now going to write in Cauchy form the equation of a spring subjected to a force of $a \sin t$ newton, where $a = 2$ N. Let us choose for states $w_1 = \ell'$ and $w_2 = \ell$. The equation can then be written in Cauchy form as

$$w'_1 = -c/m w_1 - h/m w_2 + a/m \sin(t) \tag{9.4}$$
$$w'_2 = w_1 \tag{9.5}$$

This system has a unique solution for each set of initial conditions. Let us choose to start the simulation at $t_0 = 0$, with initial conditions

$$w_1(0) = 0$$
$$w_2(0) = 0$$

We are going to use the ode23 routine to solve the differential system thus obtained. If you were to write

```
≫ doc ode23
```

you notice, in the list of inputs, the presence of a parameter called `options`
that refers to parameters. We will not worry for the time being about these parameters, which are optional, and will simply omit them. MATLAB will assign
them a reasonable value. The first step necessary to compute the dynamics of
the spring is to write a function file that describes the system, like the following file `springb.m`. We named this file springb.m for 'spring-basic version'.
We plan to modify this function later and will call the variations `springg.m`,
`springp.m`. Such a function is called, in the MATLAB jargon, the **odefile**.
An odefile is a function that, in its simplest form, accepts as inputs a scalar
(most often time; t in the following example), a column vector of states (w
in the following example) and returns a vector (wd in the following example)
that represents the derivative of w at time t. Here is the source of `springb.m`:

```
function wd = springb(t, w)

% This funtion defines the dynamics of a
% mass-spring-damper system  driven
% by an external sinusoidal force.
% Designed by Moshe Breiner for Biran and Breiner (1995).

% assign values to system constants
a   = 2.0;              % N
m   = 2.0;              % kg
c   = 1.4;              % Ns/m
h   = 0.1;              % N/m
om = 1.0;               % rad/s
% allocate space for wd
wd = zeros(size(w));
% compute derivatives
wd(1) = -c/m*w(1) - h/m*w(2) + a/m*sin(om*t);
wd(2) = w(1);
```

To see how the mass oscillates, define the initial and final time and the
initial values of the states:

```
≫ t0 = 0.0; tf = 100; w0 = [0; 0];
```

and run the simulation calling the odefile as an anonymous function:

```
≫ [t,w] = ode23(@springb, [t0, tf], w0)
t =

 0
 0.0195
 0.0391
 0.0684
 ...
100.0000
w =

 0 0
 0.0002 0.0000
 0.0008 0.0000
 0.0023 0.0001
 ...
 -0.8430 -0.0873
```

You can now plot the position, as a function of time, with the command:

```
≫ plot(t,w(:,2), 'k-')
≫ grid on;
≫ xlabel('Time, t, s')
≫ ylabel('Spring elongation, l, m')
```

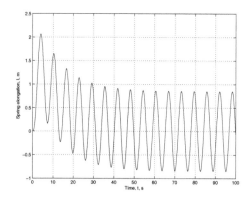

FIGURE 9.1: Simulation of mass-spring-damper system.

The result is shown in Figure 9.1. We are going now to examine some alternative commands to solve the same problem, namely:

- specifying the times of the solution;

- using alternative odesolvers;

- passing parameters to the model.

9.3.1 Specifying the times of the solution

In the previous example, the function **ode23** computed the array of times t, and correspondingly the values of the vector of the states, w. The values of t were produced by MATLAB in a way that makes the computation most efficient. Often we want to be able to specify the vector t, for example at constant steps. A possible strategy is to solve the equation as before, and then use some form of interpolation to produce w at the assigned values of time, but it is much better, and simpler, to define the vector *tspan*, for example as in the following line:

```
≫ tspan=0.0:  0.1:  100;
```

and pass it directly to **ode23**. The command to solve the differential equation is, in this case:

```
≫ [t,w] = ode23(@springb, tspan, w0)
t =

 0
 0.1000
 0.2000
 0.3000
 ...
 100.0000
w =
 0 0
 0.0049 0.0002
 0.0190 0.0013
 0.0417 0.0043
 ...
 -0.8430 -0.0873
```

The array **span** is, in our example, an arithmetic sequence, but it can be more generally any monotonic sequence of numbers.

9.3.2   Using alternative odesolvers

Your package of MATLAB contains a number of routines that solve ordinary differential equations. We are not going to list them, as their number increases with each new release. To obtain the list of odesolvers available on your installation click on `Help`. A pull-down menu will open. Choose `Product help`. In the left-hand window click on `MATLAB`, and in the right-hand window that opens choose `Functions` → `By Category` →. Scroll down to `Ordinary Differential Equations`. Each of these routines is, in principle, capable of solving any differential system, but the performance and the precision is different for each of them. Choosing the correct routine is usually not a simple task, and requires a fair knowledge of the underlying mathematical theory, which is probably beyond the interest of the occasional user. If you are interested in going deeper on this subject, please read Section 9.5; otherwise we can suggest only that you experiment with different odesolvers. The syntax of all odesolvers is identical, so if you want to submit the odefile `springb` to a new solver, like `ode45`, the syntax is (assuming `springb`, `tspan` and `w0` defined from the previous section):

```
>> [t,w] = ode45(@springb, tspan, w0)
```

9.3.3   Passing parameters to the model

In this section we want to consider a very common situation. We have the odefile `springb.m`, but we would like to examine how the solution changes if we modify one or more parameters that define the system, like the damping coefficient, c, or the parameter *om* that changes the frequency of the external force. In the following example, for the sake of simplicity, we are going to vary only one parameter, c. We have three possibilities, which we are going to examine in the following example.

Example

Solve the differential equation of the mass-spring-damper system described in the file `springb.m`, with the new values $c = 2.5\,\mathrm{N\,s\,m^{-1}}$.

Solution 1

The first solution is to edit the file `springb.m`. Change the line

```
    c = 1.4;                    % Ns/m
```

to the new line

```
    c = 2.5;                    % Ns/m
```

Call the new file `springb1.m`:

```
    function wd=springb1(t,w);
```

```
% This function defines the dynamics of a
% mass-spring-damper system, subject
% to an external sinusoidal force.
% Written by Moshe Breiner for Biran and Breiner (1995).

% --- assign values to the constants that define the system:
a = 2.0;                        % N
m = 2.0;                        % kg
c = 2.5;                        % Ns/m
h = 0.1;                        % N/m
om= 1.0;                        % rad/s
% --- allocate space for wd
wd = zeros(size(w));
% --- compute derivatives
wd(1) = -c/m*w(1) -h/m*w(2) + a/m*sin(om*t);
wd(2) = w(1);
```

You can now compare the original system, where $c = 1.4$, with the new system, where $c = 2.5$, with the commands:

```
≫ tspan=[0,100]; w0=[0; 0];
≫ % solve initial system
≫ ode23(@springb1, tspan, w0);
≫ % compare results
≫ [ t1 w1 ] = ode23(@springb1, tspan, w0)
≫ % compare results
≫ plot(t, w(:, 2), 'k-', t1, w1(:, 2), 'k--'), grid
≫ xlabel('Time, t, s')
≫ ylabel('Spring elongation, l, m')
≫ legend('w', 'w1')
```

The resulting plot is shown in Figure 9.2.

While this method works well, it is a little inconvenient because it requires the use of the editor for each new value of the parameter you want to test.

Solution 2

We can define the varying parameter as **global**. The second solution is to change in **springb.m** the line

```
c = 1.4;                        % Ns/m
```

to the new line

```
global c                        % Ns/m
```

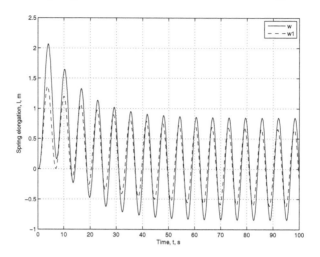

FIGURE 9.2: Plot produced by `springb1`.

Call the new file `springg.m`, for 'spring with global':

```
function wd=springg(t,w);
% This function defines the dynamics of a
% mass-spring-damper system, subject
% to an external sinusoidal force.
% Written by Moshe Breiner for Biran and Breiner (1995).

% --- assign values to the constants that define the system:
a = 2.0;                     % N
m = 2.0;                     % kg
global c;                    % Ns/m
h = 0.1;                     % N/m
om= 1.0;                     % rad/s
% --- allocate space for wd
wd = zeros(size(w));
% --- compute derivatives
wd(1) = -c/m*w(1) -h/m*w(2) + a/m*sin(om*t);
wd(2) = w(1);
```

You can now compare the original system, where $c = 1.4$, with the new system, where $c = 2.5$, with the commands:

```
t≫ span = [0,100]; w0=[0;0];
t≫ global c
t≫ % solve original system
t≫ c = 1.4; [t,w] = ode23('springg',tspan,w0);
t≫ % solve modified system
t≫ c = 2.5; [t1,w1] = ode23('springg',tspan,w0);
t≫ % compare results
t≫ plot(t,w(:,2),t1,w1(:,2)); grid
```

This solution does not require that we edit the file for each new value of the parameter, but we still need to define a global parameter, in our case c. Generally speaking, defining global parameters is a potential source of nasty bugs. Imagine what happens if some other program defines as global a different variable c! We strongly advise you to avoid this practice, except for very short programs, and employ the solution described below.

Solution 3

We can place the varying parameter in the list of the input parameters of the odefile. This alternative syntax of the odefile uses as third parameter the variable **flag**. We are not going to use this parameter, which was introduced for compatibility with Simulink, but must nevertheless put it in the list of input parameters. From the fourth place on, we are free to put as many optional parameters as we like. We therefore modify the function **springb.m** to the new file **springp.m** (spring with parameters). The changes are in the first line and the line that originally defined c which is now deleted:

```
function wd=springp(t,w,flag, c);
% This function defines the dynamics of a
% mass-spring-damper system, subject
% to an external sinusoidal force.
% Written by Moshe Breiner for Biran and Breiner (1995).

% --- assign values to the constants that define the system:
a = 2.0;                    % N
m = 2.0;                    % kg
% c now appears in the list of input parameters
h = 0.1;                    % N/m
om= 1.0;                    % rad/s
% --- allocate space for wd
wd = zeros(size(w));
% --- compute derivatives
wd(1) = -c/m*w(1) -h/m*w(2) + a/m*sin(om*t);
wd(2) = w(1);
```

The commands to run the simulation for an odefile with parameters are also slightly different:

```
≫ tspan=[0,100]; w0=[0;0];
≫ % solve original system
≫ c=1.4; [t,w]=ode23('springp',tspan,w0,[],c);
≫ % solve modified system
≫ c=2.5; [t1,w1]=ode23('springp',tspan,w0,[],c);
≫ % compare results
≫ plot(t,w(:,2),t1,w1(:,2));grid on
```

Observe the list of input parameters to **ode23**. The fourth place is occupied by a parameter, called **options**, or by a couple of empty brackets, as in the example above. In this case the optional parameters are assigned their default values. The user-defined parameter or parameters, like c in our example, start from the fifth place.

9.4 Alternative strategies to solve ordinary differential equations

In this section we are going to define what it means to solve numerically an ordinary differential equation in Cauchy form, and what tools MATLAB provides to obtain such a solution. We are fully aware that the rest of this chapter is less easy to read than the rest of the book. We have tried to simplify it, by providing most formulae for only single equations, instead of for systems of equations. Generalizing the proofs for systems of equations is just a technicality that contains no new ideas, but makes the notation a lot heavier. With this compromise, we feel more comfortable about inserting this section in a book not written for numerical analysis specialists, particularly for the following reasons:

1. It requires no more mathematical knowledge than the Taylor-McLaurin expansion.

2. It can be skipped without compromising the understanding of the following chapters.

On the other hand, if you feel uneasy, like the authors of this book, about using a routine without understanding it, we encourage you to make the effort to read this chapter to the end.

In exchange, we promise you that:

- your simulations will run much faster;

- you will be able to understand and assign correctly the values of the parameters that define the precision of the results.

We have seen in the previous section that a differential system of equations is expressed in Cauchy form as:

$$w'_{(1)}(t) = f_{(1)}(t, w)$$

$$\vdots$$

$$w'_{(n)}(t) = f_{(n)}(t, w)$$

or, in vector notation,

$$\mathbf{w}' = \mathbf{f}(t, \mathbf{w})$$

From now on, we shall consider for a while only single equations, that is the particular case when $n = 1$, and just state the results for the general case. Solving a differential equation means, by definition, producing a function Φ such that

$$\Phi'(t) = f(t, \Phi(t))$$

It can be proven that, under very modest conditions of regularity on f, for any initial condition $w_0 = w(t_0)$, there exists, in the neighborhood of the point (t_0, w_0), a unique function $\Phi(t)$ that solves the equation and such that $\Phi(t_0) = w_0$. The proof can be found in the literature. We will make our life a bit easier by assuming that f has infinitely many continuous derivatives. The theorem of existence is, unfortunately, not a constructive one. It states that a solution exists, but gives no indication on how to find it. A numerical solution to the Cauchy problem, on the other hand, is a sequence of values $[w_0, w_1, \ldots w_n]$ corresponding to an increasing sequence of time points $[t_0, t_1, \ldots, t_n]$, such that w_i approximates $\Phi(t_i)$. The points t_i are sometimes selected by the user, but it is usually preferable, for the sake of efficiency, to let the routine define them. If one needs the solution at specific points, the standard practice is to let the routine find the solution at the point it selects, and then apply some interpolation formula.

Given the initial value w_0 at time t_0, the basic algorithm must be able to produce, in correspondence of the time $t_1 = t_0 + \Delta t$, the value of $w_1 = w(t_1)$. In the process we are aware of introducing a **local error**, which in most cases decreases if we take shorter steps. This local error has two main components, the first one due to the arithmetic precision of the processor, the other one to the algorithm itself, which substitutes, as we shall see in a moment, a differential equation with an approximate algebraic expression. We can repeat the algorithm to produce, from the new initial condition w_1 corresponding to time t_1, a new point, w_2, which approximates Φ at $t_2 = t_1 + \Delta t$ (either the same Δt, in which case the algorithm is called a **fixed step algorithm**, or possibly a different Δt, in which case the algorithm is called a **variable step algorithm**). In this way we produce, step by step, a numerical solution to the Cauchy problem. But when computing w_2, in addition to the local error, we introduce a new error which was not present at step 1. In fact, the correct initial condition for the interval $[t_1, t_2]$ is $\Phi(t_1)$, which we don't know, and we start instead from the point w_1, which is just an approximation of $\Phi(t_1)$. This **cumulative error** grows, in general, as the number of steps increases.

The conceptually simplest (and probably computationally hardest) way to compute $w(t + \Delta t)$ is by Taylor expansion. Notice that, once we know $\Phi'(t) = f(t, w)$, we potentially know *all* the derivatives of the solution. In fact:

$$\Phi''(t) = \frac{d}{dt} f(t, w)$$

$$= \frac{\partial f(t, w)}{\partial t} + \frac{\partial f(t, w)}{\partial w} \cdot \frac{dw}{dt} = \frac{\partial}{\partial t} f(t, w) + \frac{\partial f(t, w)}{\partial w} f(t, w)$$

and, by taking successive derivatives of f, it is possible, in principle, to compute all the derivatives of Φ. We can then compute $\Phi(t_0 + \Delta t)$ using the first n terms of the Taylor expansion. By doing so, we contribute to the local error with a term proportional to $(\Delta t)^{n+1}$. This term is called **truncation error**, because it is the consequence of truncating the infinite Taylor expansion after a finite number of terms. For example, if we take into account the first two terms of the Taylor expansion of f, we obtain the following formula:

$$w_1 \approx \Phi(t_0 + \Delta t) \approx w_0 + \left[f + \frac{1}{2} \left(\frac{\partial f}{\partial t} + \frac{\partial f}{\partial w} f \right) \Delta t \right]_{(t_0, w_0)} \Delta t$$

with an error proportional to $(\Delta t)^3$. The term that appears in square brackets can be interpreted as the average slope of the solution in the interval $[t_0, t_0 + \Delta t]$, and we are going to refer to it as the **Taylor slope**.

Applying this method requires, in all practical cases, an incredibly long preliminary work to compute the formulae for the successive derivatives but, until the end of the last century, there was no better way to compute the numerical solution of an ordinary differential equation. The rest of this chapter will be devoted to explaining alternative methods to solve numerically a differential equation, namely **Runge–Kutta methods, predictor-corrector methods** and special methods for **stiff equations**.

9.4.1   Runge–Kutta methods

Around the turn of the century Carl Runge (German, 1856–1927) and Wilhelm Kutta (German, 1867–1944) independently developed a number of algorithms based on the observation that, instead of computing $f(t, w)$ and its derivatives at the point t_0, one can compute only the slope, at different points in the interval $[t_0, t_0 + \Delta t]$, and produce an approximation of $\Phi(t_0 + \Delta t)$ comparable to that obtained by Taylor expansion.

For example, by computing $f(t, w)$ for two different values of t and w, one can write a formula for $w(t_0 + \Delta t)$ that differs from the true solution $\Phi(t_0 + \Delta t)$ by no more than a constant times $(\Delta t)^3$, like the approximation obtained by the second-order Taylor expansion. At each step we compute

$$k_1 = f(t_0, w_0) \tag{9.6}$$
$$k_2 = f(t_0 + \alpha \Delta t, \ w_0 + \alpha k_1 \Delta t) \tag{9.7}$$

and the **average slope** for the interval,

$$k = \lambda_1 k_1 + \lambda_2 k_2$$

For analogy with the Taylor slope introduced in the previous section, we shall call this k the **Runge–Kutta slope**. We use the Runge–Kutta slope to produce $w(t_0 + \Delta t)$ as

$$w(t_0 + \Delta t) = w_0 + k \Delta t$$

As long as we choose the parameters, λ_1, λ_2, α, such that

$$\lambda_1 + \lambda_2 = 1$$
$$\alpha \lambda_2 = \frac{1}{2}$$

the Runge–Kutta slope k coincides, up to higher order terms, with the Taylor slope, as one can easily verify by writing the Taylor expansion of k. This method can be generalized to higher orders. For example, it is possible to write a fourth order Runge–Kutta algorithm that requires four calls of $f(t, w)$ at each step, and produces results accurate to a constant times $(\Delta t)^5$, like the Taylor expansion truncated after the fourth term.

Runge–Kutta methods were the most popular for the numerical solution of ordinary differential equations for more than half a century, mainly because they were easy to use with hand calculators first, and to implement on electronic computers later. Their main drawback was the lack of an intrinsic evaluation of their accuracy. The typical method of checking accuracy was to recompute the solution using twice as many points, and accept the solution if the two results were close enough. This implied, in the best case, an overhead of 200%. And, if you were unlucky, that is, if the discrepancy between the two solutions was unacceptable, you could restart the game anew with a smaller Δt.

In the fifties a new family of algorithms was developed, called predictor-correctors (see next section). These algorithms were more efficient than comparatively precise Runge–Kutta methods, and, in particular, by comparing the result of the predictor with the corrector, were able to estimate the truncation error. Eventually, in the seventies, a little modification of the Runge–Kutta method was proposed by Fehlberg (see Fehlberg, 1970). The second-order Runge–Kutta procedure, as we have seen, requires two computations of

$f(t, w)$ per integration step. Fehlberg observed that an extra computation, representing a 50% overhead, allows an estimate to be made of the factor η that appears in the computation of the truncation error formulated as

$$error \approx \eta(\Delta t)^3 \tag{9.8}$$

The situation is even better for the fourth order Runge–Kutta, which requires at each step four calls to $f(t, w)$. In this case a fifth call, representing a mere 20% overhead, produces an estimate of the truncation error. At each step we test the estimated local error and consider the integration step as successful if it is smaller than the value

$$\max(RelTol \cdot \|w\|, AbsTol)$$

where *RelTol* and *AbsTol*, the relative and absolute tolerance parameters, are constants that, as we shall see in a moment, have small default values or can alternatively be defined by the user.

In the case of a system of n equations, *RelTol* and *AbsTol* can be either vectors of length n or scalars. In the latter case that scalar value is common for all states. *RelTol* and *AbsTol* have default values, and we, in fact, used them until now. If you want to set different values, you should use the parameter *options*. This parameter is usually defined with the function **odeset**. For example, if you want to define $RelTol = 0.0001$ and $AbsTol = 0.000001$ (one tenth of their default values), the syntax is:

```
≫ options = odeset('RelTol',1e-4,'AbsTol',1e-5);
```

and this parameter, *options*, is then passed to the odesolver. So, if we want to solve the problem of the spring with these values for *RelTol* and *AbsTol* we should write, instead of the command

```
≫ c=2.5; [t1,w1]=ode23('springp',tspan,w0,[],c);
```

seen at the end of Subsection 9.3.3, a new command:

```
≫ c=2.5; [t1,w1]=ode23('springp',tspan,w0,options,c);
```

Formula 9.8 does not just satisfy our legitimate curiosity for estimating the truncation error; it makes it possible to create very efficient algorithms, by reducing the integration steps when the error is large and increasing them when the error is small.

A last note of caution. The reader should be careful not to assign to *AbsTol* and *RelTol* too small values. This would be a penny-wise, pound-foolish strategy. If the tolerances are too small, MATLAB will be forced to chose too small integration steps over a given interval; this will result in a large number of steps and possibly large errors of a different type (finite arithmetic errors, cumulative errors, etc.) that the algorithm does not take into account.

Methods based on the Runge–Kutta algorithm are ode23, ode45 and other solvers, which are not supplied with the basic package, like ode2, ode4 that come with some editions of Simulink.

9.4.2 Predictor-corrector methods

Let us consider again the problem of finding $w_{k+1} \approx \Phi(t_{k+1})$, but this time assuming that we know the past history of t and w, namely the points $[t_1, t_2, \ldots, t_k]$ and $[w_1, w_2, \ldots, w_k]$. Let us also define $\dot{w}_i = f(t_i, w_i)$, which is approximately equal to $\dot{\Phi}(t_i)$. Then

$$w_{k+1} = w_k + \int_{t_k}^{t_{k+1}} \dot{\Phi}(s)\, ds \qquad (9.9)$$

We can decide to compute the integral using some numerical routine like trapezoidal approximation, Simpson's rule (Section 11.4) or some higher order formula. To illustrate the idea, let us choose the trapezoidal approximation. Then Formula 9.9 becomes

$$w_{k+1} \approx w_k + \frac{\dot{w}_k + \dot{w}_{k+1}}{2}\Delta t = w_k + \frac{\dot{w}_k + f(t_{k+1}, w_{k+1})}{2}(t_{k+1} - t_k) \qquad (9.10)$$

So w_{k+1} can be obtained by solving the equation:

$$w_{k+1} = w_k + \frac{\dot{w}_k + f(t_{k+1}, w_{k+1})}{2}\Delta t \qquad (9.11)$$

There are various methods that find the value of w_{k+1} that solves this implicit equation. The predictor-corrector idea is to substitute the w_{k+1} that appears on the right-hand side of Equation 9.9 with a predicted value $w_{k+1}^{(p)}$ obtained by substituting the integrand in Formula 9.9 with an appropriate polynomial approximation. For example, if we decide to use parabolic approximation, we would use the polynomial of degree 2 that passes through the points $(t_{k-2}, \dot{w}_{k-2})$, $(t_{k-1}, \dot{w}_{k-1})$ and $(t_k, \dot{w}_k)$. Since this is not a book on numerical analysis but a section of a book for non-specialists and our purpose is just to illustrate the idea of the predictor-corrector, we shall just use linear approximation, and our polynomial P will be simply the line that passes through $(t_{k-1}, \dot{w}_{k-1})$ and $(t_k, \dot{w}_k)$, that is the polynomial:

$$P(s) = \dot{w}_k + \frac{\dot{w}_k - \dot{w}_{k-1}}{t_k - t_{k-1}}(s - t_k)$$

Having so defined $P(s)$, $w_{k+1}^{(p)}$ can be expressed as

$$w_{k+1}^{(p)} = w_k + \int_{t_k}^{t_{k+1}} P(s)\,ds = w_k + \int_{t_k}^{t_{k+1}} \left[\dot{w}_k + \frac{\dot{w}_k - \dot{w}_{k-1}}{t_k - t_{k-1}}(s - t_k) \right] ds$$

and finally:

$$w_{k+1}^{(p)} = w_k + \dot{w}_k(t_{k+1} - t_k) + \frac{\dot{w}_k - \dot{w}_{k-1}}{t_k - t_{k-1}} \cdot \frac{(t_{k+1} - t_k)^2}{2}$$

Now we can use this value on the right-hand side of Formula 9.11 and eventually get the new value for w_{k+1} corrected, or simply w_{k+1}, as:

$$w_{k+1} = w_k + \frac{\dot{w}_k + \dot{w}_{k+1}}{2}(t_{k+1} - t_k)$$

$$= w_k + \frac{\dot{w}_k + f(t_{k+1}, w_{k+1}^{(p)})}{2}(t_{k+1} - t_k)$$

Predictor-corrector methods are usually very efficient, because a comparison between the value of the predictor and the value of the corrector leads to a good estimate of the local error introduced at each integration cycle, without requiring an extra evaluation of the odefile as done in the Runge–Kutta–Fehlberg algorithm. As in the latter methods, this estimate can be used to determine dynamically the next integration step. Moreover, at each cycle, we can also change the *order* of the predictor (linear, parabolic, etc.) and the integration formula for the corrector (trapezoidal, Simpson, etc.). The current version of MATLAB comes with a variable-step, variable-order predictor-corrector algorithm, `ode113`.

9.4.3   Stiff systems

In previous sections, when we illustrated the ideas of Runge–Kutta and predictor-corrector odesolvers, we were concerned with the precision of the algorithm but we carefully avoided another minefield, the problem of **stability**. Stability is a qualitative property rather difficult to define. Formal definitions are complicated, and not very illustrative.

Let us agree to use the term **precision** when dealing with an algorithm that produces a solution differing only a little from the true one, and the term **instability**, when the algorithm produces a numerical solution that is very far from the true one. For example, in the latter case the error can grow to infinity or oscillate widely. In other words, we speak of instability in such cases when the error becomes so large that the numerical solution is useless for every practical purpose, as the following example shows.

To illustrate instability we shall develop a simple implementation of the second-order, fixed-step Runge–Kutta algorithm that solves differential equations, not systems. Let us call it **rk2fxeq**:

```
function [t,w]=RK2fxeq(odefun,tspan,w0);

% This is the implementation of the second order
% Runge-Kutta algorithm for functions. User defines:
% - the name of the odefile, F (a scalar function!)
% - the points of integration, tspan
% - the initial state, w0
% The outputs, t (equal to tspan), and the solution, w:
% Calling syntax : [t,w] = RK2fxeq('F',tspan,w0);
% Written by Moshe Breiner for Biran and Breiner (1995).

% --- define parameters of integration routine;
la1 = 1/4;
la2 = 3/4;
al  = 2/3;

l   = length(tspan);
t = zeros(1,1);  % allocate space for times
w = zeros(1,1);  % allocate space for the solution;
% assign initial conditions:
t(1) = tspan(1);
w(1) = w0;
% start cycle
for i= 1:(1-1);
    T = tspan(i);   % initial time of the cycle
    W = w(i);       % initial value of the cycle
    del_t = tspan(i+1)- tspan(i);
    k1 = feval(odefun,T,W);
    k2 = feval(odefun,T+al*del_t, W+al*k1*del_t);
    k  = la1*k1+la2*k2;
    t(i+1) = tspan(i+1);
    w(i+1) = W+k*del_t;
end
```

Lets us now apply this new odesolver to the simplest equation we can think of:

$$w' = -\left(\frac{1}{\tau}\right)w$$
$$w(0) = 1$$

for which we know the algebraic solution

$$\Phi(t) = e^{-t/\tau}$$

Let us assign to τ the value 0.1 second, and write the odefile **deq.m**:

```
function wd = deq(t, w)

%DEQ odefile to run with RK2fxeq
% Written by Moshe Breiner for 'MATLAB for Engineers'
tau = 0.1;        % time constant
wd  = -(1/tau)*w;
```

Let us solve this equation, at fixed, constant steps of 0.1 seconds, and find the numerical solution, using different integration steps. For example, for an appropriate integration step, like $\tau/4$, the numerical solution agrees quite well with the algebraic solution:

```
≫ tau = 0.1;
≫ w0 = 1.0;
≫ step = tau/4;
≫ tspan = 0:  step:  5;
≫ [t,w] = RK2fxeq('deq',tspan,w0); % numerical solution
≫ walg = exp(-t/tau); % algebraic solution
≫ h = plot(t, walg, 'k-', t, w, 'k:');
≫ xlabel('t')
≫ legend(h, 'Analytic solution', 'Numerical solution')
≫ title('Integration step \tau/4')
```

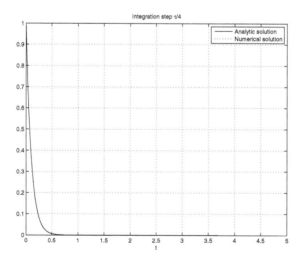

FIGURE 9.3: Small integration steps: the system is stable

The T<small>E</small>X command \tau prints the greek letter τ. The result is shown in Figure 9.3.

The reader is invited to experiment to see what happens when we increase the steps. Gradually the solution becomes less precise and, when the step is larger than 2τ, it becomes unstable:

```
>> step = 2.01*tau;
>> tspan = 0:step:5;
>> [t,w] = RK2fxeq('deq',tspan,w0);
>> walg = exp(-t/tau);
>> plot(t, walg, 'k-', t, w, 'k:'), grid
>> legend(h,'Analytic solution','Numerical solution');
>> xlabel('t, s')
>> title('Integration steps:  2.01\tau')
```

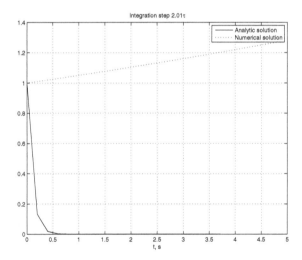

FIGURE 9.4: Large integration steps: the system is unstable

The result is shown in Figure 9.4. At this point, the reader must trust us that this phenomenon is quite typical, not only for differential equations, but for differential systems as well. Each second-order Runge–Kutta algorithm becomes unstable as soon as the integration step exceeds twice the minimum time constant of the system. Higher-order Runge–Kutta algorithms can take slightly larger steps. For example, fourth-order Runge–Kutta algorithms can take steps up to four times the minimum time constant of the system. Similar results apply to predictor-corrector algorithms. Treating this fascinating subject in more depth is beyond the scope of this book and we must refer the interested reader to specialized books. Now we have the background necessary to analyze what happens if we try to use either a Runge–Kutta or a predictor-corrector algorithm to solve a system like:

$$\dot{w}_1 = (a-2)w_1 + 2(a-1)w_2$$
$$\dot{w}_2 = (1-a)w_1 + 2(1-2a)w_2$$

with initial conditions

$$w_1(0) = 1$$
$$w_2(0) = 0$$

The solution, as the reader can easily verify, is

$$w_1 = 2e^{-t} - e^{-at}$$
$$w_2 = -e^{-t} + e^{-at}$$

that is, the solution is the sum of two exponentials, the first one with time constant $\tau_1 = 1$, and the second one with time constant $\tau_2 = 1/a$. As long as τ_1 and τ_2 are of the same order of magnitude, the system can be efficiently solved with an algorithm like ode23. For this purpose we must prepare the odefile

```
function wd = stiff(t,w,flag,a);
A =   [a-2, 2*(a-1);
       1-a, 1-2*a];
wd = A*w;
```

and run the simulation with the commands:

```
≫ tspan = [0,10];
≫ w0 = [1;0];
≫ a = 2; tic; [t,w] = ode23('stiff',tspan,w0,[],a); toc
```

So far so good, the numerical solution is correct (you are invited to compare it with the analytic solution) and the computation time reasonable. Let us now change the value of a from 2 to 2000:

```
≫ a=2000; tic; [t,w] = ode23('stiff',tspan,w0,[],a); toc
```

Now the solution takes much longer; on our computer more than 100 times longer. To add insult to injury, a simple investigation reveals that the term responsible for the extremely large number of integration cycles, and therefore for the length of the computation, is $-e^{-at}$, a component whose contribution to the solution, for such a large value of a, soon becomes insignificant. Such a system is said to be **stiff**. It is hard to give a precise definition of stiffness, since this is a qualitative, not a quantitative property. Stiff subsystems are not always easy to recognize from their mathematical formulation. A useful definition of stiffness should take into account not only the equations that define the system, but also the time frame we want to analyse. Our example is stiff if we want to study its solution in the interval $[0, 5]$ but is not stiff if we want to analyse it in the first few thousands of a second.

Runge–Kutta and predictor–corrector methods are incapable of treating efficiently stiff problems, since the length of the integration steps they can take is bounded by the time constant of the fast subsystem. To solve a stiff system we need an algorithm that is capable of taking integration steps much, much larger than the minimum time constant of the system. The first algorithm of this kind was devised by William Gear (English, born 1935). Its strategy is to solve Equation 9.11 by an implicit method, like the Newton–Raphson method shown in Section 7.5. We saw there that the solution to the equation

$$\phi(x) = 0$$

can often be obtained by starting with a good guess and then solving iteratively the linear equation:

$$0 = \phi(x_i) + \phi'(x_i)(x_{i+1} - x_i)$$

The method maintains its validity also when x is, instead of a scalar, a vector of n components, $[x_1, x_2, \ldots, x_n]$, and ϕ is, instead of a single equation, a system of equations:

$$\phi_1(x) = 0$$
$$\phi_2(x) = 0$$
$$\vdots = \vdots$$
$$\phi_n(x) = 0$$

The role of the derivative, in this case, is played by the **jacobian**, which is defined as the matrix, $\partial\phi/\partial x$, of the (i, j) component:

$$\left(\frac{\partial\phi}{\partial x}\right)_{i,j} = \frac{\partial\phi_i}{\partial x_j}$$

The equation we want to solve is then (replace w_{k+1} simply by x):

$$\phi(x) = -x + w_k + \frac{\dot{w}_k + f(t_{k+1}, x)}{2}(t_{k+1} - t_k)$$

and the jacobian of ϕ is

$$\frac{\partial \phi}{\partial x} = \left[-I_n + \frac{\partial f(t_k, x_k)}{\partial w} \cdot \frac{t_{k+1} - t_k}{2} \right]$$

where I_n denotes the n-by-n identity matrix.

The Newton–Ralphson method directs us to solve the implicit equation by iteratively computing x_{i+1} as the solution of the linear system

$$0 = \phi(x_i) + \frac{\partial \phi}{\partial x}(x_i)(x_{i+1} - x_i) \tag{9.12}$$

The first guess is not a problem. Many stiff routines use the predictor to produce a reasonable first guess, the same one used by the predictor-corrector. Also, the number of times to repeat the iteration varies from implementation to implementation. Solving the linear equation is an operation that today can be performed quite efficiently, so most time is usually spent in the computation of the jacobian. It is true that stiff algorithms allow us to take large integration steps, but if we had to perform at each step the complete computation of finding numerically the jacobian (for instance by numerical perturbation) and solving the linear system described by Equation 9.12, the computation would probably be so slow that we would be better sticking to old faithful Runge–Kutta.

In practice good implementations of this strategy compute the jacobian not at each step, but only when necessary. At each step the algorithm tries to use the previous jacobian, and only if some accuracy test fails does it compute the jacobian anew. MATLAB, in addition, lets you

- inform the solver that the jacobian has special properties, like being vectorized or sparse;

- inform the solver that the system is linear, i.e. that the jacobian remains constant and never needs to be recomputed;

- pass the jacobian to the solver, if you can compute it, for example, in closed form.

For an explanation of this, and other less common properties, we refer the reader to the manual or to the text displayed when you type

```
>> doc odeset
```

Stiff odesolvers in the recent versions of MATLAB are ode15s, ode23s, ode23t and ode23tb.

9.5 Conclusion: How to choose the odesolver

In this section you receive the rewards for having read the previous one. So, how do you choose the odesolver? There is no general rule valid in every case, and what follows is a list of suggestions derived mainly from our experience.

The first and most critical decision is whether to use a stiff solver, or a non-stiff one, like those based on Runge–Kutta and predictor-corrector methods. There is usually little advantage in choosing a stiff odesolver, unless the system is composed of subsystems whose time constants differ by at least two or three orders of magnitude. To find these time constants, it can be useful to compute the eigenvalues of the linearized system, but it is usually your engineering intuition that should suggest to you whether the system described can be broken down into subsystems of different time constants. Answering affirmatively to this question is a necessary but not sufficient condition to opt for a stiff solver. Stiff solvers take large integration steps, at the expense of introducing large errors in the description of the fast subsystem. If you are not interested in taking large steps, or if this error is unacceptable for the kind of analysis you are performing, do not choose a stiff solver.

The second factor you should take into account is that the cost of each full integration step is usually very large, so choose a stiff solver only if you think that there will be no need to recompute the jacobian at each step (this happens if your system is linear or quasi-linear). This condition can certainly be mitigated, if you can and are willing to compute the jacobian and provide it to the solver. If you decide that you are going to use a stiff solver, please go carefully through all the options offered to you by **odeset** and activate those that apply to your case. If you decide not to use a stiff solver, you are left with less critical decisions.

Generally, it can be said that a predictor-corrector is superior to a Runge–Kutta routine in the absence of discontinuous phenomena, i.e. when the polynomial extrapolation represents a good prediction. On the other hand, when the prediction is misleading, the predictor-corrector is usually inferior to a Runge–Kutta solver. Such is the case in the well-known example of simulating a ball bouncing on a table. Here the predicted trajectory of the ball is to penetrate the table. The MATLAB algorithm based on a predictor-corrector is capable of coping with a discontinuity of this kind, but only at the price of a loss of efficiency. Alternatively, you could write some extra code, to help MATLAB treat the discontinuity present when the ball hits the table. Such a phenomenon is called, in MATLAB language, an **event**, but we feel that event handling is a subject beyond the scope of this book, and we refer you to the MATLAB manual. To simulate this or similar systems, unless you want to learn how to describe events efficiently, it is probably better just to use a Runge–Kutta algorithm.

The last advice is about the order of integration. Higher-order algorithms, like ode45, are usually superior to the corresponding lower-order ones, like ode23, when describing smooth systems. Lower-order algorithms are sometimes more efficient, especially in the presence of discontinuities, or when the precision requirements are less stringent.

For more details on ODEs in MATLAB we recommend to the reader a paper written by Shampine and Reichelt (see the Bibliography).

9.6   Exercises

Exercise 9.1 Linear, time-invariant system
Using the ode23 or ode45 function, solve the following differential equation

$$\acute{\mathbf{w}} = \mathbf{A}\mathbf{w}$$

in the interval $[0, \, 4\pi]$, with initial conditions $\mathbf{W_0}$, and plot the results

$$\mathbf{A} = \begin{bmatrix} -0.5 & 1 \\ -1 & -0.5 \end{bmatrix}, \quad \mathbf{w_0} = \begin{bmatrix} 0 \\ 1 \end{bmatrix}$$

Exercise 9.2 Linear, time-varying equations
Using the ode23 or ode45 function, solve the following differential equations with the given initial conditions, in the interval $[t_0, \, t_f]$, and plot the results:

1.

$$\acute{w} + (1.2 + \sin 10t)w = 0, \, t_0 = 0, \, t_f = 5, w(t_0) = 1$$

2.

$$3\acute{w} + \frac{1}{1+t^2}w = \cos t, \, t_f = 5, \, w(t_0) = 1$$

Exercise 9.3 Foxes and rabbits
A well-known ecosystem due to Vito Volterra (Italian, 1860–1940; see, for example, Kahaner, Moler and Nash, 1989 or Giordano and Weir 1991) considers two populations of rabbits and foxes, related by the equations

$$\frac{dr}{dt} = 2r - \alpha r f$$
$$\frac{df}{dt} = -f + \alpha r f$$

where r is the number of rabbits, f, the number of foxes and the time, t, is given in years.

Using ode23, compute and plot the solution of the differential system, for $\alpha = 0.01$ and initial conditions $r_0 = 300$ and $f_0 = 150$, over the time interval $[0, 25]$.

Exercise 9.4 Pendulum

The motion of an ideal pendulum of length l is governed by the equation

$$\theta'' = -gl\sin\theta \qquad (9.13)$$

where θ is the angle of the pendulum from the vertical.

Find θ as a function of t, in the time interval $[0, 10]$, and plot θ and θ' as functions of time. Parameters: $g = 9.81\,\mathrm{m\,s^{-2}}$, $l = 0.5\,\mathrm{m}$. Initial conditions: $\theta(0) = 1\,\mathrm{radian}$, $\theta'(0) = 0\,\mathrm{rad\,s^{-1}}$. Do not be surprised if the graphs of θ and θ' look like sinusoids. The sinusoid is the solution of the linearized model obtained by substituting $\sin\theta$ by θ in Equation 9.13. The linearized model is an excellent one, as long as the angle θ and its derivative remain in the neighbourhood of 0. To see that θ and θ' are not sinusoids, repeat the simulation with a large value of $\theta'(0)$.

Exercise 9.5 Ballistics

A simple ballistic model assumes that only two forces act on the projectile, namely gravity and drag. According to this model, the equations of motion are

$$mv'_x = -Wv_x/v$$
$$mv'_y = -Wv_y/v - mg$$

Here v_x, v_y are the components of the velocity and v its magnitude. The drag W is given by the equation

$$W = c_w \frac{\rho}{2} v^2 \frac{\pi}{4} d^2$$

The symbol c_w is called the **drag coefficient** and is nearly constant as long as the flight is at subsonic speed, ρ is the air density and d, the diameter of the projectile.

(1) Plot the trajectory, using the following parameters: $g = 9.81\,\mathrm{m\,s^{-2}}$, $m = 10\,\mathrm{kg}$, $c_w = 0.2$, $\rho = 1.225\,\mathrm{kg\,m^{-3}}$, $d = 0.050\,\mathrm{m}$. The initial conditions are $x_0 = 0\,\mathrm{m}$, $y_0 = 0\,\mathrm{m}$, $v_0 = 250\,\mathrm{m\,s^{-1}}$ and the gun elevation equals $\pi/6$.

(2) If you used ode23 with the default value of tol, the graph of the trajectory is unpleasant, due to the small number of resulting points. Repeat the problem at (1), but produce a smoother graph by defining the parameter tol to be 5e–4, instead of the default tolerance, which is equal to 1e–3.

10

More graphics

10.1 Introduction

In this chapter we deal with three subjects in advanced graphics:

1. plotting at scale;

2. plotting a conic surface defined analytically, a conical section, the developed conic surface and a conical helix;

3. creating a graphical user interface, shortly a GUI.

CAD software allows the user to draw at standard scales. MATLAB is not designed for engineering graphics; therefore, it usually scales plots so as to better use the computer screen, but does not care for standard scaling. It is possible to extend the usability of MATLAB by enabling plotting at a desired scale and this chapter shows how to do it.

The powerful graphic facilities of MATLAB allow sophisticated and beautiful plots. The reader can appreciate them by running the demos that come with the product. In this chapter we show how to produce plots from analytical expressions. The examples come from the classical field of *conic sections*.

Modern software comes with friendly and powerful graphical user interfaces. MATLAB allows users to design such GUIs according to their own needs. Building such an interface requires knowledge, creativity and not a few efforts. These efforts are worthwhile for GUIs frequently used, such as for repetitive calculations, or for educational use. The latter is the case for the GUI developed in this chapter; it illustrates a simple problem in statics and is based on functions developed in Chapter 2.

10.2 Drawing at scale

In engineering it is usual to draw at some standard scale. In MATLAB plots are drawn by default at non-standard scales that do not allow easy measurements. In this section we show how to override the default option. We

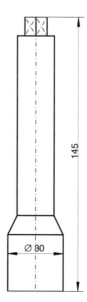

FIGURE 10.1: Machine element

acknowledge the help of Igal Yaroslavski, of Systematics Limited, for providing the solution. Let us assume that we want to draw at natural scale, that
is 1:1, the machine element shown in Figure 10.1. The displayed dimensions
are in mm. The following code does the job. We invite the reader to write it
on a file, `scaleDrawing.m`, run the file, print the drawing and check that it
is at scale. Explanations follow the code.

```
%SCALEDRAWING Draws at scale machine element

% draw something that defines limits
x1 = [ 0 1.5 1.5 1.0 1.0 0.6 0.6 0 ];
z   = [ 0 0 3.0 4.0 13.5 13.5 14.5 14.5 ];
x2 = -x1;
hp = plot(x1, z, 'k-', x2, z, 'k-');
set(hp, 'LineWidth', 1.5)
axis equal, axis off
set(gca, 'Units', 'centimeters')
set(gca, 'xlimmode', 'manual', 'ylimmode', 'manual')
set(gca, 'Position', [ 0 0 abs(diff(xlim)) abs(diff(ylim)) ] )
hold on
% complete frontal elevation
plot([ 0 0 ], [ -4 30 ], 'k-.', [ -35. 3.5 ], [ -3 -3 ], 'k-.')
set(hp, 'LineWidth', 1.5)
hp = plot([ -1.5 1.5 ], [ 3 3 ], 'k-');
```

```
set(hp, 'LineWidth', 1.5)
hp = plot([ -1 1 ], [ 4 4 ], 'k-');
set(hp, 'LineWidth', 1.5)
hp = plot([ -0.6 0.6 ], [ 13.5 13.5 ], 'k-');
set(hp, 'LineWidth', 1.5)
hp = plot([ 0 0 ], [ 13.5 14.5 ], 'k-');
set(hp, 'LineWidth', 1.5)
plot([ -0.6 0 ], [ 13.5 14.5 ], 'k-.')
plot([ -0.6 0 ], [ 14.5 13.5 ], 'k-.')
plot([ 0 0.6 ], [ 13.5 14.5 ], 'k-.')
plot([ 0 0.6 ], [ 14.5 13.5 ], 'k-.')
% some dimensioning
arrow([ 0;2], [ -1.5; 2 ], 0.1)
arrow([ 0;2], [  1.5; 2 ], 0.1)
ht = text(-0.6, 2.3, '\oslash 30');
set(ht, 'FontSize', 14)
plot([ 1.5 2.6 ], [ 0 0 ], 'k-')
plot([ 0.6 2.6 ], [ 14.5 14.5 ] ,'k-')
arrow([ 2.3; 7.25 ], [ 2.3; 0 ], 0.1)
arrow([ 2.3; 7.25 ], [ 2.3;14.50 ], 0.1)
ht = text(2.1, 7, '145', 'Rotation', 90);
set(ht, 'FontSize', 14)
hold off
```

The script file ScaleDrawing begins with a plot of the machine-element contour. This plot defines the drawing limits. Our intention is to plot on an A4 format whose height is 297 mm. Thus it is possible to plot at natural scale. MATLAB admits the unit cm; therefore, we give the dimensions in cm. The coordinates x1 belong to the right-hand half of the element, the coordinates x2, to the left-hand half. Next we assign the property 'Units' the value 'centimeters'. The command gca returns the handle of the current axes for the current figure so that the value 'centimeter' is set for this figure. In continuation we declare that the limits of the $x-$ and $y-$coordinates should be those set by us and not by the software. Next is a statement that assigns the 'Position' property the values given in a four-element array. The first value is the $x-$coordinate of the leftmost point, the second element is the $y-$value of this point. The third element states that the width of the drawing should equal the difference between the extreme $x-$values. The fourth element defines the height of the drawing.

In the script file we introduce the command \oslash; it belongs to the *TeX* typesetting system and it prints the symbol ⊘, which shows that the number 30 refers to a diameter. The text propriety Rotation says by how many degrees the text is rotated counterclockwise. Here we rotate it by 90 degrees.

10.3   The cone surface and conic sections

10.3.1   The cone surface

The theory of conic sections is a fascinating achievement of Greek geometry. Conic sections have important applications in science and engineering. Once, the study of conic sections in analytic and descriptive geometry was part of the basic education of engineers. This is no more the case in not a few technical universities. The result is that many students have difficulties understanding these matters. As in many other cases, developing simple equations and visualizing the mathematical objects can help. In this section we show how to use MATLAB for finding the intersection curve of a cone and a plane, for developing conic surfaces, and for drawing the intersection curve on the cone surface and on its development.

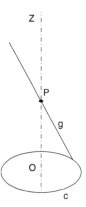

FIGURE 10.2: The definition of the cone surface

To define a general cone surface we need:

1. a fixed point, P;

2. a straight line, g, passing through P;

3. and a plane curve, c. This curve is called *directrix*.

As the line, g, moves along the curve, c, so that it passes always through the fixed point, P, it generates a *cone* surface. The line, g, is appropriately called *generatrix*. Figure 10.2 shows the particular, best known case in which the curve, c, is a circle, and the fixed point, P, lies on the perpendicular to the circle plane, in the center of the circle. In this case we talk about a *right circular cone* and the perpendicular on which the point P lies is the *axis of rotation*. In this section we deal with this kind of cones only. The general definition assumes that the generatrix is infinite. Then, it generates two *nappes* also called *sheets*. Figure 10.3 shows parts of the two nappes of a cone. To understand how open conic sections — more specifically the parabola and the hyperbola — are generated, we must consider the nappes as extending to infinity, both upwards and downwards.

α = 20 deg

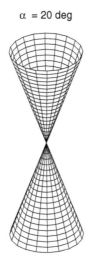

FIGURE 10.3: The two nappes of a cone

Figure 10.4 a shows the *sketch* of a cone having the half-angle at the vertex equal to α, and the height H. The frontal projection appears above, and the horizontal projection below. This arrangement of projections is common in Europe. A cartesian system of coordinates, x, y, z is attached to the cone so that the z–axis runs along the axis of rotation of the surface, positive upwards, and the origin lies in the plane of the cone base. Positive directions correspond to the default definitions of MATLAB.

All sections parallel to the base are circles and we note by r the radius of such a circle. Obviously, r is a function of the height coordinate z. In the horizontal projection we see a radius; the angle between this radius and the x–axis is marked $2\pi u$. When the *parameter* u varies from 0 to 1, the angle varies from 0 to 2π radians and the outer end of the radius describes the base circle.

Figure 10.4 b shows the same cone, but with a network of coordinates superposed on the surface. The network consists of generatrices and circles. To identify the generatrices we use the *parameter* u that runs from 0 to 1. The circles are identified by another parameter, v, that also varies from 0 to 1.

The value $v = 0$ corresponds to the vertex, the value 1 to the base circle.

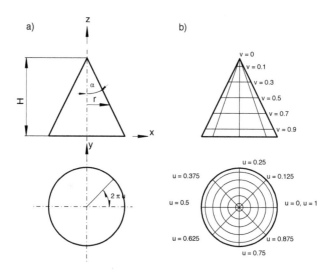

FIGURE 10.4: Coordinates and parameters of a conic surface

The 3-D cartesian coordinates defined in Figure 10.4 a are related to the parameters u, v defined in Figure 10.4 b by the equations

$$
\begin{aligned}
x &= r\cos(2\pi u) \\
y &= y_0 + r\sin(2\pi u) \\
z &= (1-v)H
\end{aligned}
\tag{10.1}
$$

We assume here that the axis of rotation has the equation $y = y_0$. With $r = Hv\tan\alpha$ we rewrite equations 10.1 as

$$
\begin{aligned}
x &= H\tan\alpha \cdot v\cos(2\pi u) \\
y &= y_0 + H\tan\alpha \cdot v\sin(2\pi u) \\
z &= (1-v)H
\end{aligned}
\tag{10.2}
$$

10.3.2   Conic sections

In Figure 10.5 we consider a cone cut by a plane perpendicular to the frontal projection plane and making an angle, β, with the horizontal projection plane. The cutting plane intersects the horizontal projection plane along the line h whose horizontal projection, h', is shown in the figure. The intersection of the cutting plane with the frontal projection plane is f; its frontal projection is marked in the figure by f''. The straight lines h and f are the *traces* of the cutting plane on the projection planes. Let the equation of the trace h be $x = d$.

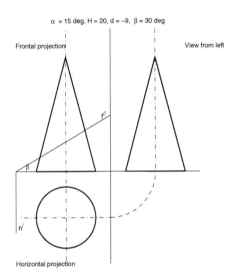

FIGURE 10.5: The sketch of a cone surface

The equation of the cutting plane is

$$z = (x - d) \tan \beta \tag{10.3}$$

The intersection of the cone with the cutting plane is a curve that belongs to the family of *conic sections*. Let the leftmost point of the conic section be characterized by the parameter v_1 and have the coordinates x_1, y_1, z_1. The radius of the base circle is $R = H \tan \alpha$. The equations of the cone surface yield

$$x_1 = -Rv_1$$
$$z_1 = (1 - v_1)H$$

while the equation of the plane gives

$$z_1 = (x_1 - d) \tan \beta$$

Substituting the value of x_1 into the equation of the plane, eliminating z_1 and solving for v_1 we obtain

$$v_1 = \frac{H + d \tan \beta}{H - R \tan \beta} \tag{10.4}$$

In a similar way we can write for the rightmost point of the conic section

$$x_2 = Rv_2$$
$$z_2 = (1 - v_2)H$$

while the equation of the plane yields

$$z_2 = (x_2 - d)\tan\beta$$

From these equations we obtain

$$v_2 = \frac{H + d\tan\beta}{H + R\tan\beta} \tag{10.5}$$

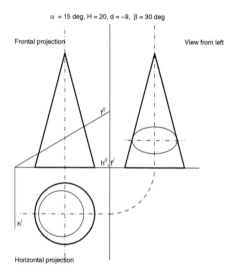

FIGURE 10.6: The sketch of the conic section

To plot the conic section we must find the current coordinates of its points, in the interval $[z_1 \; z_2]$. For each value z_c in this interval we calculate

$$v_c = 1 - z_c/H$$
$$r_c = Rv_c$$
$$x_c = \frac{z_c}{\tan\beta} + d$$
$$y_{c1} = y_0 + \sqrt{r_c^2 - x_c^2}$$
$$y_{c2} = y_0 - \sqrt{r_c^2 - x_c^2}$$

Curve of intersection

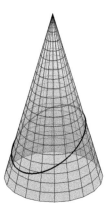

FIGURE 10.7: Axonometric view of the cone surface and of the intersection curve

We develop now a function, `consec`, that solves all the problems formulated in this section, except those described in Subsection 10.3.4. The function receives the following as input arguments

1. the half-angle at the vertex, α, in degrees;

2. the cone height, H;

3. the x-coordinate, d, of the cutting-plane trace in the horizontal projection plane;

4. the angle, β, between the cutting plane and the horizontal projection plane (xOy plane), in degrees.

The listing of the function `consec` is shown in Subsection 10.3.5. The first part of the function plots the **sketch** of the cone. *Sketch* is a concept of descriptive geometry that defines the set of two or more orthographic projections of a given object. Our function plots the projections of the cone on a frontal and on a horizontal plane. To this plot are added the **traces** of the truncating plane, that is the intersections of the cutting plane with the two projection planes. The second part of the function `consec` calculates the coordinates of the intersection curve, namely the resulting conic section. The third part of the function plots an axonometric view of the cone surface with the intersection curve on it. To make visible also what happens on the hidden side of the cone, we made the surface partially transparent by using the command

```
set(hs, 'FaceAlpha', .5)
```

where hs is the handle of the surf plot.

To produce part of the figures that illustrate this section we called the function with the command

```
≫ consec(15, 20, -9, 30)
```

Running the function until the first pause statement produced Figure 10.5. To continue we pressed Enter and obtained Figure 10.6. For the given data the intersection curve is closed; it is an ellipse, the only closed conic section (we consider the circle as a particular case of the ellipse). In this case the slope of the cutting plane is smaller than that of the generatrix, that is $\beta < \pi/2 - \alpha$. Pressing again Enter produces the plot shown in Figure 10.7, an axonometric view of the cone surface and of the intersection curve.

10.3.3   Developing the cone surface

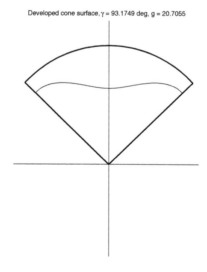

Developed cone surface, $\gamma = 93.1749$ deg, $g = 20.7055$

FIGURE 10.8: The developed cone surface

Developing the surface of a cone we obtain a circular sector with radius equal to the generatrix length, and the angle such that the length of the arc of circle delimiting the sector is equal to the perimeter of the cone base. Let the angle of the sector be Φ, and the length of the generatrix, g. Then

$$2\pi R = 2\pi H \tan \alpha = g\Phi = \sqrt{H^2 + H^2 tan\alpha^2} \cdot \Phi = \frac{H}{\cos \alpha}\Phi \qquad (10.6)$$

where the angle Φ is measured in radians.

On the developed surfaces we define a network of coordinates, ρ, ϕ, actually polar coordinates. The network consists of straight lines radiating from the center of the sector, and of arcs of circle concentric with the center of the sector. The lines correspond to constant ϕ values, and the arcs of circle, to constant ρ values. The relationships between the parameters u, v and the ρ, ϕ coordinates are given by

$$\rho = \sqrt{(vH^2 + (vH\tan\alpha)^2} = H\sqrt{1 + \tan^2\alpha} \cdot v = \frac{H}{\cos\alpha}v \tag{10.7}$$

and

$$\phi = \frac{\Phi}{2\pi}u = u\sin\alpha \tag{10.8}$$

Equations 10.7 and 10.8 can be written more concisely as one matrix equation

$$\begin{bmatrix} 0 & \frac{H}{\cos\alpha} \\ \sin\alpha & 0 \end{bmatrix}\begin{bmatrix} u \\ v \end{bmatrix} = \begin{bmatrix} \rho \\ \phi \end{bmatrix} \tag{10.9}$$

or

$$A_D\begin{bmatrix} u \\ v \end{bmatrix} = \begin{bmatrix} \rho \\ \phi \end{bmatrix} \tag{10.10}$$

To map from the developed to the cone surface we need the inverse of the matrix A_D. We invite the reader to check that this is

$$A_D^{-1} = \begin{bmatrix} 0 & \frac{1}{\sin\alpha} \\ \frac{\cos\alpha}{H} & 0 \end{bmatrix} \tag{10.11}$$

The fourth part of the function consec plots the developed surface of the cone, and the fifth part adds the intersection curve. The function consec calls twice a function con2dev that maps points on the cone surface onto points on the developed surface. The listing of the function con2dev is also given in Subsection 10.3.5. The function gives good results for small values of the angle α.

10.3.4   A helicoidal curve on the cone surface

Let us consider on the cone surface a curve generated by a point that starts on the base and goes up, at constant rate, turning on the cone until it reaches the vertex. From the base to the vertex the point turns 2π radians. The parametric equations of the curve are

$$\begin{aligned} x &= H\tan\alpha \cdot (1-t)\cos(2\pi t) \\ y &= H\tan\alpha \cdot (1-t)\sin(2\pi t) \\ z &= Ht \end{aligned} \tag{10.12}$$

where the parameter t runs from 0 to 1. We invite the reader to check that for $t = 0$ the point lies on the cone base, while for $t = 1$ it is the cone vertex. We draw the attention of the reader to the fact that, while the equations of the cone surface depend on two parameters, namely u and v, the equations of the helicoidal curve depend on one parameter only, namely t. It is so because the cone surface is a two-dimensional manifold, while the curve is a one-dimensional manifold.

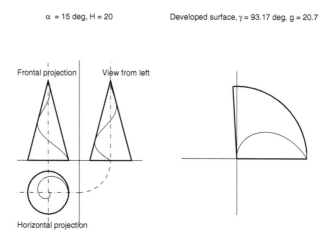

FIGURE 10.9: A helicoidal curve on the cone surface

To show the helicoidal curve on the cone surface and on its development we wrote a function `conhel` whose listing is given in Subsection 10.3.5. The first part of this function produces a `surf` plot of the cone. As in the previous subsection, the surface is made partially transparent to show what happens on the hidden part. The second part of the function calculates the cartesian coordinates of the conical helix, as functions of the parameter t, and plots the curve on the cone surface, as shown in Figure 10.9. The third part of the function plots the sketch, that is the orthographic projections of the cone and of the helicoidal curve. We made no effort to distinguish between visible and hidden lines in the left-hand part of the figure; we invite the reader to solve this as an exercise. The fourth part of the function `conhel` calculates the data of the developed surface and plots it. The fifth and last part of the function plots the helicoidal curve on the developed cone surface.

10.3.5   The listing of functions developed in this section

In the following listings we use the TEX command \alpha to print the Greek letter α, the command \beta to print the Greek letter β, and the command

\gamma to print the Greek letter γ.

Listing of function consec

```
%CONSEC    Conic section and development
% consec(alpha, H, d, beta) plots the sketch of the cone and the
% truncating plane defined by the input parameters
%          alpha     semiangle at cone vertex, degrees
%          H         cone height
%          d         x-coordinate of horizontal trace of cutting plane,
%          beta      slope of cutting plane, degrees
%          Developed by Adrian Biran, 2008

function     consec(alpha, H, d, beta)


ui     = 0:360;          % fine angular parameter, degrees
tana = tand(alpha);      % tangent of half-angle at cone vertex
tanb = tand(beta);       % tangent of cutting plane slope
R       = H*tana;        % radius of cone base
ls      = 1.5*R;         % offset of side axis
clf                      % make sure the screen is clean
% define frame
d1      = min(d, -1.5*R);
plot([ d1   4.5*R ], [ 0 0 ], 'k-', [ 1.5*R  1.5*R ], [ -3*R  1.25*H ], 'k-')
t = [ '\alpha  = ' num2str(alpha)  ' deg, H = ' num2str(H)]
t = [ t ', d = ' num2str(d)  ', \beta = ' num2str(beta)  ' deg' ]);
title(t);
axis equal, axis off
text(d1, 1.1*H, 'Frontal projection')
text(d1, -3*R, 'Horizontal projection')
text(4*R, 1.1*H, 'View from left')


hold on
% complete frame
plot([ 0 0 ], [ -3*R  1.25*H ], 'k-.', [ -1.5*R  1.5*R ], [ - 1.5*R  -1.5*R ], 'k-.')
y0 = -1.5*R;             % y-coordinate of cone axis
% show connection between horizonal and side projections
t1   = 270: 360;
r1   = 1.5*R;
x1   = 1.5*R + r1*cosd(t1);
y1   = r1*sind(t1);
plot(x1, y1, 'b--')          % ligne de rappel
plot([ 3*R 3*R ], [0 1.25*H ], 'k-.')

% horizontal projection
x    = R*cosd(ui);
y    = y0 + R*sind(ui);
hp = plot(x, y, 'k-');
set(hp, 'LineWidth', 1.5)
plot([ d d ], [ -2*R 0 ], 'k-') % horizontal trace of cutting plane
text(0.95*d, -1.8*R, 'h^/')


% frontal projection
hp = plot([ -R 0 R -R ], [ 0 H 0 0 ], 'k-');
set(hp, 'LineWidth', 1.5)
d2   = 1.5*R - d;
if d2*tanb <= H
    plot([ d 1.5*R ], [ 0 d2*tanb ], 'k-') % frontal trace of cutting plane
else
    plot([ d R ], [ 0 (R -d)*tanb ], 'k-') % frontal trace of cutting plane
end
text(1.2*R, d2*tanb, 'f^{//}')
rb = R/3;
text((d + rb), rb/2, '\beta')
```

```matlab
% side projection
hp = plot([ 2*R 3*R 4*R 2*R ], [ 0 H 0 0 ], 'k-');
set(hp, 'LineWidth', 1.5)

pause
%%%%%%%%%%%%% calculation of the conic section %%%%%%%%%%%%%%%
% find v parameter of leftmost and rigthmost points of conic
% first check if the truncating plane intersects the cone at left
if d < -R
        v1  = (H + d*tanb)/(H - R*tanb);
else
        v1 = 1;
end
        z1  = (1 - v1)*H;
        v2  = (H + d*tanb)/(H + R*tanb);
        z2  = (1 - v2)*H;
% define range of z
zc    = z1: (z2 - z1)/180: z2;
vc    = 1 - zc/H;               % v parameter
rc    = R*vc;                   % variable radius of circular section
xc    = zc/tanb + d;            % x-coordinate on intersection curve
yc1   = y0 + (rc.^2 - xc.^2).^(1/2); % y-coordinate on intersection line
yc2 = y0 - (rc.^2 - xc.^2).^(1/2);
plot(xc, yc1, 'r-', xc, yc2, 'r-') % horizontal projection
plot((ls - yc1), zc, 'r-', (ls - yc2), zc, 'r-') % side projection
zm    = (z1 + z2)/2;           % middle height of intersection curve
plot([ 2*R 4*R ], [ zm zm ], 'k-.') % conic axis in side  projection


hold off
pause
%%%%%%%%%%%%%%%%%%%%%%% SURFACE PLOT %%%%%%%%%%%%%%%%%%%%%%%
clf                              % clear figure for new plot
n         = 20;                  % plotting resolution
u         = (0:n)/n;             % u parameter
v         = (n: -1: 0)/n;        % v parameter
theta = 2*pi*u;                  % angle corresponding to u parameter
xs         = R*cos(theta);       % x coordinate on base circle
ys         = R*sin(theta);       % y coordinate on base circle
xxs        = v'*xs;              % (n+1)-by-(n+1) array of x-coordinates on cone surface
yys        = y0 + v'*ys;         % (n+1)-by-(n+1) array of y-coordinates on cone surface
zs         =  (1 - v)*H;         % (n+1)-by-(n+1) array of z-coordinates on cone surface
zzs        = zs'*ones(size(v));

hs = surf(xxs, yys, zzs);        % plot cone surface
axis equal, axis off
colormap(autumn)
set(hs, 'FaceAlpha', .5)         % make surface transparent to view details on hidden side
hold on
hp1 = plot3(xc, yc1, zc, 'k-', xc, yc2, zc, 'k-'); % plot intersection curve
set(hp1, 'LineWidth', 2)
title('Curve of intersection')
% to print decomment one of the following
% print -depsc ConeParabolaC.eps
% print -deps ConeParabola.eps
% print -djpeg ConeParabola.jpg
pause

%%%%%%%%%%%%%%%%%%%%%%% DEVELOPED SURFACE %%%%%%%%%%%%%%%%%%%%%%
clf                              % clear screen for new plot
R15  = 1.5*R;
% copy initial frame to make sure dimensions are the same as in sketch
plot([ d1 4.5*R ], [ 0 0 ], 'k-', [ R15 R15 ], [ -3*R 1.25*H ], 'k-')
g        = H/cosd(alpha);        % generatrix length
Phi = 360*sind(alpha);           % angle of circular sector, deg
Phi1 = (180 - Phi)/2;            % initial angle for symmetrical plot
title([ 'Developed cone surface, \gamma = ' num2str(Phi) ' deg, g = ' num2str(g) ])
axis equal, axis off
```

```
hold on
td      = Phi1: Phi/90: (Phi1 + Phi);  % parameter of circular sector
xd      = R15 + g*cosd(td);
yd      = g*sind(td);
hp      = plot(xd, yd, 'k-', [ R15 xd(1) ], [ 0 yd(1) ], 'k-', [ R15 xd(end) ],...
            [ 0 yd(end) ], 'k-');
set(hp, 'LineWidth', 1.5)
%%%%%%%%%% PLOT INTERSECTION CURVE ON DEVELOPED SURFACE %%%%%%%%%%%%%
 % call function that maps from cone to developed surface
[ RHO1 PHID1 ]   = con2dev(H, alpha, y0, xc, yc1, zc);
XD1                       = R15 + RHO1.*sin(PHID1);
YD1                       = RHO1.*cos(PHID1);
hp3                       = plot(XD1, YD1, 'r-');
% call again function for the other half of the intersection curve
[ RHO2 PHID2 ]   = con2dev(H, alpha, y0, xc, yc2, zc);
XD2                       =  R15 + RHO2.*sin(PHID2);
YD2                       = RHO2.*cos(PHID1);
hp4                       = plot(XD2, YD2, 'r-');

hold off
% to print decomment one of the following
% print -deps ConeDevelop.eps
% print -djpeg ConeDevelop.jpg
```

Listing of function con2dev

```
%CON2DEV  maps from the cone to the developed surface
%               Input arguments:
%                      H        - cone height
%                      alpha - half-angle at cone vertex, degrees
%                      y0   - y-coordinate of axis of  rotation
%                      X, Y, Z  - coordinates of curve on conic surface
%               Output arguments:
%                      rho    - polar radius
%                      phid   - polar angle, radians
%          Developed by Adrian Biran, 2008

function     [ rho phid ] = con2dev(H, alpha, y0, X, Y, Z)

cosa = cosd(alpha);
rho    = (H - Z)/cosa;
u          = atan2((Y - y0), X);
phid  = sind(alpha)*u;
```

Listing of function conhel

```
%CONHEL Conic helix and its development
% conhel(alpha, H) plots helicoidal curve on cone surface defined
% by the input parameters
%       alpha              semiangle at cone vertex, degrees
%       H                  cone height
% The pitch of the curve equals H.
% Developed by Adrian Biran, 2008

function     conhel(alpha, H)

tana = tand(alpha);
R        = H*tana;     % radius of cone base
R15      = 1.5*R;
y0       = -R15;       % y-coordinate of cylinder axis
ls       = R15;        % offset of side axis

clf                    % make sure the screen is clear

%%%%%%%%%%%%%%%%%%%%%%%%%% SURFACE PLOT %%%%%%%%%%%%%%%%%%%%%%%%%%
```

```
n          = 20;              % plotting resolution
u          = (0:n)/n;         % u parameter
v          = (n: -1: 0)/n;    % v parameter
theta = 2*pi*u;               % angle corresponding to u parameter
xs         = R*cos(theta);    % x-coordinates on cone basis
ys         = R*sin(theta);    % y-coordinates on cone basis
xxs        = v'*xs;           % (n+1)-by-(n+1) array of x-coordinates on cone surface
yys        = y0 + v'*ys;      % (n+1)-by-(n+1) array of y-coordinates on cone surface
zs         = (1 - v)*H;       % z-coordinates on cone basis
zzs        = zs'*ones(size(v)); % (n+1)-by-(n+1) array of z-coordinates on cone surface

hs = surf(xxs, yys, zzs);
axis equal, axis off
colormap(autumn)
set(hs, 'FaceAlpha', .5)      % make surface transparent to see on hidden face
hold on

%%%%%%%%%%%%%%%%%%%%%%% A CONIC HELIX %%%%%%%%%%%%%%%%%%%%%%%
t     = 0: 0.05: 1;
xh  = R*(1 -t).*cos(2*pi*t);
yh  = R*(1 -t).*sin(2*pi*t) + y0;
zh  = H*t;
hp1 = plot3(xh, yh, zh, 'b-');
set(hp1, 'LineWidth', 2)
title('A conic helix')

print  -depsc  ConeHelix1C.eps
print  -deps ConeHelix1.eps
print  -djpeg ConeHelix1.jpg

hold off
pause
%%%%%%%%%%%%%%%%% ORTHOGRAPHIC PROJECTIONS %%%%%%%%%%%%%%%%%
clf
subplot(1, 2, 1)
% define frame
R15  = 1.5*R;
plot([ -R15  4.5*R ], [ 0 0 ], 'k-', [R15  R15 ], [ -3*R  1.25*H ], 'k-')
title([ '\alpha  = ' num2str(alpha) ' deg, H = ' num2str(H)  ]);
axis equal, axis off
text(-R15, 1.1*H, 'Frontal projection')
text(-R15, -3*R, 'Horizontal projection')
text(2.6*R, 1.1*H, 'View from left')

hold on
% complete frame
plot([ 0 0 ], [ -3*R  1.25*H ], 'k-.', [ -R15  R15 ], [ -R15  -R15], 'k-.')
t1  = 270: 360;
r1  = 1.5*R;
x1  = 1.5*R + r1*cosd(t1);
y1  = r1*sind(t1);
plot(x1, y1, 'b--')                       % ligne de rappel
plot([ 3*R 3*R ], [0 1.25*H ], 'k-.')

% horizontal projection
ui = 0: 360;
x     = R*cosd(ui);
y     = y0 + R*sind(ui);
hp = plot(x, y, 'k-');
set(hp, 'LineWidth', 1.5)
plot(xh, yh, 'r-')

% frontal projection
hp = plot([ -R 0 R -R ], [ 0 H 0 0 ], 'k-');
set(hp, 'LineWidth', 1.5)
plot(xh, zh, 'r-')
```

```
% side projection
hp = plot([ 2*R 3*R 4*R 2*R ], [ 0 H 0 0 ], 'k-');
set(hp, 'LineWidth', 1.5)
plot((ls - yh), zh, 'r-')

%%%%%%%%%%%%%%%%%%%%%% DEVELOPED SURFACE %%%%%%%%%%%%%%%%%%%%%%
subplot(1, 2, 2)
% reuse frames to ensure the same cone dimensions
plot([ -R15   4.5*R ], [ 0 0 ], 'k-', [ R15   R15 ], [ -3*R   1.25*H ], 'k-')
g      = H/cosd(alpha);         % generatrix length
Phi = 360*sind(alpha);          % angle of circular sector, deg
PP     = fix(Phi*100)/100       % round off to two decimal digits
gg     = fix(g*100)/100;        % round off to two decimal digits
title([ 'Developed surface, \gamma = ' num2str(PP) ' deg, g = ' num2str(gg) ])
axis equal, axis off
hold on
td     = 0: Phi/90: Phi;        % parameter of circular sector
xd     = R15 + g*cosd(td);
yd     = g*sind(td);
hp     = plot(xd, yd, 'k-', [ R15 xd(1) ], [ 0 yd(1) ], 'k-', [ R15 xd(end) ],...
              [ 0 yd(end) ], 'k-');
set(hp, 'LineWidth', 1.5)
%%%%%%%%%%%%%%%%% DEVELOPMENT OF THE CONIC HELIX %%%%%%%%%%%%%%%%%%%%%
% mapping from parameter t
phi = 2*pi*t*sind(alpha);
rho = g*(1 - t);
XD     = R15 + rho.*cos(phi);
YD     = rho.*sin(phi);;
hp5    = plot(XD, YD, 'r-');

hold off
% to print decomment one of the following
% print -depsc  ConeHelix2C.eps
% print -deps ConeHelix2.eps
% print -djpeg ConeHelix2.jpg
```

10.4 GUIs - graphical user interfaces

Graphical user interfaces are common features of today's software and, as the user may have learned by now, MATLAB and its various toolboxes are no exception. In addition, MATLAB allows the user to customize several built-in interfaces and use them in user-written script files and functions. We mean the *predefined dialog boxes* and we give some examples in Subsection 6.3.4. There is more in MATLAB, and the basic package enables the user to develop sophisticated graphical interfaces; we call them *graphical user interfaces*, shortly *GUIs*. The development of GUIs is not a trivial matter; it requires knowledge, creativity, and not a few efforts. Therefore, it is worth developing a GUI only for repetitive tasks. Such tasks include calculations performed often, but also educational applications in which the user can better visualize a phenomenon and learn the influence of various parameters by playing interactively with them.

In this section we develop a rather simple GUI, just to show the reader another powerful feature of MATLAB. It is beyond the scope of this book to

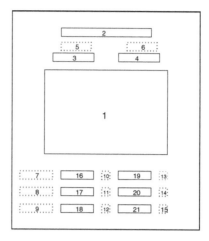

FIGURE 10.10: The sketch of the GUI

treat the subject in more detail. The interested reader can refer to dedicated books, such as Smith (2006). There are two ways of developing a GUI in MATLAB, the first *programatically*, the second with the help of a specially built interface called *GUIDE*. We chose here the first way because thus we can explain the various elements of a GUI. Developing a GUI is a matter of *design*. Before beginning the work we must have a good idea of what we want to see in the final product. This is what appears in Figure 10.12. In the middle of the figure we want the drawing of the two-bar system, as generated by the function Stat1FigFun developed in Example 2.5. At the top of the figure we place a *slider*. Pulling the slider to the left or to the right should change the horizontal position of the hanging weight and the values of the variables a and b.. Below the slider we put two *text* boxes that mark the variables a and b, and one row below two *edit* boxes to display the values of the above variables. Below the drawing we place six *text* boxes that display the $x-$ and $y-$components of the forces **R1** and **R2**, and the magnitudes of these forces. On the left of the group of six text boxes we place three other text boxes to show the names of the displayed variables, on the right of each one of the six text boxes, other boxes display the letter N, the symbol of the force unit newton.

Having the final image in mind, we sketch, at scale, the *layout* of the GUI. The sketch suiting Figure 10.12 is shown in Figure 10.10. We like to draw such sketches on millimetric paper, or at least on square-grid paper. We use solid lines to show the elements that should appear as individual windows, and dotted lines to mark the elements that should appear as character strings without an own window.

Now we are ready to **dimension** the various elements of the GUI. Several units can be used, we choose *normalized* units. This means, for example, that each horizontal value is given as a fraction of the horizontal extension of the GUI figure. Thus, we define the position of the drawing, element 1 in Figure 10.10, by the array [0.174 0.346 0.652 0.385]. The first element says that the lower, left corner of the drawing is placed at 0.174 times the GUI width, from the left of the GUI. The second number, 0.346, places the lower corners of the drawing at a distance equal to 0.346 times the GUI height, from the bottom of the GUI. The third number is the width of the drawing as a fraction of the GUI width, and the fourth, the height of the drawing given as a part of the GUI height.

We write the code in several steps, adding in each step more features. In the first step we define the general template of the GUI; it consists of

- a line that defines the GUI as a function

- a set of comments

- a blank line

- a line that defines the figure object, allocates it a handle, and hides the figure while it is under construction;

- the definitions of the GUI components;

- a set of commands that name the GUI and makes it visible;

- the word **end.**

The first component is a graphics object. We create it with the command **axes**. Mind the 'e', to distinguish from the command **axis**. The GUI displays the user-selected plot in an **axes** object. All other elements are *user interface control objects* created with commands like

handle = uicontrol('PropertyName',PropertyValue,...)

From all the user interface controls available in MATLAB, in this GUI we use only the following ones:

Static text labels These are static text boxes that display text without any possibility of user intervention. In Figure 10.10 all elements shown in dotted frames are of this type, that is the elements numbered 5-15.

A slider Sliders accept numeric input, within a specific range, by enabling the user to move a sliding bar. Users move the bar by pressing the mouse button and dragging the pointer over the bar, or by clicking in the trough or on an arrow. This is the component marked 2 in Figure 10.10.

Edit text boxes Editable text fields enable users to enter or modify text values. We use editable text when we want text as input. Such are the elements marked 3 and 4 in Figure 10.10.

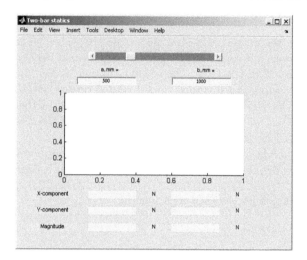

FIGURE 10.11: The layout of the GUI `TwoBarStatics`

Below is the code developed in the first step. Write it to a file called TwoBarStatics.m. To run the file write `TwoBarStatics` at the command line. The result is shown in Figure 10.11.

```
     function    TwoBarStatics
%TWOBARSTATICS  GUI for the statics of a system consisting of
%               a weight of 50 N hanging from the common point
%               of two bars whose upper ends are connected to
%               the ceiling by articulations.
%               Use the slider to change the position
%               of the weight relative to the upper
%               articulations of the two bars.
%               Written by Adrian Biran, June 2009.

    % Initialize and hide the GUI during its construction
    f = figure('Visible', 'off');

    %%%%%%  Construct the components  %%%%%%%%%%%%
    haxes = axes('Units', 'normalized',...
                 'Position', [ 0.174 0.346 0.652 0.385 ]);
    % text boxes
    htext1   = uicontrol('Style', 'Text',...
                         'String', 'a, mm = ',...
                         'Units', 'normalized',...
                         'Position', [ 0.26 0.82 0.16 0.038 ]);
    htext2   = uicontrol('Style', 'Text',...
                         'String', 'b, mm = ',...
                         'Units', 'normalized',...
                         'Position', [ 0.609 0.82 0.16 0.038 ]);
```

```
htext3   = uicontrol('Style', 'Text',...
                     'String', 'X-component',...
                     'Units', 'normalized',...
                     'Position', [ 0.043 0.23 0.175 0.038 ]);
htext4   = uicontrol('Style', 'Text',...
                     'String', 'Y-component',...
                     'Units', 'normalized',...
                     'Position', [ 0.043 0.154 0.175 0.038 ]);
htext5   = uicontrol('Style', 'Text',...
                     'String', 'Magnitude',...
                     'Units', 'normalized',...
                     'Position', [ 0.043 0.077 0.175 0.038 ]);
htext6   = uicontrol('Style', 'Text',...
                     'String', 'N',...
                     'Units', 'normalized',...
                     'Position', [ 0.478 0.23 0.043 0.038 ]);
htext7   = uicontrol('Style', 'Text',...
                     'String', 'N',...
                     'Units', 'normalized',...
                     'Position', [ 0.478 0.154 0.043 0.038 ]);
htext8   = uicontrol('Style', 'Text',...
                     'String', 'N',...
                     'Units', 'normalized',...
                     'Position', [ 0.478 0.077 0.043 0.038 ]);
htext9   = uicontrol('Style', 'Text',...
                     'String', 'N',...
                     'Units', 'normalized',...
                     'Position', [ 0.783 0.23 0.043 0.038 ]);
htext10 = uicontrol('Style', 'Text',...
                     'String', 'N',...
                     'Units', 'normalized',...
                     'Position', [ 0.783 0.154 0.043 0.038 ]);
htext11 = uicontrol('Style', 'Text',...
                     'String', 'N',...
                     'Units', 'normalized',...
                     'Position', [ 0.783 0.077 0.043 0.038 ]);
%slider control
hslide  = uicontrol('Style', 'slider',...
                     'BackgroundColor', [ 0.9 0.1 0.0 ],...
                     'Min', 200,...
                     'Max', 1300,...
                     'SliderStep', [ 0.01 0.1 ],...
                     'Value', 500,...
                     'Units', 'normalized',...
                     'Position', [ 0.26 0.885 0.489 0.038 ]);
% edit text boxes above figure
hedit1 = uicontrol('Style', 'edit',...
                     'BackgroundColor', 'y',...
                     'String', num2str(get(hslide, 'Value')),...
```

```
                      'Units', 'normalized',...
                      'Position', [ 0.217 0.769 0.218 0.038 ]);

    hedit2 = uicontrol('Style', 'edit',...
                      'BackgroundColor', 'y',...
                      'String', num2str(1500 - get(hslide, 'Value')),...
                      'Units', 'normalized',...
                      'Position', [ 0.565 0.769 0.218 0.038 ]);
    % text boxes below the figure
              % left
    hedit3 = uicontrol('Style', 'text',...
                      'BackgroundColor', 'y',...
                      'String', [ ],...
                      'Units', 'normalized',...
                      'Position', [ 0.261 0.231 0.174 0.038 ]);
    hedit4 = uicontrol('Style', 'text',...
                      'BackgroundColor', 'y',...
                      'String', [ ],...
                      'Units', 'normalized',...
                      'Position', [ 0.261 0.154 0.174 0.038 ]);
    hedit5 = uicontrol('Style', 'text',...
                      'BackgroundColor', 'y',...
                      'String', [ ] ,...
                      'Units', 'normalized',...
                      'Position', [ 0.261 0.077 0.174 0.038 ]);
              % right
    hedit6 = uicontrol('Style', 'text',...
                      'BackgroundColor', 'y',...
                      'String', [ ],...
                      'Units', 'normalized',...
                      'Position', [ 0.565 0.231 0.174 0.038 ]);
    hedit7 = uicontrol('Style', 'text',...
                      'BackgroundColor', 'y',...
                      'String', [ ],...
                      'Units', 'normalized',...
                      'Position', [ 0.565 0.154 0.174 0.038 ]);
    hedit8 = uicontrol('Style', 'text',...
                      'BackgroundColor', 'y',...
                      'String', [ ],...
                      'Units', 'normalized',...
                      'Position', [ 0.565 0.077 0.174 0.038 ]);

    %%%%%%%  Name and make the GUI visible %%%%%%%%%%%%%
    set(f, 'NumberTitle', 'off',...
                  'Name', 'Two-bar statics');
    set(f, 'Visible', 'on')
end
```

In the definition of the slider, with the handle hslide, we gave the property

BachgroundColor the value [0.9 0.1 0.0], an array of the intensities of
the three color components red, green, blue *RGB components*. To the property Value we assign the *value* 500 thus initializing the hanging point at $a =$
500. To the BachgroundColor property of the edit boxes we assign the value
'y', the short notation for yellow. We use the function set to assign values to
the properties of the figure object. The first argument of set is the handle of
the object, *f*. Assigning the value *off* to the property NumberTitle prevents
the printing of the title *Figure* and of a number in the heading of the figure
window. Instead of this we force the printing of a name.

Up to now we designed the layout of the GUI and obtained a static figure. Call the GUI as written until now. You should see what appears in
Figure 10.11. To bring the GUI to life we must call the graph and add **callbacks**. The first job is done by the simple command

[r1 r2 R1 R2] = Stat1FigFun(500);

The input argument 500 initializes the distance *a*, the output arguments
describe the characteristics of the force vectors **R1** and **R2**.

A *callback* is a function that executes when we perform various actions in
the GUI, in our example moving the slider or changing the text in an edit
box. More generally, we can use callbacks to execute MATLAB commands, or
M-files. In our GUI we want to define callback functions for three of its components, the slider, with handle hslide, and the edit boxes, with handles hedit1
and hedit2. Therefore, in the definition of the slider uicontrol we must insert
the line 'Callback', @slider_callback, in the definition of the first edit
uicontrol the line 'Callback', @edittext_callback, and in the definition of
the second edit uicontrol the line 'Callback', @edittext_callback2. The
reader may note that, in this case, we define callbacks as *anonymous functions*.

Separately we write *nested functions* that implement the callbacks. These
callbacks have access to component handles and initialized data because they
are nested at a lower level. Before the function definitions we write comments.
The definition of the slider callback, for example, is

    function slider_callback(hObject, eventdata)

The argument hObject is the handle of the component for which the event is
triggered. In our case, the first object is the slider, and the event is the change
of the slider position by means of mouse clicks. The second object is the edit
box of the variable *a* and the event is composed of a change of text followed
by a mouse click or the pressing of Enter. As to the argument eventdata,
at least one of the examples given by the MATLAB documentation mentions

eventdata reserved - to be defined in a future version of MATLAB

As the calculation of the two forces, R_1, R_2, of their components and the
display in text boxes can be triggered by the slider or by one of the two edit
boxes, we write a function, PlotCalc, which appears at the end of the function
TwoBarStatics and can be called from any one the three controls mentioned
above. This function carries on the following actions:

- accepts as input the argument `val` with which it is called in one of the uicontrols;

- calls the function `Stat1FigFun` with the input argument `val` and four output arguments;

- as the default four-digit display is inadequate for the variables a and b, it rounds the number a to zero decimal digits, using the MATLAB function `fix`, and converts it to a string for display in the text box with handle `hedit1`;

- converts the default, four-letter display of the variables *r1, r2, **R1, R2*** into three-digit displays;

- updates the displays in the text boxes with handles `hedit3` to `hedit8`.

We are going to explain now how we have written the code of the callbacks, while the complete code is given after the explanation. The function `slider_callback` performs the following actions:

- reads the value of the variable that defines the slider position;

- converts this variable to a string to be displayed in the first edit box;

- calculates the variable b and converts it to a string for display in the text box with handle `hedit2`;

- calls the function `PlotCalc`.

The second callback function is `edittext_callback`. The input arguments are the same as those of the first callback. This second function performs the following actions

- stores the current value of the slider, for the case that the reader entered an incorrect value in the edit box;

- reads the value entered in the edit box;

- checks if this value is a numeric scalar, and in the affirmative if it lies within the range of values declared for the slider;

- if the test succeeded, sets the slider to the value entered in the edit box;

- updates the display of the edit box with handle `hedit2`;

- if the text entered in the edit box was inadequate it displays an error message in that box and resets the slider to the previous value.

The final, complete code of the GUI is listed below.

```
        function      TwoBarStatics
%TWOBARSTATICS    GUI for the statics of a system consisting
%                    of a weight of 50 N hanging from the common
%                    point of two bars whose upper ends are
%                    connected to the ceiling by articulations.
%                    Use the slider to change the position
%                    of the weight relative to the upper
%                    articulations of the two bars.
%                    Written by Adrian Biran for
%                    Taylor & Francis, June 2009.

% Initialize and hide the GUI during its construction
f = figure('Visible', 'off');

%%%%%%  Construct the components %%%%%%%%%%%%%%%
haxes = axes('Units', 'normalized',...
               'Position', [ 0.174 0.346 0.652 0.385 ]);
% text boxes
htext1 = uicontrol('Style', 'Text',...
                    'String', 'a = ',...
                    'Units', 'normalized',...
                    'Position', [ 0.26 0.82 0.16 0.038 ]);
htext2 = uicontrol('Style', 'Text',...
                    'String', 'b = ',...
                    'Units', 'normalized',...
                    'Position', [ 0.609 0.82 0.16 0.038 ]);
htext3 = uicontrol('Style', 'Text',...
                    'String', 'X-component',...
                    'Units', 'normalized',...
                    'Position', [ 0.043 0.23 0.175 0.038 ]);
htext4 = uicontrol('Style', 'Text',...
                    'String', 'Y-component',...
                    'Units', 'normalized',...
                    'Position', [ 0.043 0.154 0.175 0.038 ]);
htext5 = uicontrol('Style', 'Text',...
                    'String', 'Magnitude',...
                    'Units', 'normalized',...
                    'Position', [ 0.043 0.077 0.175 0.038 ]);
htext6 = uicontrol('Style', 'Text',...
                    'String', 'N',...
                    'Units', 'normalized',...
                    'Position', [ 0.478 0.23 0.043 0.038 ]);
htext7 = uicontrol('Style', 'Text',...
                    'String', 'N',...
                    'Units', 'normalized',...
                    'Position', [ 0.478 0.154 0.043 0.038 ]);
htext8 = uicontrol('Style', 'Text',...
                    'String', 'N',...
                    'Units', 'normalized',...
```

```
                          'Position', [ 0.478 0.077 0.043 0.038 ]);
htext9  = uicontrol('Style', 'Text',...
                          'String', 'N',...
                          'Units', 'normalized',...
                          'Position', [ 0.783 0.23 0.043 0.038 ]);
htext10 = uicontrol('Style', 'Text',...
                          'String', 'N',...
                          'Units', 'normalized',...
                          'Position', [ 0.783 0.154 0.043 0.038 ]);
htext11 = uicontrol('Style', 'Text',...
                          'String', 'N',...
                          'Units', 'normalized',...
                          'Position', [ 0.783 0.077 0.043 0.038 ]);
%slider control
hslide  = uicontrol('Style', 'slider',...
                          'BackgroundColor', [ 0.9 0.1 0.0 ],...
                          'Min', 200,...
                          'Max', 1300,...
                          'SliderStep', [ 0.01 0.1 ],...
                          'Value', 500,...
                          'Units', 'normalized',...
                          'Position', [ 0.26 0.885 0.489 0.038 ],...
                          'Callback', @slider_callback);
% edit text boxes above figure
hedit1 = uicontrol('Style', 'edit',...
                          'BackgroundColor', 'y',...
                          'String', num2str(get(hslide, 'Value')),...
                          'Units', 'normalized',...
                          'Position', [ 0.217 0.769 0.218 0.038 ],...
                          'Callback', @edittext_callback);
hedit2 = uicontrol('Style', 'edit',...
                          'BackgroundColor', 'y',...
                          'String', num2str(1500 - get(hslide, 'Value')),...
                          'Units', 'normalized',...
                          'Position', [ 0.565 0.769 0.218 0.038 ],...
                          'Callback', @edittext_callback2);
% text boxes below the figure
                          % left
hedit3 = uicontrol('Style', 'text',...
                          'BackgroundColor', 'y',...
                          'String', [ ],...
                          'Units', 'normalized',...
                          'Position', [ 0.261 0.231 0.174 0.038 ]);
hedit4 = uicontrol('Style', 'text',...
                          'BackgroundColor', 'y',...
                          'String', [ ],...
                          'Units', 'normalized',...
                          'Position', [ 0.261 0.154 0.174 0.038 ]);
hedit5 = uicontrol('Style', 'text',...
```

```
                          'BackgroundColor', 'y',...
                          'String', [ ] ,...
                          'Units', 'normalized',...
                          'Position', [ 0.261 0.077 0.174 0.038 ]);
                             % right
hedit6 = uicontrol('Style', 'text',...
                          'BackgroundColor', 'y',...
                          'String', [ ],...
                          'Units', 'normalized',...
                          'Position', [ 0.565 0.231 0.174 0.038 ]);
hedit7 = uicontrol('Style', 'text',...
                          'BackgroundColor', 'y',...
                          'String', [ ],...
                          'Units', 'normalized',...
                          'Position', [ 0.565 0.154 0.174 0.038 ]);
hedit8 = uicontrol('Style', 'text',...
                          'BackgroundColor', 'y',...
                          'String', [ ],...
                          'Units', 'normalized',...
                          'Position', [ 0.565 0.077 0.174 0.038 ]);
[ r1 r2 R1 R2 ] = Stat1FigFun(500);
%%%%%%%  Name and make the GUI visible %%%%%%%%%%%%%%
set(f, 'NumberTitle', 'off',...
                 'Name', 'Two-bar statics');
set(f, 'Visible', 'on')
%%%%%%%%%% First nested functions %%%%%%%%%%%%%%%%%%
% Set the value of the first edit text box String property
% to the value of the slider and the value of the second
% edit text box to 1500 - the value of the slider
function slider_callback(hObject, eventdata)
        val = get(hObject, 'Value');
        set(hedit1, 'String', num2str(fix(val)));
        set(hedit2, 'String', num2str(1500 - fix(val)));
        [ r1 r2 R1 R2 ]  = PlotCalc(val);
end
% Set the slider value to the number the user types in
% the first edit text or display an error message.
function edittext_callback(hObject,eventdata)
    val = str2double(get(hslide, 'Value'));
    previous_val = val;
    val = str2double(get(hObject,'String'));
    % Determine whether val is a number between the
    % slider's Min and Max. If it is, set the slider Value.
    if isnumeric(val) && length(val) == 1 && ...
            val >= get(hslide, 'Min') && ...
            val <= get(hslide, 'Max')
            set(hslide, 'Value', val);
            set(hedit2, 'String', num2str(1500 - val));
            [ r1 r2 R1 R2 ]  = PlotCalc(val);
```

```
        else
                set(hObject,'String', 'Invalid entry ');
                 val = previous_val;
        end
    end
% Set the value of the second edit text box String property
% to 1500 - the value of the slider and the value of the first
% edit text box to the value of the slider
    function slider_callback2(hObject, eventdata)
            val = get(hObject, 'Value');
            set(hedit2, 'String', num2str(fix(val)));
            val1 = 1500 -val;
            set(hedit1, 'String', num2str(fix(val1)));
            [ r1 r2 R1 R2 ]  = PlotCalc(val1);
    end
% Set the slider value to the number the user types in
% the second edit text or display an error message.
    function edittext_callback2(hObject,eventdata)
        val = str2double(get(hslide, 'Value'));
        previous_val = val;
        val = str2double(get(hObject,'String'));
        % Determine whether val is a number between the
        % slider's Min and Max. If it is, set the slider Value.
        if isnumeric(val) && length(val) == 1 && ...
                val >= get(hslide, 'Min') && ...
                val <= get(hslide, 'Max')
                val1 = 1500 -val;
                set(hslide, 'Value', val1);
                set(hedit1, 'String', num2str(val1));
                [ r1 r2 R1 R2 ]  = PlotCalc(val1);
        else
                set(hObject,'String', 'Invalid entry ');
                 val = previous_val;
        end
    end
    function [ r1 r2 R1 R2 ] = PlotCalc(val)
            [ r1 r2 R1 R2 ] = Stat1FigFun(val);
            nd              = 3; % number of displayed decimal digits
            p               = 10^nd;
            r1              = fix(p*r1)/p;
            r2              = fix(p*r2)/p;
            R1              = fix(p*R1)/p;
            R2              = fix(p*R2)/p;
            set(hedit3, 'String', R1(1));
            set(hedit4, 'String', R1(2));
            set(hedit5, 'String', r1);
            set(hedit6, 'String', R2(1));
            set(hedit7, 'String', R2(2));
```

```
                    set(hedit8, 'String', r2);
        end
end
```

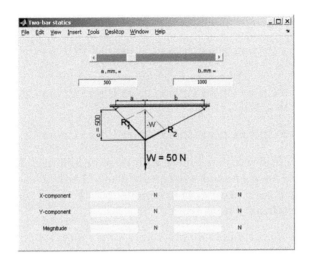

FIGURE 10.12: The final aspect of the GUI `TwoBarStatics`

10.5 Summary

In this chapter we introduced the following commands

axis off suppresses the axes of coordinates.

\oslash prints the symbol ⊘.

FaceAlpha set the transparency property of plot faces.

gca returns the handle of the current figure axes.

Rotate text property that defines by how many degree the text is rotated counterclockwise.

10.6   Exercises

Exercise 10.1 Conic sections 1

We return to the developments in Subsection 10.3.2. For the values exemplified there we obtained an ellipse. Other possibilities are

Parabola If the cutting plane is exactly parallel to a generatrix, that is if $\beta = \pi/2 - \alpha$.

Hyperbola If the slope of the cutting plane is greater than that of the generatrices, that is if $\beta > \pi/2 - \alpha$.

Two lines If the cutting plane passes through the cone vertex.

We invite the reader to experiment with these possibilities. It may be necessary to change the position of the cutting plane, that is to change the value of the third input argument, d. For example, run the program developed in Section 10.3.2 with the parameters

$$\alpha = 15^\circ, \ H = 20, \ d = -4, \ \beta = 75^\circ.$$

Which conic section did you obtain?

Let us make a few comments on the three possibilities listed in this exercise.

When the cutting plane is parallel to a generatrix, it cuts only one cone nappe. Thus we obtain a curve with one branch, that comes from infinity, makes a turn and returns to infinity in the same direction from which it came. The only conic section that behaves so is the parabola.

When the slope of the cutting plane is greater than that of the generatrices, the plane cuts both cone nappes. We obtain a curve with two branches, each one coming from infinity and returning to infinity. The only conic section that behaves so is the hyperbola.

When the cutting plane passes through the vertex, it cuts the cone surface along two of its generatrices. These are, obviously, straight lines.

There are two additional possibilities, we may call them *degenerate*: the intersection of the plane with the cone surface is a *point* or *one line*. Can you see how this may happen?

The circle is a special case of the ellipse. Then the cone sections are the circle, the ellipse, the parabola, the hyperbola, two lines and the degenerate conics that are the point and one line. All these curves can be described in cartesian coordinates by second-order equations, and they are the only curves described by second-order equations.

Exercise 10.2 Conical helix

In Subsection 10.3.4 we considered a helicoidal curve that turns once around the cone while rising from the base to the vertex. The height a point of the curve rises while it turns 2π radians is called *pitch*; let us note it by p. In Subsection 10.3.4 $p = H$. Modify the function *conhel* to draw a conical helix with the pitch $p = H/2$.

11

An introduction to Simulink®

11.1 What is simulation?

Today, probably the best-known engineering use of differential equations is in modelling real-life phenomena, engineering systems and industrial processes. Running such a model on a computer produces a **simulation** of the phenomenon, engineering system or industrial process. Originally, the word *simulation*, and others related to the same root, such as *simulate* or *simulator*, had other meanings, mainly with negative connotations. Thus, we can read in a Latin dictionary (Kidd, 1993), 'simulatio ... pretence, shamming, hypocrisy', or 'simulo -are ... imitate, represent, impersonate, pretend, counterfeit'. In such a sense we meet the imperative *simulate* in Mozart's *Don Giovanni* (scene 22, certainly attributable to the librettist Lorenzo da Ponte). Recent dictionaries include the original meanings, but add the new meaning that interests us. For example, in *Webster's Ninth New Collegiate Dictionary* (1990) we read, 'simulation .. 1: the act or process of simulating :FEIGNING 2: a sham object: COUNTERFEIT 3 a: the imitative representation of the functioning of one system or process by means of the functioning of another < a computer ˜ of an industrial process >. ...'

A first use of simulation in engineering is in checking and improving the functioning of a system under design. Running a computer model costs less than building a prototype and experimenting with it. On the other hand, the computer model is usually based on approximations. Therefore, in advanced stages of a project it may be necessary to build a real-life prototype, compare the results of experiments with those yielded by simulation, and calibrate, or even correct, the computer model.

There are also cases in which experiments are impossible, either because of prohibitive costs, or because of the dangers involved. Then, the only possibility is simulation. Possible examples are the sinking of a ship and the malfunctioning of a power plant, especially a nuclear one.

The classic way of building a computer model for simulation consists of the following phases:

1. building a model of physical, chemical, and/or other laws;

2. the translation of the model of laws into a system of algebraic, differen-

tial and/or difference equations;

3. the translation of the equations into computer code.

It is possible to carry on the third phase in MATLAB. However, there is a MATLAB toolbox that greatly facilitates the second and the third phase; this is **Simulink**. While MATLAB is a *procedural* language that describes **how** to carry on the calculation, Simulink shows **what** should be calculated. In procedural languages, known also as *imperative* languages, the sequence of commands is written exactly in the order in which they should be executed. In Simulink the system to be simulated is described by block diagrams. The user has to drag icons from libraries of blocks, place them in a model window, and connect the blocks according to their functional relationships. The blocks available in the various Simulink libraries spare the work of programming special functions, such as step, ramp or wave inputs, backlash, saturation, or detection of zero-crossing, and especially integration. These are only a few examples out of the many functions provided by the package. Block diagrams may also help in gaining more insight into the process we want to simulate.

The software translates the block diagram into MATLAB code. The model can be run, modified, or stopped interactively from the graphical interface that displays the block diagram.

One important point is that there exists a student's edition of MATLAB that includes Simulink. Another point is that there are many toolboxes, provided by the MathWorks and third-party suppliers, that greatly enhance the possibilities of Simulink.

In this chapter the reader will find a short introduction to Simulink. An extended tutorial would require a full book, possibly more than one book. To explain the good qualities of Simulink and convince the reader to use them, we begin with examples that have analytical solutions. We encourage the reader to implement these solutions in MATLAB and compare the results with those yielded by simulation in Simulink. To introduce Simulink in a smoother way, we begin with an example that does not imply integration. We continue with examples of first order, linear differential equations, next with second-order, linear differential equations, and end up with non-linear differential equations. We give examples both in mechanical and in electrical engineering.

11.2   Beats

The phenomenon of *beats* is caused by the superposition of two waves of slightly different frequencies. A well-known example is the sound produced by striking two adjacent piano keys. Let us consider, for example, a pair of corresponding *A* and *B-flat* keys. The former — known in French, Italian and Spanish as *La*, and in German as *A* — has the frequency of 440 Hz, the latter — known in French as *Si bémol*, in Italian as *Si bemolle*, and in German as *H* — has the frequency of 466 Hz. As known, in higher octaves these frequencies are multiplied by powers of 2. So far for the physical model and now let us translate it into a mathematical model. We assume that the two vibrations have the same amplitude, equal to 1, and then we represent the *A* tone by the equation

$$A = \sin(2 \times 440\pi t) \tag{11.1}$$

and the *B-flat* tone by the equation

$$B = \sin(2 \times 466\pi t) \tag{11.2}$$

The beats are produced by the superposition of the two tones, that is

$$C = A + B = \sin(2 \times 440\pi t) + \sin(2 \times 466\pi t) \tag{11.3}$$

The sum of the two sine functions can be converted into a product by a classical formula of trigonometry

$$C = 2 \times \cos \frac{2\pi(466 - 440)t}{2} \sin \frac{2\pi(466 + 440)t}{2} = 2\cos(26\pi t)\sin(906\pi t) \tag{11.4}$$

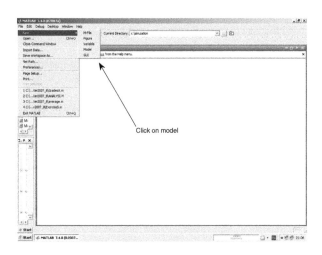

FIGURE 11.1: To open a new model window

The result is a basic acoustic vibration of 453 Hz whose amplitude varies with the frequency of 13 Hz. We invite the reader to check this result by adding the two acoustic waves in MATLAB.

Now we implement the model on a computer by using Simulink. To do so we must open the Simulink **library browser** and a **model window**. Usually these actions are carried out in the above order. We have found that the reverse order can be more convenient for arranging the screen. Accordingly, on the toolbar open the `File` menu, click on `New` to open a submenu, and click on `Model` as shown in Figure 11.1. A new window appears; dock it on the right-hand side of the screen as shown in Figure 11.2. Next, type `simulink` at the command prompt, or click the Simulink icon, as shown in Figure 11.3. The library browser opens as shown in Figure 11.4.

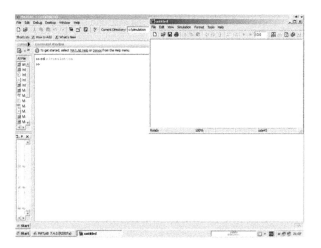

FIGURE 11.2: The new model window

To build our model we have to drag *blocks* from the appropriate libraries to the model window and connect them with lines that show how the *signals* are transmitted from one block to another. For our model we need:

1. two blocks that generate the acoustical vibrations described by Equations 11.1 and 11.2;

2. one block that performs the intended mathematical operation on the signals issued by the blocks described above, in our case the sum of the two signals, and

3. a block that displays the result of the operation, in our case the beats.

The blocks that generate the two sine waves are **sources** of signals. Therefore, in the `Library browser` we double click on `Sources`. In the window that opens we look for `Sine wave` and click on the icon. Holding down the left mouse button we drag the icon into the model window. After placing it

FIGURE 11.3: The Simulink icon

in the appropriate position, as shown in Figure 11.5, we release the mouse button. For the second tone we drag again the **Sine wave** icon and place it in the model window as shown in Figure 11.6

Now we need the block that represents the addition. In the left column of the library browser we look for **Commonly used blocks** and double click on it. The corresponding library opens. We find there the **Sum** icon and drag it into the model window as shown in Figure 11.7.

At this stage we need a block that displays the result of the mathematical operation. Blocks that receive the final signals are stored in the **Sinks** library. We return to the left column of the library browser, look for the **Sinks** icon and double click on it. The **Sinks** library window opens and we seek in it the **Scope** icon; we drag it into the model window, as in Figure 11.8.

To complete the block diagram we must show the paths followed by the signals. All blocks are provided with ports. More specifically sources have output ports, sinks input ports, and many other blocks have both input and output ports. Output ports are represented by > signs pointing outwards from the block, while input ports are represented by > signs pointing towards the block they belong to.

Let us begin by connecting the first sine block with the sum block. We bring the mouse over the output port of the first sine block, press the left mouse button and holding it down we drag a line till the left-hand input port of the **Sum** block, as seen in Figure 11.9. While we drag the mouse pointer between the two blocks we draw a dotted line. When we release the button, a solid line appears; it ends with an arrow that indicates the direction of the signal flow. In a similar way we connect the output port of the second sine wave to the bottom input port of the **Sum** block, and the output port of the latter block to

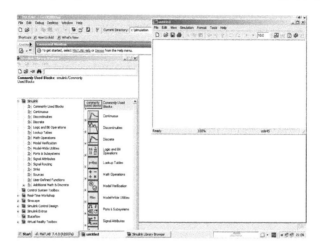

FIGURE 11.4: Opening the Simulink library browser

the input port of the **Scope** block. The model looks now as in Figure 11.10.

At some stage it is wise to save our work and give it a name. To do so open the **File** menu and click on *Save as*. Save the file in an appropriate directory, for instance **Simulation**. Let us choose the file name **Beats**; the file type will be **mdl**, an obvious abbreviation of *model*.

We can enhance readability by giving significative names to the various blocks and signals. For example, let us click on the label 'Sine wave' and add a '1' to read 'Sine Wave 1'. Next, let us click on the initial label 'Sine wave 1' of the second source block and change '1' to '2' to read 'Sine wave 2'. Double click on the line that connects the first sine-wave block to the sum block. In the text box that opens write 'A tone'. Double click on the line that connects the second wave-sine block to the sum block and write in the opening text box 'B-flat tone'. Finally, double click on the line that leads from the sum block to the scope block and write in the text box 'Superposed waves'.

Now, that we have drawn the block diagram, we must define the *parameters* of each block. To begin double click on the first sine icon. A dialogue box opens as in Figure 11.11. We accept the default values with the exception of the frequency that we define as **2*pi*440** rad/sec. MATLAB works as an interpreter; we can enter mathematical expressions like the one shown here. There is no need to input the calculated value; MATLAB will perform the operation. Click on **OK** to apply the choice and close the dialogue box.

Now we double click on the icon of the second wave source. Again we leave all default values untouched, with the exception of the frequency that we change to **2*pi*466** rad/sec. Click on **OK** to apply the choice and close the dialogue box (see Figure 11.12).

Figure 11.13 shows the dialogue box of the **Sum** block. Under **List of signs** appear two '+' signs; it means that two signals are added. In other

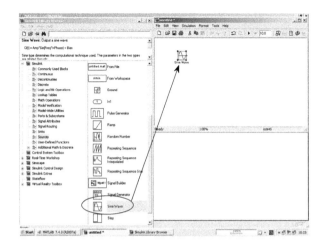

FIGURE 11.5: Dragging a `Sine wave` block from the `Sources` library

examples we are going to see that they can be changed. In this example we leave the default parameters untouched.

To see the parameters of the `Scope` block we must have the scope opened. If it does not appear, double click on its icon in the model window. Then click on the parameters icon marked by an ellipse in Figure ??. Figure 11.14 shows the dialogue box that opens. We leave the default values as they are.

Before running the model we must fix the run parameters. To do so click on `Simulation` in the tool bar of the model window. A pull-down menu with two items opens as in Figure 11.15. Click on `Configuration parameters` to open the dialogue box shown in Figure 11.16. Accept the default values except the three parameters indicated here:

Stop time to be set to 0.1,

Solver to be set to `discrete`,

Fixed-step to be set to 0.0001.

The values indicated above suit the frequencies of the sine tones; they ensure a meaningful plot. General information about the configuration parameters can be found on the Mathworks site for Simulink documentation, under the title `Solver panel`.

Our model is now ready and we can run the simulation. Return to `Simulation` on the toolbar and in the pull-down menu click on `Start`. The end of the run is announced by a sound. If the scope is not open, double click on it. The scope window opens and Figure 11.17 shows the plot of the resulting beats. We in-

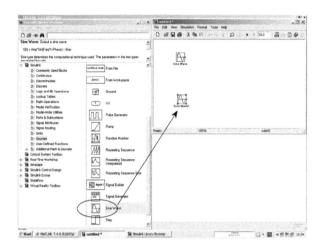

FIGURE 11.6: Dragging a second `Sine wave` block from the `Sources` library

vite the reader to check on the graph that the frequency of the basic oscillation and the period of the envelope correspond to those predicted by Equation 11.4.

11.3   A model of the momentum law

In this example we show how to simulate a first-order, linear differential equation derived from elementary laws of mechanics. The physical model is that of a mass accelerated by a force that increases linearly. This is the case of all accelerating means of transportation. To keep the model as simple as possible we do not assume a limit of the force increase. We will return to this point at the end of this section.

And now to the mathematical model. One formulation of Newton's second law states that the external force acting on a body equals the rate of change of the *momentum*:

$$F = \frac{d(mv)}{dt} \tag{11.5}$$

where F is the external force, m the body mass, v the velocity of the body centre of mass, and t the time variable. The quantity mv is called *momentum*. Assuming a law of evolution of the force F, we can calculate the behaviour of the velocity v. To give an interesting example that shows how easily Simulink treats piecewise continuous functions, let F be a *ramp* function that starts

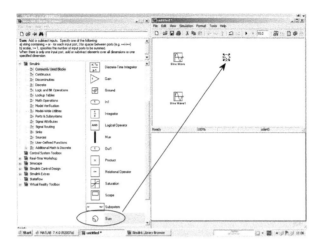

FIGURE 11.7: Dragging a `Sum` block from the *Commonly used blocks* library

from 0 N (newton), at the time $t = 0.5$ s, and increases indefinitely with the slope 0.5. Considering the mass, m, constant, the solution of Equation 11.5 is

$$v = \frac{1}{m} \int F(t)dt \tag{11.6}$$

We have written $F(t)$ to show that this force is a function of time. We further express $F(t)$ as a ramp function by writing

$$F = F_0 r(t) \tag{11.7}$$

where F_0 is a constant that we assume here as equal to 2 N, and $r(t)$ is a ramp function provided in a Simulink library. Multiplying F by $1/m$ we obtain the integrand of Equation 11.6. The integration of this quantity yields the velocity, v.

Equations 11.5 to 11.7 constitute our mathematical model. To implement it on the computer we build the Simulink block diagram corresponding to the equations described above and plot the velocity, v, against the time, t. Again, we do this in three steps:

1. we drag blocks from the `Simulink Library Browser` and place them in the block diagram;

2. we connect the blocks according to their relationships;

3. we assign values to the parameters of the various blocks.

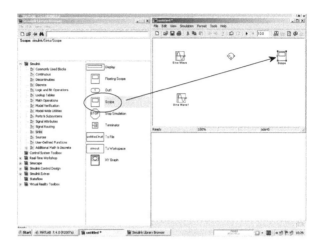

FIGURE 11.8: Dragging a `Scope` block from the *Sinks* library

We start by opening MATLAB and typing `simulink` in the command line. The browser of Simulink libraries opens. To open a window in which the Simulink block diagram will be built, click on `File` in the toolbar, then on `New` to open a pull-down menu and in this menu click on `Model`. Alternatively, to open the new model window click in the **toolbar of the library browser** on the leftmost icon that looks like an empty page.

In the library browser click on `Commonly Used Blocks` and in the new browser that opens look for the `Constant` block. Point the mouse on this block and keeping the left button pressed down drag the block into the model window to the place shown in Figure 11.18. Release the mouse button. To complete the diagram, as shown in Figure 11.18, drag a `Ramp` block from the `Sources` library, a second `Constant` block, two `Product` blocks, and an `Integrator` block from the `Commonly Used Blocks` library, and a `To Workspace` block from the `Sinks` library. This completes the first phase.

Now we must connect the blocks. We proceed as in the previous example. As to 'cosmetics', it is easy to move the blocks for aligning them, and to 'shape' the lines until we obtain a nice-looking, easy-to-understand block diagram. Figure 11.18 shows the completed block diagram.

We go now to phase 3 in which we fix the parameters of the various blocks. Double click on the first `Constant` block. A dialogue box opens. Set the `Constant value` to 2 and leave for the other parameters the default values that appear. Thus, the line `Interpret vector parameters 1-D` should be checked, that is the symbol ✓ should be there. The `Sampling mode` is set to `Sample based`, and the `Sample time` to *inf*.

Double click now on the `Ramp`. In the dialogue box that opens set the

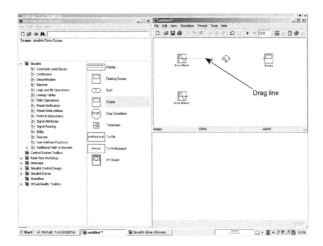

FIGURE 11.9: Dragging a connecting line between two blocks

Slope to 0.5 and the **Start time** to 0.5. Do not change the default values 0 for **Initial output**, and -1 **Interpret vector parameters as**.

In the **Product** block the **Number of inputs** should be 2. Leave the default values **Element-wise** for **Multiplication** and -1 for the **Sample time**.

In the **Constant 1** block set the **Constant value** to 0.5. This is the inverse of the mass assumed by us, that is $1/2N^{-1}$. The other values are the same as in the first constant block.

In the **To Workspace block** set **Save format** to **Array**.

We are ready to run the model. Take care that the configuration parameters include start time 0, stop time 10.0, **Type variable step** and **Solver ode45 (Dormand-Prince)**. This time the output is sent to the workspace and we can check that we have in it the arrays **simout** and **tout**. Therefore, we can plot one against the other with the commands

```
≫ plot(tout, simout), grid
≫ xlabel('Time, s')
≫ ylabel('Speed, m/s')
```

and obtain the graph shown in Figure 11.19.

The assumption of a ramp input that continues to increase without limit is certainly unrealistic. In real-life systems the increase must stop at some moment. We leave to an exercise the task of modeling an input signal that starts to increase linearly at a given moment, and remains constant from another given moment on. As a hint, build a simple Simulink model in which you add the signals produced by two ramp blocks with the parameters

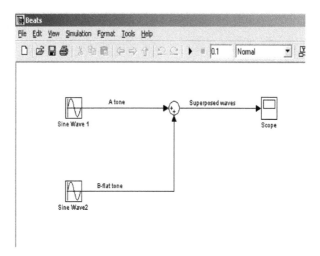

FIGURE 11.10: The completed block diagram

	Slope	Start time	Initial output
Ramp 1	1	0	0
Ramp 2	-1	5	0

Run the model for 10 s.

11.4 Capacitor discharge

A simple example in electrical engineering that can be modeled by a first-order, linear differential equation is that of a capacitor discharge over a resistor. The physical model is shown in Figure 11.20, where C is the capacitor, R a resistor, and E a DC source. We are going to use the same letters for the values of these electrical components, respectively capacitance, resistance, and voltage. A switch can be brought to position SA or SB. If the switch is set as in Figure 11.20, the capacitor is charged until the voltage across it is equal to that of the DC source. When the switch is brought to the position SB, the capacitor discharges through the resistor producing a current whose intensity, i, decays to zero while the stored energy is dissipated.

The mathematical model of the discharge process is constituted by four equations. First, the initial charge, Q, of the capacitor is equal to

$$Q = EC \tag{11.8}$$

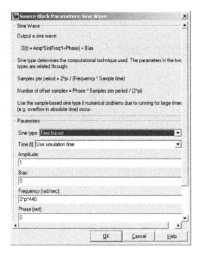

FIGURE 11.11: The parameters of the first `Sine wave` block

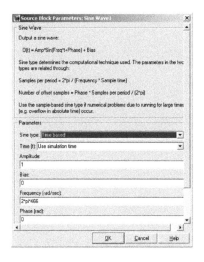

FIGURE 11.12: The parameters of the second `Sine wave` block

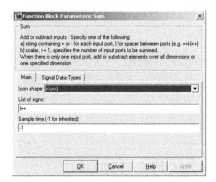

FIGURE 11.13: The parameters of the `Sum` block

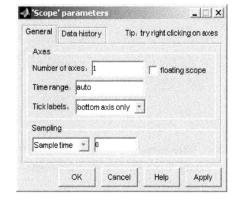

FIGURE 11.14: The parameters of the `Scope` block

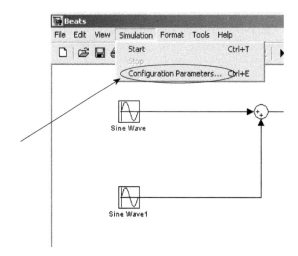

FIGURE 11.15: The dialogue box of simulation parameters

FIGURE 11.16: The parameters of the simulation run

As the charge of the capacitor, Q, and the voltage across it, V, varies with time, we write

$$V = \frac{Q}{C} \tag{11.9}$$

The current produced by the discharge is equal to the rate of change of the charge. As this charge gets smaller during the process, we use a minus sign when writing

$$i = -\frac{dQ}{dt} \tag{11.10}$$

Finally, Ohm's law gives

$$i = \frac{V}{R} \tag{11.11}$$

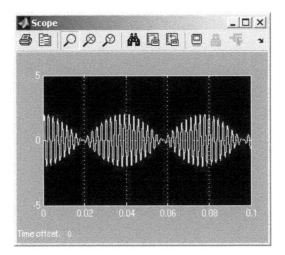

FIGURE 11.17: Beats resulting from the superposition of the A and B-flat tones

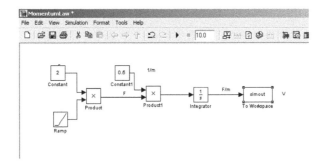

FIGURE 11.18: The Simulink model of the momentum-law

From the four equations we derive

$$\frac{dV}{dt} = -\frac{1}{CR}V \tag{11.12}$$

that we rewrite as

$$\frac{dV}{V} = -\frac{1}{CR}dt \tag{11.13}$$

Simple integration yields

$$\log V = -\frac{t}{CR} + k \tag{11.14}$$

where k is a constant of integration. To define it we remark that at the time $t = 0$, $V = E$; therefore, the final result is

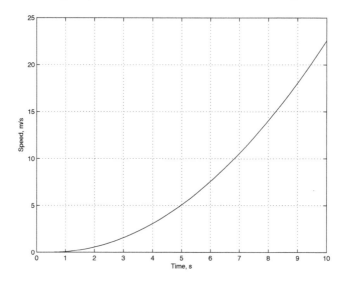

FIGURE 11.19: The output of the momentum-law model

$$V = Ee^{-t/CR} \tag{11.15}$$

where CR is called the *time constant* of the process and we note it by τ.

Let us assume now that the capacitance is $C = 250\ \mu$ F, the resistance $R = 20$ k Ω, and the source potential $E = 12$ V. The resulting time constant is $\tau = 5$ s. We invite the reader to plot the result in MATLAB and keep the plot for comparison with that yielded in Simulink. Use a time scale of at least three time constants. Add to the plot the tangent in the origin. It is obtained by connecting the point with coordinates $[\ 0;\ E\ ]$, with the point $[\ \tau;\ 0\ ]$. Derive the proof of this statement for yourself.

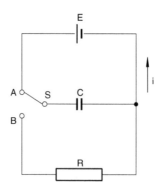

FIGURE 11.20: Capacitor discharge over a resistor

To implement the mathematical model in Simulink we start from Equation 11.12. We store the values of C and R in separate `Constant` blocks dragged from the `Commonly used blocks` library. In this way it is possible to play independently with their values.

We need another `Constant` block to store the value 1. We multiply C by R in a `Product` block dragged from the `Commonly Used Blocks` library. A `Divide` block, dragged from the `Math operation` library, will be used to calculate the inverse of the time constant, $\tau = 1/CR$. We prepare a `Product` block, taken from the `Commonly Used Blocks` library, to multiply V by the $1/CR$ value.

The output of the `Product` block is the derivative dV/dT that we input to an `Integrator` block to obtain the voltage, V. The integrator block is chosen from the `Continuous` library. It is marked by $\frac{1}{s}$, which means integration in operational calculus (remember the Laplace transform). The voltage signal, V, is supplied to a `Scope` block obtained from the `Sinks` library.

We also have to supply the voltage, affected by the minus sign, to the product block, as it appears in the right-hand side of Equation 11.12. To multiply the V signal by -1 we drag a `Gain` block from the `Commonly used blocks` library. The icon of the `Gain` block is a triangle pointing rightwards. To change the orientation click on `Format` in the toolbar of the model window. In the pull-down menu that opens click on `Flip`. To bring the V signal to the input port of the `Gain` block we have to draw a *branch line* from the line that connects the integrator to the scope. To do so hold the `Ctrl` key, point the mouse to the V line and drag a line downwards, click and continue to draw the line towards the input port of the `Gain` block.

The completed model is shown in Figure 11.21.

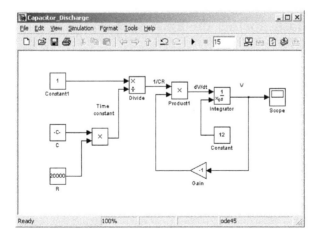

FIGURE 11.21: The Simulink model of capacitor discharge

If the default axes of the Scope do not show all the graph, look in the toolbar of the scope for the icon looking like a binocular and click on it. Thus the axes are resized to show the graph well. We see the resulting display in Figure 11.22.

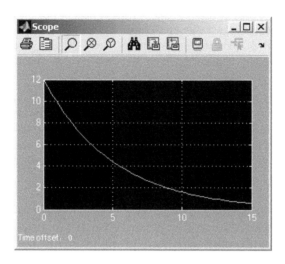

FIGURE 11.22: The voltage across the discharging capacitor

11.5 A mass–spring–dashpot system

In this section we describe a simple mechanical system that is modeled by a second-order, linear equation. Figure 11.23 shows the physical model. A mass, m, is supported by a spring characterized by the *spring constant* k. The motion is damped by a *dashpot* that produces *viscous* friction. If the mass is initially displaced a distance x_0, and pushed with an initial velocity $\dot{x}_0$, after being left free it oscillates freely with an amplitude that decays to zero. We measure the displacement of the mass m in the direction of the x-axis shown at the right of the mass. The forces acting on the mass are:

1. inertia;

2. the force of the extended or compressed spring;

3. the viscous friction generated by the dashpot.

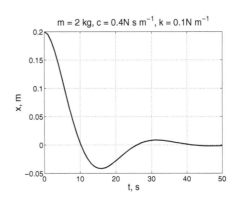

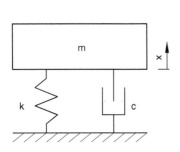

FIGURE 11.23: A mass-spring-dashpot system

FIGURE 11.24: Damped oscillations of the mass-spring-dashpot system

The mathematical model is obtained by writing that, in the absence of an external force acting on the mass, the sum of the three forces mentioned above is zero

$$inertia\ force\ +\ spring\ force\ + friction\ force\ =\ 0$$

We assume the following simplifications:

Linear spring - the amplitude of the oscillations is limited to the domain within which the force developed by the spring is proportional to its extension or compression. We write this force as kx.

Linear damping - the damping force is proportional to the velocity of the mass, $\dot{x}$. We write this force as $c\dot{x}$.

The resulting equation is

$$m\ddot{x} + c\dot{x} + kx = 0 \tag{11.16}$$

The corresponding *characteristic equation* is

$$ms^2 + cs + k = 0 \tag{11.17}$$

with the solutions

$$s_1 = \frac{-c + \sqrt{c^2 - 4mk}}{2m}, \quad s_2 = \frac{-c - \sqrt{c^2 - 4mk}}{2m} \tag{11.18}$$

The general solution of Equation 11.16 is

$$x(t) = Ae^{s_1 t} + Be^{s_2 t} \tag{11.19}$$

To find the constants A and B we use the initial conditions defined above, that is

$$x(0) = A + B = x_0$$
$$\dot{x}(0) = s_1 A + s_2 B = \dot{x}_0$$

To give an example let us assume the values

$$m = 2\text{kg}, \quad c = 0.4\text{N sm}^{-1}, \quad k = 0.1\text{N m}^{-1}$$

and the initial conditions $x_0 = 0.2$ m, and $\dot{x}_0 = 0$ m s^{-1}. The solutions of the characteristic equation are

$$s_1 = -0.1 + 0.2i, \quad s_2 = -0.1 - 0.2i$$

and the constants of the solution are

$$A = 0.1 - 0.05i, \quad B = 0.1 + 0.05$$

Carrying on the calculations in MATLAB produces the result shown in Figure 11.24. The oscillations are damped. Keep the graph for comparison with the results obtained in Simulink.

For numerical calculation of the solution of a second-order differential equation it is convenient to present it as a system of two first-order differential equations. We say then that we work in the *state space*. It also can be convenient for building the block diagram in Simulink. We define two *state variables*

$$x_1 = \dot{x}, \ x_2 = x,$$

and rewrite Equation 11.16 as

$$\dot{x}_1 = -\frac{c}{m}x_1 - \frac{k}{m}x_2 \qquad (11.20)$$
$$\dot{x}_2 = x_1$$

A good practice is to begin the block diagram with the elements that define the highest derivative. In our case this means to build first the model of the first of Equations 11.20. We start with a Sum block dragged from the Commonly used blocks library. In the model we click on this block and then in the toolbar we click on Format. In the menu that opens we click first on Rotate block and next on Flip block. When the block looks as in Figure 11.25, we double click on it and in the parameter dialogue box we change the list of signs to '- -'.

To calculate the two terms of the first of Equations 11.20 we need two Product blocks dragged from the Commonly used blocks library, and two

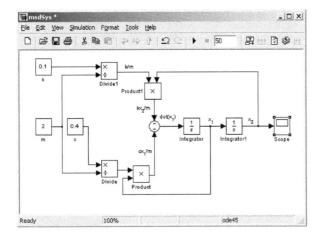

FIGURE 11.25: The Simulink model of the mass-spring-dashpot system

Divide blocks dragged from the Math operations library. To store the values of k, m and c we use three Constant blocks taken from the Commonly used blocks library. The output of the Sum block is the acceleration $\ddot{x}_1$; we lead it to the input of an Integrator block to obtain the speed x_1. We use the latter as input to a second Integrator block that yields the displacement x. To visualize the displacement we use a Scope found in the Sinks library. We connect the blocks as explained in the previous sections. Recall that to derive a branch line, from an existing connection line, we use the mouse while holding down the Ctrl key.

In the MATLAB calculations run for this example we used the initial values $x_0 = 0.2$ m, and $\dot{x}_0 = 0$ m s^{-1}. To reuse them we leave in the parameter dialogue box of the first Integrator block the value 0, while in the box of the second Integrator we set the initial condition to 0.2. We click on Simulation in the toolbar, and in the menu that opens we click on Configuration parameters. We take care that the following values are set

Start time	0.0
Stop time	50
Type	Variable step
Solver	ode45

Next we click on Start in the Simulation menu. The result appears on the Scope display. To see it as in Figure 11.26 click on the binocular icon in the toolbar of the Scope.

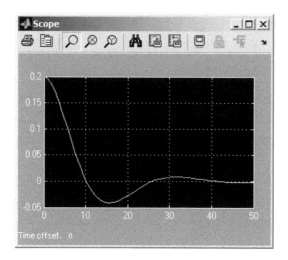

FIGURE 11.26: Mass-spring-dashpot system — Simulation in Simulink

11.6   A series RLC circuit

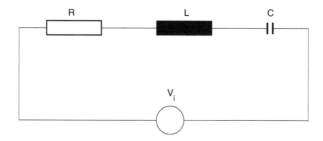

FIGURE 11.27: A series RLC circuit

In this section we show how to model and simulate in MATLAB an alternating-current circuit containing a resistor, an inductor and a capacitor connected in series. We present the mathematical model as an equation containing a variable, its first derivative and its integral. Actually this is an *integro-differential equation.*To find the canonical solution the equation is usually transformed into a second-order differential equation. We do not have to do this for modeling in Simulink; we can use directly the integro-differential equation.

Figure 11.27 shows a resistor, an inductor and a capacitor connected in

series with an ac source. According to one of Kirchoff's laws the voltage produced by the ac source is equal to the sum of the voltage drops across each
component

$$V_i = V_R + V_L + V_C \tag{11.21}$$

where V_i is the voltage supplied by the ac source, V_R is the voltage drop across
the resistor, V_L, that across the inductor, and V_C, that across the capacitor.
Noting the current by i, and detailing the voltage drop across each component
we obtain

$$V_i = Ri + L\frac{di}{dt} + \frac{1}{C}\int_{t0}^{t} i \, dt \tag{11.22}$$

As usual, the values of the various components are noted by R for the resistance, L for the inductance, and C for the capacitance. For illustration we
choose the values $R = 70 \ \Omega$, $L = 200\mu H$, $C = 200$ pF. The corresponding
undamped resonant frequency is

$$\omega_0 = \frac{1}{\sqrt{LC}} = \frac{1}{\sqrt{200 \times 10^{-6} \times 200 \times 10^{-12}}} = 5000000 \text{ rads}^{-1} \tag{11.23}$$

and the resonant frequency $f = 759.8 \times 10^3$, that is approximately 760 kHz.
The *damping coefficient* is

$$\alpha = \frac{R}{2L} = \frac{70}{2 \times 200 \times 10^{-6}} = 1.75 \times 10^5$$

and the *damping ratio*

$$\zeta = \frac{\alpha}{\omega_0} = \frac{R}{2} \cdot \sqrt{\frac{C}{L}} = 0.035$$

The *natural frequency* is given by

$$\omega_n = \sqrt{\omega_0^2 - \alpha^2} = 4.9969 \times 10^6 \text{ rad s}^{-1}$$

As $\zeta < 1$ and $\alpha < \omega_0$, the system is *underdamped*.
We assume that the impressed voltage is sinusoidal with an amplitude V_0,
and radian frequency ω.

In the preceding section we recommended the reader to begin the block
diagram by modeling the highest derivative. In this example the highest (and
only) derivative is a factor of the inductor voltage. We isolate it from Equation 11.22

$$\frac{di}{dt} = \frac{1}{L}\{V_i - Ri - \frac{1}{C}\int i \, dt\} \tag{11.24}$$

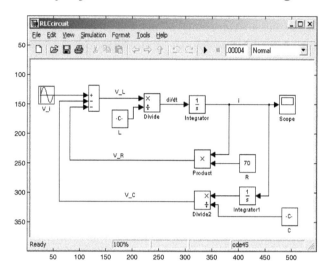

FIGURE 11.28: The Simulink model of the series RLC circuit

and begin by modeling this expression. See Figure 11.28 to follow the explanations.

From the `Commonly used blocks` library we drag a `Sum` block. To shape it
according to our needs we double click on the block and in the dialogue box
that opens we set:

Icon shape rectangular

List of signs + - -

If the block looks too small it is possible to double click on it and extend
its size. We continue by dragging a `Divide` block from `Math operations`,
a `Constant` and an `Integrator` block from `Commonly used blocks`, and a
`Scope` block from `Sinks`. On a return branch we need a `Product` block that
we drag from the `Commonly used blocks` library. Being on a return branch
we have to change the block orientation. To do so we click on the toolbar on
`Format` and in the menu that opens we click on `Flip block`. In that block we
must multiply by R, a value supplied by a constant block dragged again from
`Commonly used blocks`. This block also must be 'flipped'. To calculate the
capacitor voltage, V_C, we need another `Integrator` block, a `Divide` block,
and a `Constant` block to store the capacitance, C. The orientation of all these
blocks must be changed by clicking on `Flip block` in the `Format` menu. Finally, to supply the input voltage, V_i, we drag a `Sine wave` block from the
`Sources` library. Run this model and open the scope block. To see the plot
of the decaying oscillations click on the binocular icon that means *autoscale*.

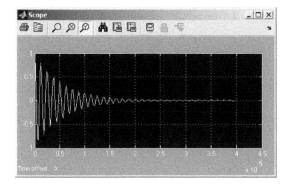

FIGURE 11.29: The result of the simulation of the series RLC circuit

Figure 11.29 shows an example run with the data described above. It is possible to count the number of oscillations in a chosen time interval and see that it corresponds to the predicted one.

11.7 The pendulum

11.7.1 The mathematical and the physical pendulum

Until now we have given examples of only linear differential equations. In this section the mathematical model is a non-linear equation and we are going to show that Simulink allows us to deal with it as simply as we did with linear equations. Step by step we will make the model more complex and let the reader see how to adapt the Simulink scheme to the mathematical model. The subject of this section, the **pendulum**, is extensively treated in the technical literature, be it in textbooks or on the web.

Figure 11.30 shows an idealized pendulum consisting of a weightless rod and a mass, m. For the sake of the figure the mass is represented by a sphere, but we consider it as concentrated in one point. The pendulum rotates freely around a fixed point represented in the figure by a black circle. What we obtain with these assumptions is not a real-life pendulum, but a simplified model that can be treated by simpler mathematics and yields good insight of some phenomena. Let us call this model *mathematical pendulum*, as opposed to a *physical pendulum*. For the moment we neglect frictional forces and consider that only two forces act on the mass:

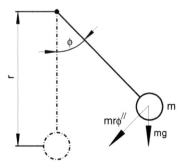

FIGURE 11.30: The mathematical pendulum

inertia equal to $-mr\ddot{\phi}$;

gravity equal to mg.

We project these forces on the tangent to the trajectory of the mass, that is on the direction of the inertia force, and write that their sum is equal to zero:

$$-mr\ddot{\phi} - mg\sin\phi = 0$$

an equation that can be rewritten as

$$\ddot{\phi} + \frac{g}{r}\sin\phi = 0 \qquad (11.25)$$

We are going to simulate this equation, but we first discuss a linearized model based on an assumption that allows elegant mathematical treatments, namely that the angle ϕ is small. Then $\sin\phi \approx \phi$. Noting the *natural frequency* by

$$\omega_0 = \sqrt{\frac{g}{r}}$$

we obtain

$$\ddot{\phi} + \omega_0^2\phi = 0 \qquad (11.26)$$

This is the *linearized equation* of the mathematical pendulum. We invite the reader to check that the *steady-state solution* can be represented by

$$\phi = \phi_0\cos(\omega_0 t + \epsilon) \qquad (11.27)$$

To find the *amplitude*, ϕ, and the *phase*, ϵ, we consider the *initial conditions* $\phi(0)$, $\dot{\phi}(0)$. The physical interpretation of these conditions is simple: we assume that at the start of the process the pendulum was deflected by an angle $\phi(0)$ and launched from there with the speed $\dot{\phi}(0)$. Then, we can write that for $t = 0$

$$\phi(0) = \phi_0 \cos \epsilon \tag{11.28}$$

$$\dot{\phi}(0) = -\omega_0 \phi_0 \sin \epsilon \tag{11.29}$$

Dividing Equation 11.29 by Equation 11.28 yields

$$\tan \epsilon = -\frac{1}{\omega_0} \frac{\dot{\phi}(0)}{\phi(0)} \tag{11.30}$$

Next we isolate the sines and the cosine of ϵ in Equations 11.28 and 11.29, raise both sides of the resulting equations to the power 2 and add them. The result is

$$\phi_0^2 = \phi(0)^2 + \frac{\dot{\phi}^2(0)}{\omega_0^2} \tag{11.31}$$

For a first experiment let us assume the pendulum length $r = 1$ m, and the initial conditions $\phi(0) = \pi/90$, $\dot{\phi}(0) = 0$. For these values the linearized equation yields the radian frequency $\omega_0 = 3.1321$ rad $\cdot$ s^{-1}, the frequency $f = \omega_0/(2 * \pi) = 0.4985$ Hz, and the *natural period* $T = 1/f = 2.0061$ s. The amplitude of the steady-state motion is $\phi_0 = \phi(0) = \pi/90$, that is 2^0. Simulating the linearized model in MATLAB is a trivial job. As to the nonlinear equation, we are going to simulate it in Simulink.

We start by isolating the highest derivative in Equation 11.25

$$\ddot{\phi} = -\frac{g}{r} \sin \phi \tag{11.32}$$

This immediately shows that the model needs two integrator blocks. We arrange them as shown in Figure 11.31. To compare the output of the non-linear model to that of the linear one, we send the result of the simulation, ϕ, to a **To Workspace** block. To obtain the sine of the angle ϕ we drag a **Trigonometric function** block from the **Math operations library**. The default option of this block is **sine** and we do not change it. To divide the angle ϕ by the pendulum length, r, we use a **Divide** block. The value r is stored in a **Constant** block. Finally, we use a **Gain** block to multiply by $-g$. Thus, we obtain the right-hand side of Equation 11.32. This is the input of the first integrator block.

Let us use another technique for connecting the blocks. Click on the **Gain** block, then press down the **Ctrl** key and click on the first integrator. The connection line appears immediately. Continue for the other connections. Next annotate the signals as shown in Figure 11.31.

We still must fix a few parameters. Double click on the **Gain** block and fix the gain to *-9.81*. For the first integration block fix the initial condition to *0*, for the second integration block fix the initial condition to *pi/90*. Open the dialogue box of the **To workspace** block and change the **Variable name**

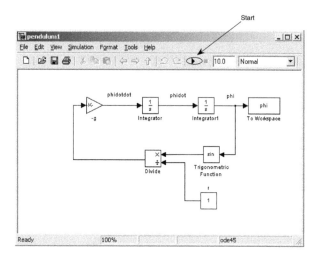

FIGURE 11.31: The model of the nonlinear pendulum

to *phi* and the `Save format` to *Array*. In the `Constant` block representing *r* the value should be *1*. Pulling down the `Simulation` menu and opening the `Configuration parameters` you may find the simulation time set by default to 10 seconds. This suits well the oscillation period of about 2 seconds found above.

We can start the simulation as we did in the other examples, that is by pulling down the `Simulation` menu in the toolbar and clicking on `Start`. To try another option, click in the toolbar on the arrow marked `Start` in Figure 11.31.

The variable ϕ and the simulation time, *tout*, are stored now in the workspace, as we can check by looking into the *workspace* window. To compare the result with that predicted by the linearized model (Equation 11.26) we use the following commands:

```
≫ omega0 = sqrt(9.81);
≫ phi0 = pi/90
phi0 =
   0.0349
≫ t = 0:  0.1:  10;
≫ y = phi0*cos(omega0*t);
≫ plot(tout, phi, 'k-', t, y, 'r+'), grid
≫ xlabel('t, seconds')
≫legend('Simulink output', 'MATLAB sampling')
```

The resulting plot is shown in Figure 11.32. The two models yield the same

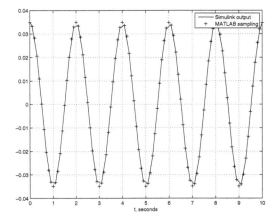

FIGURE 11.32: Comparing the output of the nonlinear model with that of the linearized model

result because $2°$ is, indeed, a small angle for which the approximate model holds well.

Let us change, now, the initial condition and assume $\phi(0) = \pi/4$. This is certainly not a small angle. Proceeding as before we obtain the plot shown in Figure 11.33. The result yielded by the nonlinear model differs from that yielded by the linear one. The amplitude, ϕ_0, is still that predicted by Equation 11.31, but the frequency is different.

Exercise 11.1 Mathematical pendulum

Run the Simulink model of the mathematical pendulum developed in the preceding section and set the initial conditions to $\phi(0) = \pi/3$, $\dot{\phi}(0) = 0.1$. Compare the results with those predicted by the linear model. This time you will distinguish differences both in amplitude and in frequency.

11.7.2 The phase plane

Let us return to the linear model. From Equation 11.27 we derive

$$\phi = \phi_0 \cos(\omega_0 t + \varepsilon)$$
$$\dot{\phi} = -\phi_0 \omega_0 \sin(\omega_0 t + \varepsilon)$$

and

$$\frac{\phi^2}{\phi_0^2} + \frac{\dot{\phi}^2}{(\phi_0 \omega_0)^2} = 1 \qquad (11.33)$$

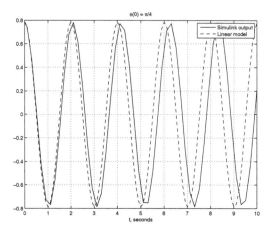

FIGURE 11.33: The output of the pendulum model for $\phi(0) = \pi/4$, $\dot{\phi}(0) = 0$.

In the ϕ, $\dot{\phi}$ plane this is the equation of an ellipse with semi-axes $a = \phi_0$ and $b = \phi_0\omega_0$. The plot of $\dot{\phi}$ as function of ϕ is called **phase plane** or **phase portrait**. This representation is a powerful tool for the study of dynamic systems. Equation 11.31 shows that the *trajectories* plotted in the phase plane depend on the initial conditions. In Figure 11.34 the upper graph shows the trajectories for constant $\dot{\phi}(0) = 0.1$ and $\phi(0)$ varying from 0 to $\pi/2$ rad, in steps of $\pi/10$. The bottom graph shows the trajectories corresponding to constant $\phi(0) = 0.1$ rad, and $\dot{\phi}(0)$ varying from 0 t0 1 rad s^{-1} in steps of 0.1 rad s^{-1}.

The area enclosed by any ellipse is equal to π times the product of the semi-axes. For the ellipses in the phase plane introduced above the area is

$$S = \pi\omega_0\phi_0^2$$

Substituting the value of ϕ_0^2 given by Equation 11.31 we obtain

$$S = \pi\omega_0(\frac{r}{g}\dot{\phi}^2(0) + \phi^2(0))$$

The quantity between parentheses is proportional to the total energy imparted to the pendulum by the initial conditions. To see this multiply the two terms between parentheses by $mrg/2$. In conclusion, the ellipses in the phase plane are curves of constant energy.

Now let us draw the phase portrait of the non-linear model. To store the angular speed, $\dot{\phi}$, in the workspace, we modify the model shown in Figure 11.31 by adding a To workspace block. Open the dialogue box of the second To workspace block and change the Variable name to dotphi and the Save format to *Array*. The modified model is shown in Figure 11.35.

After running the model enter the following MATLAB commands:

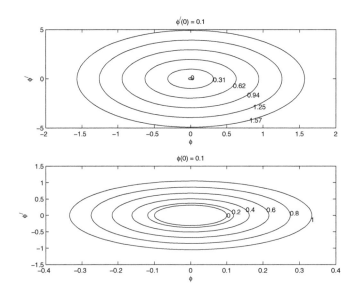

FIGURE 11.34: The phase plane of the linear pendulum. In the upper figure curves are marked by the initial condition $\phi(0)$, in the lower by the initial condition $\dot{\phi}(0)$.

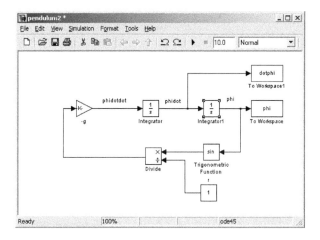

FIGURE 11.35: Pendulum model with angular speed output

```
≫ phi0 = pi/90
 phi0 =
   0.0349
≫ omega0 = sqrt(9.81)
omega0 =
   3.1321
≫ t = 0:  0.1:  10;
≫ y = phi0*cos(omega0*t);
≫ plot(y, doty, 'r--', phi, dotphi, 'b-'), grid
≫ title('\phi(0) = \pi/90, d \phi/dt(0) = 0')
≫ xlabel('phi, rad')
≫ ylabel('d\phi/dt, rad s^-1')
≫ legend('Linear model', 'Nonlinear model')
```

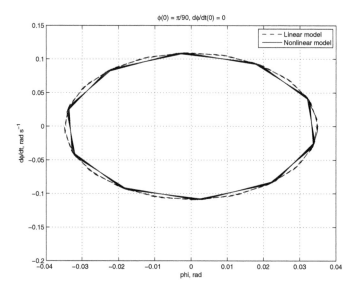

FIGURE 11.36: The phase plane of the pendulum models

The T_EX command \phi prints the Greek letter ϕ. In Figure 11.36 we can compare the phase portraits of the linear and non-linear models, for the initial conditions listed in the title of the plot.

11.7.3 Running the simulation from a script file

In Figure 11.36 we obtained the phase portrait for one set of initial conditions. We may be interested in plotting the phase plane for several values of the initial conditions. To do so we must run the simulation iteratively, for example in a FOR loop. The software package includes commands for running the model and changing its parameters from the command line. It is convenient to write these commands on a script file, as exemplified by the file **RunSim** shown below. In this application we run the model **pendulum2** for constant $\phi(0) = \pi/90$ rad and six values of $\dot{\phi}(0)$ varying from 0 to 1 rad s^{-1}. The value $\phi(0)$ is set in the dialogue box of the block called **Integrator1**.

```
%RUNSIM Runs iterations of the non-linear pendulum simulation
%        for constant initial angle and variable initial
%        angular speed.

dotphi0 = 0;                    % Initial angular velocity
for k = 1:6
    dotx0     = num2str(dotphi0);
    set_param('pendulum2/Integrator', 'Initial', dotx0)
    sim('pendulum2', 5)
    plot(phi, dotphi, 'k-')
    hold on
    text(phi(2*k), dotphi(2*k), dotx0)
    dotphi0 = dotphi0 + 0.5;
end
title('\phi(0) = \pi/2')
xlabel('\phi')
ylabel('\phi^/')
hold off
```

We begin the file by initializing $\dot{\phi}(0)$ as 0. Next we open a FOR loop. The first line of the loop converts the value of the initial condition, $\dot{\phi}(0)$, to a character string, as required by the command that follows. The syntax of the command set_param is

```
    set_parameter('object', 'parameter', value)
```

In our example the string 'pendulum2/Integrator' identifies the model and the bock in which we want to change a parameter. The string 'Initial' identifies the parameter that we want to change. The value of this parameter is specified by the string dotx0. The command for running the model is sim and its syntax is

```
    sim(model, timespan)
```

where `timespan` is the duration of the simulation run.

The other commands in the loop plot the curve yielded by the current iteration and increment $\dot{\phi}(0)$ by 0.2 rad s^{-1}. The resulting plot is shown in Figure 11.37.

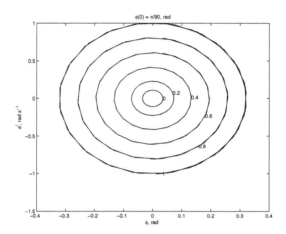

FIGURE 11.37: The phase plane of the non-linear pendulum for constant $\phi(0)$ and variable $\dot{\phi}(0)$

Alternatively, we may run the model for $\dot{\phi}(0)$ constant and variable $\phi(0)$. To do this use the following script file **RunSim1** in which $\phi(0)$ varies from 0 to π. This time the command **set_param** is applied to the initial condition of the **Integrator1** block. The initial condition $\dot{\phi}(0) = 0.1$ is set in the parameter dialog box of the **Integrator** block. The result appears in Figure 11.38.

```
%RUNSIM1  Runs the pendulum simulation from this script file
%         for constant initial angular velocity and variable
%         initial angle.

phi0 = 0; % Initial angle
for k = 1:11
    x0    = num2str(phi0, 3);
    set_param('pendulum2/Integrator1', 'Initial', x0)
    sim('pendulum2', 6)
    plot(phi, dotphi, 'k-')
    hold on
    text(phi(4*k), dotphi(4*k), x0)
    phi0 = phi0 + pi/10;
end
```

```
title('\phi^/(0) = 0.1')
xlabel('\phi, rad')
ylabel('\phi^/, rad s^{-1}')
hold off
```

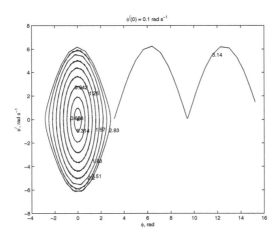

FIGURE 11.38: The phase plane of the non-linear pendulum for constant $\dot{\phi}(0)$ and variable $\phi(0)$

11.8 Exercises

Exercise 11.2 Capacitor discharge

Add a `To workspace` block to the model described in Section 11.4 so as to store the t and V data. Calculate separately the same data by using Equation 11.15. Plot the data obtained from this calculation together with the data obtained from the Simulink model. Compare the results.

Exercise 11.3 Capacitor recharge

Based on the discharge model described in Section 11.4 develop a model of *capacitor recharge*. More specifically, assume that the initial voltage across the capacitor is 0. Closing the switch in Figure 11.20 to position SA starts a charging process that continues until the voltage across the capacitor equals that of the DC source. The `Scope` block should again display the voltage against time.

12

Applications in the frequency domain

12.1 Introduction

In this chapter we give a few examples of the use of MATLAB in analysis and calculations in the frequency domain. We explain the notion of *signal* and introduce the reader to the representation of signals, their decomposition in the frequency domain and the notion of power spectrum. We end the chapter by an example of *Bode diagram*, a helpful tool in the analysis and design of control systems. The treatment employs the basic MATLAB software, with the exception of one simple example in which we call two functions belonging to the Signal Processing Toolbox.

12.2 Signals

The word *signal* has several meanings; the one that interests us is defined in Webster (1990) as 'a detectable physical quantity or impulse (as a voltage, current, or magnetic field strength) by which messages or information can be transmitted'. For example, the desired information can be a temperature, and the signal a voltage proportional to this temperature.

Many books refer to signal variation as a function of time. Measuring the sea elevation at one point would generate such a signal; it represents the evolution of the sea wave at that point. We may, however, be interested in signals that vary spatially. Examples of these are the roughness of a surface, and the grey intensity across an image. It is also possible to find signals that are functions of both time and space.

Many signals are continuous: the air temperature, the sea elevation at one point, the surface roughness. Other signals are discrete, for example letters transmitted in Morse code. Signals may be discrete because they are obtained by **sampling** continuous information, for example atmospheric temperature and pressure as transmitted at regular intervals by radio. The computer can process only discrete signals; therefore, signals intended for this purpose, unless discrete by nature, must be obtained by sampling.

To be processed by a computer, most signals can be conveniently represented by either one-, two- and less often multi-dimensional arrays. Signals produced by sensors, such as thermometers and tachometers, produce one-dimensional arrays of real numbers when sampled at constant intervals. The digitization of an image produces a two-dimensional array, each entry representing, for example, the grey level of a picture element. Three-dimensional arrays appear naturally if we want to describe the temperature of a room. We can think of dividing the room into little boxes, using three indexes for the three spatial dimensions. A four-dimensional array is the obvious way to store data, if we want to study how the temperature of the same room varies with time. Such arrays and matrices are usually manipulated in various ways; they can be filtered to attenuate noise, can be modulated, and can be processed to enhance the contrast of images, or to compress the dynamic range. While such operations can be performed in most cases either in the time or in the frequency domain, selecting the correct domain and the correct algorithm often results in a considerable increase of efficiency. MATLAB provides very fast and precise algorithms for switching back and forth from the time to the frequency domain.

In this chapter we touch only one problem of signal processing, the *discrete fourier transform*, shortly DFT, and the power spectrum. The basic MATLAB software provides tools for their treatment. More, specific tools for signal processing are provided by the Signal Processing Toolbox^{TM}.

A signal $f(t)$ of a continuous variable t is said to be **periodic of period T** if, for each t, $f(t+T) = f(t)$. If there is a minimum real positive period, this period is called the **fundamental period** of f.

Similarly, a digital signal $d = [\ldots, d(-1), d(0), d(1), d(2), \ldots]$ is said to be **periodic of period P** if, for each integer k, $d(k + P) = d(k)$.

If we sample a periodic function f of real period T at constant sampling intervals T_s, and T_s is a submultiple of T, say $T_s = T/N$, the sampled version d of f is a periodic digital signal of period N. Recall that, by definition of the sampled version, $d(k) = f(k \cdot T_s)$ and, therefore,

$$
\begin{aligned}
d(k + N) &= f((k + N) \cdot T_s) \\
&= f((k + N) \cdot T/N) \\
&= f(k \cdot T/N + T) \\
&= f(k \cdot T/N) \\
&= f(k \cdot T_s) \\
&= d(k)
\end{aligned}
$$

A periodic signal of period N is uniquely represented by any N consecutive terms, for example the terms with indices 1 to N, and can, therefore, be conveniently represented in MATLAB by a vector x of length N. The vector x

is called the **time domain representation** of the signal d, and is simply defined by the array of components x(h) = d(h), for $h = 1$ to N.

From the time domain representation x of d, using the periodicity of the signal, it is easy to reconstruct the signal itself, by means of the relation $d(h) = x(k)$, for the only k that simultaneously satisfies the following two conditions:

1. $1 \leq k \leq N$

2. $h - k$ is divisible by N.

With a slight abuse of notation, we shall identify the signal d with its time domain representation x.

Each entry $x(k)$ of the signal x can be either a real or a complex number. Actually, as we shall see in a moment, there is little advantage in considering real signals separately, either from the computational or the theoretical point of view, and it is probably more convenient to study only the general case when each $f(k)$ is complex, and consider real signals as a particular case, when all imaginary parts just happen to be equal to zero.

For a continuous signal defined as a function of time, $f(t)$, its **Fourier transform** is

$$F(j\omega) = \int_{-\infty}^{\infty} f(t)e^{-j\omega t}\,dt$$

where j is the imaginary unit. The **inverse Fourier transform** is given by

$$f(t) = F^{-1}(j\omega) = \frac{1}{2\pi} \int_{-\infty}^{\infty} F(j\omega)e^{j\omega t}\,d\omega$$

(see, for example, Dorf and Bishop, 2008). For a given function, system, or phenomena, the Fourier transform relates a representation in the **time domain** to one in the **frequency domain**.

For discrete signals, the **discrete Fourier transform**, **DFT**, of an array x of length N is another array, X, of the same length N. As shown above, the Fourier transform is invertible, its inverse usually being denoted as **IFT**. The mathematical definitions of the DFT and IFT functions will be given later in this section.

The vector X is also called the **frequency domain representation** of the signal x.

IFT and DFT are implemented in MATLAB very efficiently, using the algorithm known as the **Fast Fourier Transform**, or **FFT**. The Fast Fourier Transform is an algorithm that became popular after a famous article by Cooley and Tukey (1965). This algorithm requires approximately $N \log(N)$ arithmetic operations to compute the DFT, in comparison to approximately N^2 operations required by the routines in use until then.

Fourier transforms have two main families of applications. The first is transparent to the user. Many operations on a signal are much faster when implemented in the frequency domain. The second use of the DFT is to identify the frequency components of a signal, as we shall see in the next section.

12.3   A short introduction to the DFT

The reader is certainly familiar with the fact that in the N-dimensional space R^N (or C^N), a vector is the sum of its projections on the coordinate axes. This property is sometimes called the **decomposition property** of the vectors in R^N (or C^N).

In C^N, for example, we have a **natural basis**, denoted by the symbols $i_1, i_2, \ldots, i_N$, and defined by

$$i_1 = (1, 0, 0, \ldots, 0)$$
$$i_2 = (0, 1, 0, \ldots, 0)$$
$$i_3 = (0, 0, 1, \ldots, 0)$$
$$\vdots =$$
$$i_N = (0, 0, 0, \ldots, 1)$$

C^N, moreover, has a **dot product**, denoted by the symbol $\cdot$ and defined as follows. If

$$x = [x_1, x_2, \ldots, x_N]$$
$$y = [y_1, y_2, \ldots, y_N]$$

their dot product is

$$x \cdot y = \sum_{h=1}^{N} x_h \overline{y_h}$$

where $\overline{y_h}$ means the *complex conjugate* of y_h. The particular definition of the dot product in R^N and its MATLAB implementation are introduced in Subsection 2.1.8. We also recall that the projection of the vector x on a non-zero vector y is the vector

$$x_y = (x \cdot y) \frac{y}{y \cdot y}$$

This vector is also called the **component** of x in the direction of y.

A basis of R^N (or C^N) is said to be **orthogonal** with respect to a dot product if the dot product of any two distinct elements of the basis is equal to 0. For example, the vectors $i_1, i_2, \ldots, i_N$ just defined are an **orthogonal basis** of R^N (or C^N).

The decomposition property can be generalized into the following theorem:

For any orthogonal basis of R^N (or C^N), a vector is equal to the sum of its components in the direction of the vectors of the basis.

In other words, the decomposition property holds not only for the natural basis, but also for *any* orthogonal basis.

It is well known that the family of N vectors, each of length N, $e_m = [e_m(h)]$, defined as

$$e_m(h) = \exp\left(2\pi i \frac{(m-1)(h-1)}{N}\right) \qquad 1 \le m \le N$$

forms an orthogonal basis for C^N with respect to the dot product previously defined. Moreover, for each h, $e_h \cdot e_h = N$. The proof of these two facts can be found in virtually any book on signal processing. We will verify this result using MATLAB in a particular case, say for $N = 16$.

In the matrix $\mathbf{E}$ we are about to create, the m-th row represents the vector e_m. Type the following MATLAB commands:

```
≫ N = 16;
≫ for m = 1:N
        for n = 1:N
                E(m,n) = exp(2*pi*i*(m-1)*(n-1)/N);
        end
end
```

Let us now construct the matrix $\mathbf{D}$ that has the dot product of e_h and e_k in position (h, k). Verify that D is given in MATLAB by the expression

```
≫ D = E*E'
```

and is equal to `N*eye(N)`. The command `eye(N)` produces the $N - by - N$ `identity matrix`.

The discrete Fourier transform, X, of a vector x of length N, is defined as $X(h) = x \cdot e_h$. Therefore, $X(h)$ represents, up to a multiplicative coefficient, the magnitude of the projection of x in the direction of e_h. The operation of reconstructing x from X, called the *inverse Fourier transform*, is performed in MATLAB by the `ifft` function. As a consequence of the decomposition property, `ift` performs the algebraic operation

$$x = \sum_{h=1}^{N} \frac{X(h)}{N} e_h \qquad (12.1)$$

EXAMPLE 12.1 Inverse Fourier transform

Let us verify Equation 12.1 for a random vector of 128 terms.

```
≫ N = 128;
≫ x = rand(1,N);
≫ X = fft(x);
≫ t = (0:(N-1))/N;
≫ for h=1:N
        yy(h,:)  = X(h)/N*exp(2*pi*sqrt(-1)*(h-1)*t);
   end
≫ y = sum(yy);
```

Now compare x and y, either graphically

```
≫ plot(1:N, x, 1:N, y)
```

or numerically

```
≫ max(abs(x-y))
```

In the latter case, instead of the expected zero you will obtain a very small number. As explained in Chapter 5, this is due to the computer representation of numbers.

```
≫ P2 = fft(eye(N, N))';
```

12.4   The power spectrum

Let us consider a spring characterized by its constant, k. Within its range of linearity, a displacement x corresponds to a force $F = kx$, and a work

$$w = \int_0^x F(\xi)d\xi = \frac{k}{2}x^2$$

Let us consider now a resistor whose resistance is R. If a dc current of intensity i passes through that resistor, the voltage drop equals Ri and the dissipated power is Ri^2. In the two cases mentioned here, like in many other engineering systems or physical phenomena, the produced, or consumed power or energy are proportional to the square of a variable. This variable can be a function of time, not rarely a periodic function of time. The latter is the case of a spring that is part of an oscillating, linear mass-damper-system.

MATLAB index convention

If x is the discrete version of a periodic signal
of period T sampled at frequency $f_s = N/T$ Hz
and X is its Fourier transform,

1. $X(1)$ is associated with the DC component of the signal

2. For $b \leq N/2 + 1 X(b)$ is associated with the frequency $\frac{b-1}{T} = \frac{b-1}{N} f_s$ Hz

3. For $f \leq f_s/2$ The frequency f Hz   is associated with the bin $b = \frac{N \cdot f}{f_s}$

FIGURE 12.1: Relationship between the frequency components of a signal and its DFT

There are oscillatory systems whose response consists in a superposition of periodic functions, each having its own frequency. Then, to each component corresponds a power, or energy, proportional to the square of the amplitude at that frequency. When analyzing such phenomena or systems, we may be interested in the distribution of power, or energy, as a function of frequencies. This section is a short introduction to this subject.

Let x be the time-domain representation of a periodic signal of period T, sampled at intervals $T_s = T/N$, and X its discrete Fourier transform.

The base vector e_h defined in the previous section is the sampled version of the continuous signal

$$\exp\left(2\pi i \frac{h-1}{T} t\right) \tag{12.2}$$

of frequency $(h-1)/T$ Hz.

The vector $(X(h)/N)e_h$ is the projection of x in direction of e_h. For $h > 1$, this vector is often referred to as the **component** of x of frequency $(h-1)/T$ Hz. The h-th entry of the vector X, that is, $X(h)$, is sometimes called **the h-th bin of X**, especially in the signal processing community.

The average of x is $X(1)/N$ and is sometimes called the **DC component** of the signal x, especially by electrical engineers; $(X(2)/N)e_2$ and $(X(N)/N)e_N$, which are associated to the period T, are called the **fundamental components** of the signal.

Figure 12.1 shows the relationship between the components of $X = \text{fft}(x)$ and the frequency components of x. If you are familiar with the literature on the subject, please notice that MATLAB notation is slightly different from that used in many books.

For each $h = 1, 2, \ldots, N-1$, the terms $X(1+h)$ and $X(1+N-h)$ are called **conjugate terms**. Exercise 16.1 justifies this definition.

The array of elements $|X(h)|^2/N$ is called the **power spectrum** of x, as for many phenomena the square of the signal, or a multiple of it, represents its power. If this is the case, the decomposition theorem has an important physical interpretation.

The square of the magnitude of the vector x, by definition equal to

$$x \cdot x = \sum_{h=1}^{N} |x_h|^2$$

can be computed using Equation 12.1:

$$x \cdot x = \sum_{h=1}^{N} \frac{|X_h|^2}{N} \tag{12.3}$$

as all the mixed products $\langle e_h | e_k \rangle$ are equal to 0 when $h \neq k$. The term $|X_h|^2/N$ represents the power of the component of x of frequency $f = (h - 1)f_s/N$. So Equation 12.3 states that *the power of a signal is equal to the sum of the powers of its individual components.* This is one of the forms of Parseval's theorem (Marc Antoine, 1755–1836).

EXAMPLE 12.2 Extracting the spectral frequencies of a signal

We are going to sum two signals of different frequencies and amplitudes, and see how each contributes to the power spectrum. The only difficulty in this example is to keep track of which bin of the power corresponds to which frequency, and how the magnitude of the power relates to the amplitude of the signal. We begin by creating a signal x_1 of amplitude $a_1 = 7$ and frequency $f_1 = 16\,\text{Hz}$ and a second signal x_2 of amplitude $a_2 = 3$ and frequency $f_2 = 48\,\text{Hz}$, sampled at $128\,\text{Hz}$, and call x their sum:

```
≫ N = 512;              % number of points
≫ b = 1:N;              % bins
≫ Ts = 1/128;           % sampling interval in seconds
≫ fs = 1/Ts;            % sampling frequency in hertz
≫ ts = Ts*(b-1);        % sampling instants
≫ a1 = 7; f1 = 16;
≫ x1 = a1*sin(2*pi*f1*ts); % 1st signal
≫ a2 = 3; f2 = 48;
≫ x2 = a2*sin(2*pi*f2*ts); % 2nd signal
≫ x = x1+x2;
```

If you want to see the resulting signal (Figure 12.2) enter the commands:

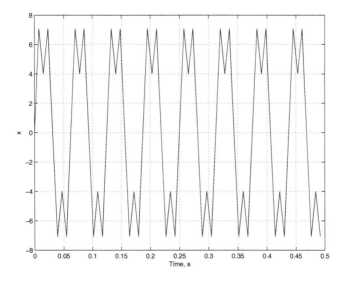

FIGURE 12.2: The sum of two sinusoids

```
≫ plot(ts(1:  N/8), x(1:  N/8), 'k-')
≫ grid
≫ xlabel('Time, s'), ylabel('x')
```

To enhance readability, we plotted only one eighth of the signal. We can now construct and plot the power spectrum:

```
≫ X = fft(x);            % DFT of x
≫ pwr = X.*conj(X)/N;    % power of the signal
≫ frs = (b-1)/N*fs;      % frequencies
≫ subplot(2,1,1);
≫ plot(b, pwr, 'k-')     % plot power spectrum
≫ grid; xlabel('bin number'); ylabel('Power');
≫ subplot(2,1,2);
≫ plot(frs, pwr);
≫ grid on; xlabel('Frequency, Hz'); ylabel('Power');
≫ axis([ 1 frs(N) 0 8000 ])
```

The resulting plot appears in the upper graph of Figure 12.3. The signal x_1 contributes to the power at bin 65, associated with the frequency $f_1 = 16\,\text{Hz}$ ($65 = 1 + 512 \times 16/128$, see Figure 12.1) and at bin 449, the latter because the conjugate of the number found in bin $65 = 1 + 64$ is stored in bin $449 = 1 + 512 - 64$.

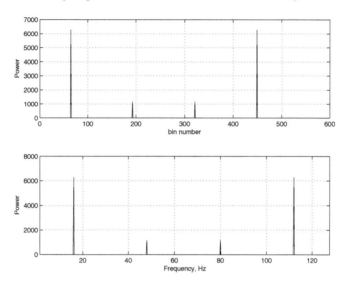

FIGURE 12.3: Power spectrum of the signal $x = x1 + x2$

The power spectrum is shown in the lower graph of Figure 12.3. The reader is invited to verify that `pwr(65)` is equal to $(a_1/2)^2 N$. Similarly, the second sinusoidal signal x_2 contributes its power to bins 193 and 321, and `pwr(193)` = `pwr(321)` is equal to $(a_2/2)^2 N$.

The Signal Processing Toolbox includes powerful functions for spectral analysis. Their use requires a deeper knowledge of the subject than what we assume for the basic reader of this book. Therefore, we are giving here only one example. If you have in the workspace the sum, x, of the two signals calculated in this example, type

```
≫ Fs = 128;
≫h = spectrum.welch;
≫Hpsd = psd(h, x, 'Fs', Fs);
≫plot(Hpsd)
```

The resulting plot is shown in Figure 12.4. The reader interested in the applications of MATLAB to signal processing can refer to one of the many specialized books published in this domain. To find an updated list refer to http://www.mathworks.com. On the upper bar of the main page click on **Products and Services**. On the right-hand side of the page that opens look for **MATLAB & Simulink Books** and click **View all books**.

EXAMPLE 12.3 Identification of the frequencies that contain most power

In this example we are going to decompose a triangular signal of period $T = 5$ seconds and peak-to-peak amplitude 1 into its frequency components,

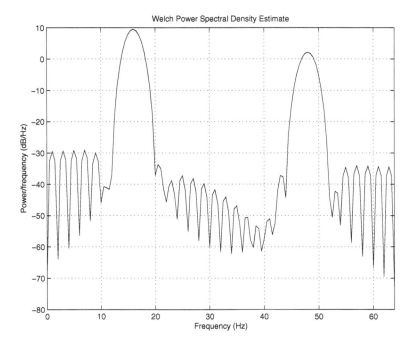

FIGURE 12.4: Power spectrum of the signal $x = x1 + x2$; plot obtained with functions belonging to the Signal Processing Toolbox

using 512 sampling points. We are interested in finding what percentage of the total power is contained in a signal obtained from the original, by neglecting all components except the four most significant ones. We also want to know how close the approximating signal is to the original. First of all, we are going to construct the discrete version x of the signal, by sampling it at 512 equally spaced points:

```
≫ T = 5;
≫ N = 512;
≫ t = linspace(0, T, N+1);
≫t = t(1:N);
≫ x1 = 2*t/T-1/2;
≫x2 = 2*(T-t)/T-1/2;
≫ x = min(x1,x2);         % the triangular signal
≫ plot(t ,x, 'k-')
```

The resulting plot shows the triangular signal (solid line in Figure 12.5). Now we construct its power spectrum:

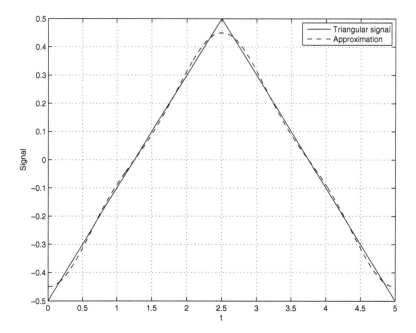

FIGURE 12.5: Approximation of a triangular signal

```
≫ b = 1:N;                 % bin sequential number
≫ X = fft(x);
≫ Ts = T/N; fs = N/T;      % sampling interval and frequency
≫ frs = (b-1)/T;           % equal to (b-1)/N*fs;
≫ pow = X.*conj(X)/N;
```

To check our results so far, we can verify Parseval's equality. The following two numbers should be equal, up to the numerical precision of the computer:

```
≫ [sum(pow) norm(x)^2]
ans =
    42.6680    42.6680
```

It is easy to identify the frequencies that contain the largest part of the power, using the **sort** function which returns the elements of pow in increasing order:

```
≫ [spow, spos] = sort(pow);
```

Let us find the indices of the four frequencies that contain most power:

```
>> m = 4; spos(N: -1:   (N-m+1))
```

We see that these frequencies are contained in bins 512, 2, 510 and 4. Let us now construct the approximating signal:

```
>> X4 = zeros(X); % allocate space for approximated X
>> h = [512 2 510 4];
>> X4(h) = X(h); % copy the bins that contain most power
```

The percentage of the power contained in the four most significant terms is then given by

```
>> perc = 100*(norm(X4)/norm(X))^2
```

In conclusion, 99.7698% of the power is contained in just four terms, those corresponding to the fundamental frequency 0.2 Hz, associated with bin number 2, its conjugate frequency, associated with bin 512, the second harmonic, 0.6 Hz, associated with bin 4, and its conjugate, associated with bin 510.

The following lines complete the plot in Figure 12.5. We can see how closely the original triangular signal is approximated.

```
>> x4 = ifft(X4);
>> hold on
>> plot(t, x4, 'r--')
>> grid on
>> xlabel('t');
>> ylabel('Signal')
>>legend('Triangular signal', 'Approximation')
```

12.5   Trigonometric expansion of a signal

The purpose of this section is to show how the discrete version of a real signal, periodic of period T, and sampled at intervals $T_s = T/N$, can be expressed as a linear combination of sines and cosines in the form

$$x = \sum_{h=1}^{N} \left(A_h \cos\left(2\pi(h-1)\frac{t}{T} \right) + B_h \sin\left(2\pi(h-1)\frac{t}{T} \right) \right) \qquad (12.4)$$

for each t multiple of T_s. We have seen that, if X denotes the Fourier transform of x, Equation 12.1 implies

$$x = \sum_{h=1}^{N} \frac{X(h)}{N} \exp\left(2\pi i(h-1)\frac{t}{T}\right) \tag{12.5}$$

Using Euler's identity, and calling R and I, respectively, the real and imaginary parts of X, Equation 12.5 can be rewritten as

$$x = \sum_{h=1}^{N} \left(\frac{R_h}{N}\cos\left(2\pi(h-1)\frac{t}{T}\right) - \frac{I_h}{N}\sin\left(2\pi(h-1)\frac{t}{T}\right)\right)$$
$$+i\sum_{h=1}^{N} \left(\frac{R_h}{N}\sin\left(2\pi(h-1)\frac{t}{T}\right) + \frac{I_h}{N}\cos\left(2\pi(h-1)\frac{t}{T}\right)\right) \tag{12.6}$$

This identity holds for each x, but can be further simplified when x is real. In this case we know *a priori* that the imaginary part of Equation 12.6 must vanish, proving Identity 12.4 for

$$A_h = R_h/N$$
$$B_h = -I_h/N$$

and each h from 1 to N.

Formula 12.4 is called the **trigonometric expansion** of x.

EXAMPLE 12.4 Trigonometric decomposition of a signal

We now want to decompose the triangular signal x defined in Example 12.3 into its trigonometric components and check the results.

```
≫ T = 5;
≫ N = 512;
≫ t = linspace(0, T, N+1); t = t(1:N);
≫ x1 = 2*t/T - 1/2 ; x2 = 2*(T-t)/T - 1/2;
≫ x = min(x1, x2); % the triangular signal
≫ plot(t, x, 'k-')
```

The plot appears in Figure 12.6. Let us now compute the coefficients of the sines and the cosines:

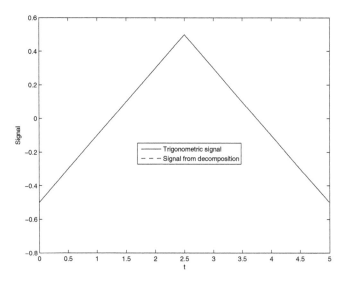

FIGURE 12.6: Trigonometric decomposition of a triangular signal

```
≫ X = fft(x);
≫ A = real(X)/N;    % cosine coefficients
≫ B = -imag(X)/N;   % sine coefficients
≫ sumcos = zeros(N,N);
≫ sumsin = zeros(N,N);
≫ for h=1:N
        sumcos(h,:)  = A(h)*cos(2*pi*(h-1)*t/T);
        sumsin(h,:)  = B(h)*sin(2*pi*(h-1)*t/T);
end
≫ y = sum(sumcos + sumsin);
```

We can check the results by comparing x and y, either graphically

```
≫ hold on
≫ plot(t,y, 'r--')
≫ legend('Trigonometric signal',...
'Signal from decomposition')
```

or numerically

```
≫ max(abs(x-y))
```

In our version of MATLAB the result was 1.2331e-013, which is zero within the computer representation of numbers.

12.6   High frequency signals and aliasing

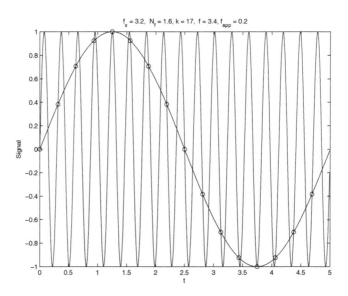

FIGURE 12.7: A high frequency signal sampled as a low frequency one

The frequency response of a digital filter depends on the sampling rate. Half the sampling frequency is called the **Nyquist** frequency and some MAT-LAB functions are expressed in terms of **normalized frequency**, that is, by definition, the frequency in Hz divided by the Nyquist frequency. So, for example, if the sampling rate is 1000 Hz, the frequency of 50 Hz will correspond to the normalized frequency $50/(1000/2) = 0.1$, and 150 Hz will correspond to $150/(1000/2) = 0.3$.

It is interesting to investigate what happens when we sample at constant intervals T_s a continuous periodic signal of frequency higher than the Nyquist frequency $N_f = 1/(2T_s)$. As we shall see in a moment, the sampled version of such a signal is identical to that of another signal of lower frequency. This phenomenon is called **aliasing**, from '*alias*', which comes from a Latin root meaning 'other'. With the help of MATLAB, we are going to show in an intuitive, if not rigorous, fashion, why aliasing occurs.

To fix the ideas, let us choose a time interval T of 5 seconds, $N = 16$ samples per period, and indicate the sampling interval with $T_s = T/N$ and the sampling frequency with $f_s = 1/T_s$.

A continuous periodic signal of period T has as its fundamental period a submultiple of T, say T/k, for a suitable k. Let us indicate its frequency, k/T, with the letter f. Such signals are, for instance, $\sin(2\pi ft)$ and $\cos(2\pi ft)$. The frequency f can always be written as

$$f = f_{app} + nf_s$$

where n is an integer and $0 \le |f_{app}| < N_f$. It is then easy to verify that, at each t multiple of T_s, say $t = hT_s$, $\sin(2\pi ft) = \sin(2\pi f_{app}t)$. In fact,

$$
\begin{aligned}
\sin(2\pi ft) &= \sin(2\pi(f_{app} + nf_s)t) \\
&= \sin(2\pi(f_{app} + nf_s)hT_s) \\
&= \sin(2\pi f_{app}hT_s + 2\pi nf_shT_s) \\
&= \sin(2\pi f_{app}hT_s + 2\pi nh) \\
&= \sin(2\pi f_{app}t)
\end{aligned}
$$

So the signal $x = \sin(2\pi ft)$, of frequency f, when sampled at frequency f_s, is indistinguishable from the signal $x_l = \sin(2\pi f_{app}t)$ of a lower frequency f_{app}.

MATLAB allows us to produce an illustrative plot of aliasing. Prepare the following M-file, `Aliasing.m`:

```
function Aliasing(k)

%ALIASING High frequency signals and aliasing

T = 5;                          % fundamental period
Np = 512;                       % number of plotting points
t = linspace(0, T, (Np+1));
t = t(1:Np);                    % fine resolution time
N = 16;                         % number of sampling points
Ts = T/N;                       % sampling interval
fs = 1/Ts;                      % sampling frequency
ts = Ts*(0: (N - 1));           % sampling instants
Nf = 1/(2*Ts);                  % Nyquist frequency
f = k/T;                        % frequency of continuous signal
x = sin(2*pi*f*t);              % high-resolution signal
xs = sin(2*pi*f*ts);            % sampling-resolution signal
% find fapp, such that f = n*fs*fapp
n = round(f/fs);
fapp = f - n*fs;
xa = sin(2*pi*fapp*t);
plot(t, x, 'k-', t, xa, 'r-', ts, xs, 'o');
xlabel('t')
ylabel('Signal')
str1 = [ 'f_s = ', num2str(fs), ',   N_f = ', num2str(Nf) ];
```

```
str2 = [ 'k = ', num2str(k), ',   f = ', num2str(f)  ];
 str3 = [ 'f_{app} = ' , num2str(fapp) ];
 str = [ str1, ', ',  str2, ', ',  str3 ];
 title(str);
```

and run the file with the command

```
≫ k = 17; alias
```

You may compare the results in Figure 12.7. In the file `Aliasing.m` we have used the command `round`; it rounds the elements of `f/fs` to the nearest integers.

12.7   Bode plot

When analyzing the behavior of a system, or designing a control system in the frequency domain, a great deal of information on performance and stability can be obtained by examining its **Bode diagram**. This diagram is composed of two curves, the $x-$axes of both representing frequencies, usually on a logarithmic scale. The $y-$axis of the first curve represents the magnitude of the transfer function, in **decibels**, and the $y-$axis of the second curve, the phase, usually in degrees. There are certain advantages in plotting against a logarithmic scale of frequencies, or frequency ratios, and using a scale in decibels for the magnitude (sometimes called *amplification factor*).

The decibel, shortly dB, is defined as ten times the decimal logarithm of the ratio of two powers. As powers in harmonic motions are proportional to the square of amplitudes, for two amplitudes, a, B, we can write

$$10 \log_1 0 \frac{|A|^2}{|B|^2} = 20 \log_1 0 \frac{|A|}{|B|}$$

An interesting explanation for the use of decibels can be found in Broch (1984), Appendix F. As to the logarithmic scale, it allows us to 'expand' the graph in the region close to zero. The logarithmic scale also allows us to extend the graph towards larger frequencies without excessively enlarging the figure.

In Subsection 7.8.2 we analyzed a mechanical system consisting in a mass, a damper and a spring, and represented its frequency response in Figure 7.8. In this section we refer again to the same system, but want to display its frequency response in a classical Bode diagram. This time we use the *transfer function* of the System. The Control System Toolbox^{TM} of the MathWorks provides specialized tools for doing the job. In this example we show how to produce the diagram using only the facilities of the basic MATLAB software. The transfer function of the system is

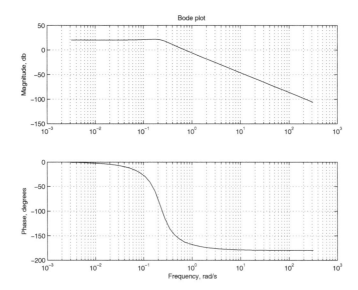

FIGURE 12.8: Bode diagram of mass-damper-spring system

$$T = \frac{1}{ms^2 + cs + k} = \frac{1}{2s^2 + 0.4s + 0.1}$$

To calculate and plot the Bode diagram write a file, `MyBode.m`, whose contents are shown below. The numerator of the transfer function is noted `pnum`,
and the denominator, `pden`. The logarithmic scale of frequencies is obtained
with the command `logspace`, and the logarithmic x−axis, with the command
`semilogx`.

```
%MYBODE Bode diagram of mechanical mass-spring-damper system

% define system parameters
m = 2;                 % mass, kg
c = 0.4;               % damping, Ns/m
k = 0.1;               % spring constant, N/m
disp('Own frequency')
sqrt(k/m)

% define transfer function
pnum =  [ 0 0 1 ];     % numerator
pden = [ m   c k ];    % denominator
% calculate magnitudes, decibels
om = logspace(-2.5, 2.5); % frequency scale, rad/s
s              = i*om;
```

```
ol          = polyval(pnum, s)./polyval(pden, s);
olmag = 20*log10(abs(ol));
% calculate phase, degrees
olpha = 180/pi*angle(ol);
% plot magnitude curve
subplot(2, 1, 1)
    semilogx(om, olmag, 'k-')
    grid
    ylabel('Magnitude, db')
    title('Bode plot')
% plot phase curve
subplot(2, 1, 2)
    semilogx(om, olpha, 'k-')
    grid
    xlabel('Frequency, rad/s')
    ylabel('Phase, degrees')
```

The resulting plot is shown in Figure 12.8.

Before concluding this section, we mention that the Control System Toolbox and the Signal Processing Toolbox contain a function, `mag2db`, for converting magnitudes to decibels, and another function, `db2mag`, for the inverse conversion. For example, check that the result of `20*log10(0.4` equals, indeed, `mag2db(0.4)`.

12.8   Summary

The commands introduced in this chapter include

fft - above X = fft(x) calculates the discrete Fourier transform (DFT) of x using the fast Fourier algorithm.

ifft - above x4 = ifft(X4) returns in x4 the inverse Fourier transform of X4.

logspace - above om = logspace(-2.5, 2.5) generates a row vector, om, containing 50 logarithmically spaced frequency values ranging from $10^{-2.5}$ to $10^2.5$.

psd - a function that belongs to the Signal Processing Toolbox; it produces a *power spectral density*, or PSD, estimate.

round - round(x) rounds the elements of x to the nearest integers.

semilogx - produces a logarithmic $x-$axis.

sort - above [spow, spos] = sort(pow) returns in spow the elements of pow sorted in increasing order, and in spos, their indices.

spectrum.welch - a function that belongs to the Signal Processing Toolbox; it produces a power-estimate object by the Welch method.

12.9 Exercises

Exercise 12.1 Ratios to decibels

Starting from the definition of the decibel find the number of dB corresponding to the ratios 0.5, 10, and 25. If you have access to the Control System Toolbox or the Signal Processing Toolbox, repeat the exercise with the function mag2db.

Exercise 12.2 Decibels to ratios

Starting from the definition of the decibel find the ratios corresponding to the 10, 68, and 120 dB. If you have access to the Control System Toolbox or the Signal Processing Toolbox, repeat the exercise with the function db2mag.

Answers to selected exercises

Chapter 2

Exercise 2.8

For example, $\overline{P_1 P_2} = \overline{Q_1 Q_2} = 2.2361$.

Exercise 2.9.

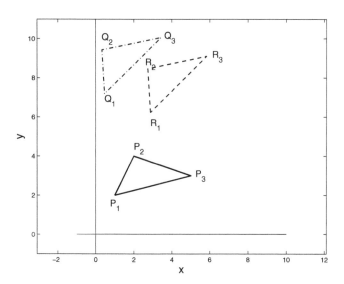

FIGURE 1: $P_1 P_2 P_3 P_1$, initial. $Q_1 Q_2 Q_3 Q_1$, translation first. $R_1 R_2 R_3 R_1$, rotation first

Exercise 2.10. The results, rounded up to two digits, are shown in Table 1.

Table 1: Weight data of container ship

		Mass, t	VCG, m	LCG, m
1	Lightship	12165	10.53	92.12
2	Deadweight	25670	10.98	103.46
3	Full load	37835	10.83	99.82

Chapter 3

Exercise 3.2

A1 = [1 1; 0 l1]; B1 = F*[1; l2]; X1 = A1\B1, R1y = X1(1) R2y = X1(2) A2 = [cosd(alpha) cosd(beta); sind(alpha) sind(beta)]; B2 = [0; F]; X2 = A2\B2, T1 = X2(1), T2 = X2(2) T1*cosd(alpha), T2*cosd(beta)

Exercise 3.3

Define P1 = [1 1] and calculate P2 = conv(P1, P1), P3 = conv(P1, P2.

Chapter 4

Exercise 4.2. The area is 1.8333.

Chapter 5

Exercise 5.5 While 0.01 and 0.02 do not belong to the set of computer numbers, 0.03125, equal to 2^{-5}, is a computer number.

Chapter 6

Exercise 6.1

weight = 55446.12 m · kg · s^{-2}.

Exercise 6.5

```
function r = ohm2S(p)
% calculates conductance corresponding to given resistance
    if (p.unit ==  'ohm')
        Value = 1/p.value;
        Unit  = 'S';
        r     = ElQuant1(Value, Unit);
    else
        errordlg('Argument not a resistance', ...
                 'Input error')
    end   % of conditional construct
end         % of ohm2S
```

```
≫ R1 = ElQuant1(2, 'ohm');
≫ R2 = ElQuant1(3, 'ohm');
Req = S2ohm(ohm2S(R1) + ohm2S(R2))
Req =
 1.2 ohm
```

Chapter 7

Exercise7.7

```
≫ cos(z1 + z2) - cos(z1)*cos(z2) + sin(z1)*sin(z2)
ans =
   3.5527e-015 +3.1086e-015i
≫ sin(z1 + z2) - sin(z1)*cos(z2) - cos(z1)*sin(z2)
ans =
  -2.6645e-015 +5.3291e-015i
```

Chapter 8

Exercise 8.2.
  $L = 22.1035$

Chapter 9

Exercise 9.2

1) Write the following function to a file derv2a.m

```
function wd = derv2a(t, w)

% example of linear differential equation

wd  = -(1.2 + sin(10*t))*w;
```

and produce the solution with the commands

```
≫ h = @derv2a;
≫ [ t, w ] = ode23(h, [ t0 tf ], w0);
≫>> plot(t, w, 'k-'), grid
```

2) Write the following function to a file derv2b.m

```
function wd = derv2b(t, w)

% example of linear differential equation
% in the form required by ODE3 and ODE45

wd = (cos(t) - w/(1 + t^2))/3;
```

and produce the solution with the commands

```
≫ tspan= [ 0 5 ]; w0 = 1;
≫ h = @derv2b;
≫ [ t, w ] = ode23(h, tspan, w0);
≫ plot(t, w, 'k-'), grid
```

Chapter 10

Exercise 10.1

The conic section is a parabola, as shown in Figure 2.

Curve of intersection

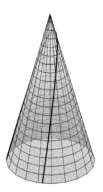

FIGURE 2: The conic section is a parabola

Chapter 11

Exercise 11.1.
    See Figure 3 on next page.

Chapter 12

Exercise 12.2.

```
≫ r = 10^(68/20)
r =
   2.5119e+003
≫ db2mag(68)
ans =
   2.5119e+003
```

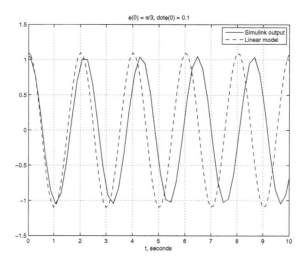

FIGURE 3: Comparing the nonlinear and the linear model of the pendulum

Bibliography

Abate, M. 1996. *Geometria*, Milano, McGraw-Hill.

Anonymous. 1992. *What every computer scientist should know about floating point-arithmetics*, Revision A,
http://docs.sun.com/app/docs/doc/800-7895/6hos0aou1?a =view.

Anonymous. 2000. *The NIST reference constants, units and uncertainty*, Physics Laboratory of NIST, http:/physics.nist.gov/cuu/Units.

Anton, H., and Rorres, C. 2005. *Elementary linear algebra – Application version*, Anton Textbooks, Inc.

Banchoff, T, and Wermer, J. 1983. *Linear algebra through geometry*, New York: Springer-Verlag, Undergraduate Texts in Mathematics.

Biran, A. and Breiner, M. 1995. *MATLAB for engineers*, Harlow, England: Addison-Wesley.

Biran, A. and Breiner, M. 1999. *MATLAB 5 for engineers*, Harlow, England: Addison-Wesley.

Biran, A. and Breiner, M. 2002. *MATLAB 6 for engineers*, Harlow, England: Pearson Education.

Biran, A. and Breiner, M. 2009. *MATLAB pour l'ingénieur – Version 7*, 2nd ed., Paris: Pearson Education France.

Broch, J.T. 1984. *Mechanical vibration and shock measurements*, 2nd ed., 4th reprint, Glostrup: K. Larsen & Søn A/S.

Cooley, J.B., and Tukey, J.W. 1965. *An algorithm for the machine computation of complex Fourier series*, Mathematics of Computation, 19, April, pp. 297-301.

Colonna, J-F. 2006. *Les calculs flottants sont-il fiables?*.
http://www.lactamme.polytechnique.fr.

Corge, C. 1975. *Éléments d'informatique — Informatique et démarche de l'esprit*, Paris: Larousse.

Davis, M. 2000. *The universal computer — The road from Leibniz to Turing*, New York: W.W. Norton and Company.

Dorf, R.C., and Bishop, R.H. 1992. *Modern control systems*, 11th ed, Upper Saddle River, NJ: Pearson - Prentice Hall.

Duncan, M. 2000. *Applied geometry for computer graphics and CAD*, London: Springer.

Emerson, W.H. 2008. *On quantity calculus and units of measurement*, Metrologia, vol. 45, pp. 134-8.

GAO (General Accounting Office). 1992. *GAO report : Patriot missile defense — System failure at Dhahran Saudi Arabia*, http://www.fas.org/spp/starwars/gao/im92026.htm.

Gerald, C.F. and Wheatley, P.O. 1994. *Applied numerical analysis*, 2nd ed., Reading-MA, Addison-Wesley.

Giordano, R.R., and Weir, M.D., 1991. *Differential equations – A modeling approach*, Reading, MA: Addison-Wesley.

Harel, D. 2003. *Computers Ltd. — What they really can't do*, revised paperback edition, Oxford University Press. German edition 2002, Italian 2002, Chinese 2003, Hebrew 2004.

Harel, D., with Feldman, Y. 2004. *Algorithmics: The spirit of computing*, Reading, MA: Addison Wesley. Hebrew edition 1991, Dutch 1989, Polish 1992, 2001.

Hofstadter, D.R. 1979. *Gödel, Escher, Bach: an eternal golden braid*, New York: Vintage Books.

Hultquist, P.F. 1988. *Numerical methods for engineers and computer scientists*, Menlo Park, CA: The Benjamin/Cummins Publishing Company.

Institute of Electrical and Electronics Engineers. 1985. *IEEE standard for binary floating-point arithmetic* (ANSI/IEEE Std 754-1985), New York: IEEE. New edition 754-2008, *IEEE Standard for Floating-Point Arithmetic*. See also Wikipedia, the entry *IEEE 754-2008*.

Kahaner, D., Moler, C., and Nash, S. 1989. *Numerical methods and software*, Englewood Cliffs, NJ: Prentice Hall.

Penrose, R. 2004. *The road to reality – A complete guide to the laws of the universe*, London: Jonathan Cape.

Schneider, P.J. and Eberly, D.H. 2003. *Geometric tools for computer graphics*, Amsterdam: Morgan Kaufmann Publishers.

Shampine, L.F., and Reichelt, M.W. (no date indicated) *The MATLAB ODE suite*, available in PDF form at

```
http://www.mathworks.com/access/helpdesk/help/pdf_doc/
otherdocs/ode_suite.pdf
```

Skeel, R. 1992. *Roundoff error and the Patriot missile*, extract from SIAM News, July 1992, Vol. 25, No. 4, p.11, http://www.siam.org/siamnews/general/patriot.thm.

Smith, S. 2006, *MATLAB advanced GUI development*, Indianapolis: Dog Ear Publishing.

Spiegel, M., and Liu, J. 2001. *Mathematical handbook of formulas and tables*, Schaum's Easy Outline of Mathematical Handbooks, New York: McGraw-Hill.

Standish, T.A. 1994. *Data structures, algorithms, and software principles*, Reading, MA: Addison-Wesley.

Taggart, R., editor 1980. *Ship design and construction*, N.Y: SNAME.

Toich, S. and Roberts, E. 1998. *The Patriot missile failure in Dhahran: Is software to blame?*,
http://shelley.toich.net/rojects/CS201/patriot.html.

Webster 1990. *Webster's ninth new collegiate dictionary*, Springfield, MA: Meriam-Webster Inc.

Wood, A. 1999. *Introduction to numerical analysis*, Harlow, England: Addison-Wesley.

Documentation provided by The Mathworks^{TM} can be downloaded from
http://www.mathworks.com/access/helpdesk/help/techdoc/
We recommend especially:

MATLAB®   7 - Getting Started Guide

ControlSystemToolbox^{TM} 8 - Getting Started Guide

Simulink®   7 - Getting Started Guide

SymbolicMathToolbox^{TM} 5 - User's Guide

Index

Milton Keynes UK
Ingram Content Group UK Ltd.
UKHW031138141024
449569UK00024B/1227